Introductory Physics for the Life Sciences

This textbook provides an accessible introduction to physics for undergraduate students in the life sciences, including those majoring in all branches of biology, biochemistry, and psychology, and students working on pre-professional programs such as pre-medical, pre-dental, and physical therapy. The text is geared for the algebra-based physics course, often named College Physics in the United States.

The order of topics studied are such that most of the problems in the text can be solved with the methods of Statics or Dynamics. That is, they require a free-body diagram, the application of Newton's Laws, and any necessary kinematics. Constructing the text with a standardized problem-solving methodology simplifies this aspect of the course and allows students to focus on the application of physics to the study of biological systems. Along the way, students apply these techniques to find the tension in a tendon, the sedimentation rate of red blood cells in hemoglobin, the torques and forces on a bacterium employing a flagellum to propel itself through a viscous fluid, and the terminal velocity of a protein moving in a Gel Electrophoresis device.

This is part one of a two-volume set; volume 2 introduces students to the conserved quantities and applies these problem-solving techniques to topics in Thermodynamics, Electrical Circuits, Optics, and Atomic and Nuclear Physics, always with continued focus on biological applications.

Key features:

- Organized and centered around analysis techniques, not traditional Mechanics and E & M
- Presents a unified approach, in a different order, meaning that the same laboratories, equipment, and demonstrations can be used when teaching the course
- Demonstrates to students that the analysis and concepts they are learning are critical to the understanding of biological systems

David V. Guerra, Ph.D., is a Professor of Physics at Saint Anselm College (SAC), United States of America. During his time at SAC, he has taught many of the courses offered by the department, including the physics course for biology majors, and has developed and taught courses in Laser Physics and Remote Sensing. At SAC, he has conducted Remote Sensing research both in instrument development and in data analysis. Professor Guerra designed, built, and operated a novel lidar (laser radar) that utilized a holographic optical element (HOE) as its primary optics. This work was done in coordination with NASA-GSFC and resulted in a series of publications and the successful development of a lidar system that was flown by NASA using the (HOE) technology. Professor Guerra has also conducted research in the analysis of remote sensing data in the investigation of natural systems. As part of a National Science Foundation grant, Professor Guerra conducted research with student researchers and faculty from other universities from across New Hampshire to study natural systems throughout their state. He continues his collaborative remote sensing work investigating relationships between environmental conditions and other natural systems. Professor Guerra has also done work in Physics education research ranging from the development of new laboratory experiences to new pedagogies and even contributing chapters to a high school physics textbook.

Introductory Physics for the Life Sciences

(Volume One)

Mechanics

David V. Guerra

CRC Press is an imprint of the
Taylor & Francis Group, an **informa** business

Front cover image: the Author

First edition published 2023
by CRC Press
6000 Broken Sound Parkway NW, Suite 300, Boca Raton, FL 33487-2742

and by CRC Press
4 Park Square, Milton Park, Abingdon, Oxon, OX14 4RN

CRC Press is an imprint of Taylor & Francis Group, LLC.

ISBN: 978-1-032-30040-5 (hbk)
ISBN: 978-1-032-31106-7 (pbk)
ISBN: 978-1-003-30806-5 (ebk)

DOI: 10.1201/9781003308065

Typeset in Times
by codeMantra

Contents

1 Introduction

1.1 VOLUME 1: INTRODUCTION

This introductory, algebra-based, physics textbook was written specifically for students interested in the life sciences. This, the first of two volumes of this textbook, is constructed around the concepts and practices of statics and dynamics as applied to biological systems, such as the tension in a tendon, the sedimentation rate of red blood cells in hemoglobin, the torques and forces on a bacterium employing a flagellum to propel itself through a viscous fluid, and the terminal velocity of a protein moving in a gel electrophoresis device.

This volume begins with an introduction to the basic concepts of forces as they are applied to the study of an object at rest. These static arrangements provide an opportunity to develop a foundational framework through a phenomenological understanding of forces. In addition, by focusing on the forces instead of the equations of kinematic at the beginning of the volume, the study of statics provides an opportunity to begin routinizing of the predominant problem-solving techniques in the volume.

The material is next scaffolded upon this problem-solving technique, as it is applied to the forces of nature. After the application of vectors to the study of physical forces is firmly established, the concepts and expressions of kinematics are introduced to provide the language necessary to explain the motion associated with a system that includes unbalanced forces. At this point in the text, it is clear that physics is more about understanding the forces associated with a system and a process based on concepts, and less about finding the correct equation to plug into to find an answer. This emphasis on the cause and effect approach to the analysis is applied to other forces including the buoyant force, the viscous force, and the restoring force associated with vibrations and waves.

Throughout this volume, a standard problem-solving process is emphasized, so that the focus of the studies can move from the frustration associated with the problem-solving process in physics, to an understanding of the importance of mechanics in the analysis of biological systems and the devices used to study these systems. In addition, to stimulate the curiosity of the reader and to provide context through an example of the application of the topics to be studied, each chapter begins with a chapter question that is answered at the end of the chapter by employing the techniques and concepts learned throughout the chapter. Throughout the volume, the techniques applied in solving mechanics problems, both static and dynamic, will become routine, and it will become apparent how these methods can be applied to better understand systems important to the study of biology.

1.2 CHAPTER 1: MATTER, UNITS, AND VECTORS

1.2.1 Introduction

This chapter begins with a brief overview of the structure of matter to set the vocabulary that is employed throughout the volume. It is not critical for all the details included in the discussion of matter to be internalized, but students should be familiar with the terminology presented in the section. Next, the units used throughout the text are presented, and again this section provides the groundwork needed to employ units successfully throughout the textbook. The most significant portion of this chapter is the presentation of concepts and techniques used to analyze vectors. In this section, the concept of a vector is presented along with the operations needed to analyze vectors. By the end of this chapter, you should be able to break a vector apart into its components and add two or more vectors together to find a resultant vector.

DOI: 10.1201/9781003308065-1

1.2.2 Matter

The starting point of the study of physics begins with a definition of matter, the stuff from which we, and everything around us, is composed. The basic building blocks of matter are atomic nuclei and electrons, because they cannot be subdivided by "ordinary" means. For this discussion, "ordinary" means are processes that do not involve nuclear reactions. This means that if an object is smashed, burned, or even dumped into acid, the object may look completely different, but the matter from which the object is made will still be electrons and the original type of atomic nuclei. In fact, the nuclei of the atoms that make up your body were not made when you were born but have existed for billions of years with some dating back to the Big Bang, others formed in the core of stars, and others resulting from supernovae (exploding stars). The concept of the atom is not new; the ancient Greeks hypothesized that matter had a limit to the size to which it could be broken down. They named this limit the atom.

Today, it is understood that atoms are composed of three types of particles: protons, neutrons, and electrons,, and they are often pictured as described in Figure 1.1. The dense conglomerate at the center of the atom is known as the nucleus, and it comprises particles known as protons and neutrons. One of the fundamental characteristics of matter is electrical charge, which will be further discussed in Chapter 5 in the context of electrical forces. At this point in the text, electrical charge will simply be considered a label of a characteristic of matter. For example, protons are labeled as a positive (+) charge and neutrons are electrically neutral, meaning they have no net electrical charge. Electrons that exist outside the nucleus have the same magnitude of charge as the proton, but the charge of an electron is negative (–).

Even though the electrons have the same magnitude of charge as the protons, they are much less massive. In fact, the mass of a neutron is roughly the same as the mass of a proton but roughly 2,000 times the mass of an electron. The nucleus is extremely dense, such that the radii of the electron orbitals are approximately 100,000 times the radius of the nucleus. So, most of the mass of atoms is located in a tiny percentage of the atom's volume.

Even though the nucleus is tiny, the distinguishing characteristic of an atom is the number of protons in its nucleus. All of the amazing varieties of objects on Earth, including living things like us, are made up of 92 different types of atoms. These 92 different atoms have 1–92 protons in their nucleus. They are commonly referred to as the natural elements. Some of the most common on Earth are hydrogen, carbon, nitrogen, and oxygen. The elements are commonly arranged, based on their chemical properties, in a chart called the periodic table. In high-energy experiments, physicists have made additional types of atoms with more protons, bringing the total number of elements up to 118, but these additional atoms are unstable, meaning they decay very quickly.

The labeling on the periodic table is explained in the insert on Figure 1.2. Each element has a label made up of letters, for example oxygen is O and helium is He. The integer above the label ranging from 1 to 118 is the number of protons in the nucleus.

In practice, when one or a few elements are listed in a chemical equation, the individual text symbols are given. If one chooses to include that number, it is written as a subscript to the left of the

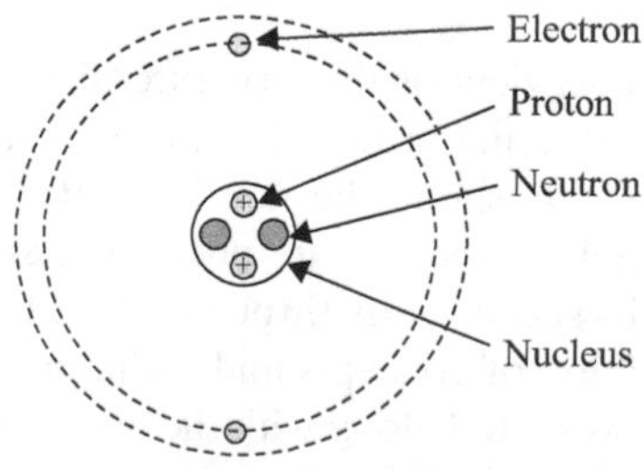

FIGURE 1.1 Atomic structure based on the Bohr model of the atom.

Periodic Table of the Elements

Atomic Number → 1; H ← Symbol; Name → Hydrogen; 1.008 ← Atomic Weight

1 IA	2 IIA	3 IIIB	4 IVB	5 VB	6 VIB	7 VIIB	8 VIIIB	9 VIIIB	10 VIIIB	11 IB	12 IIB	13 IIIA	14 IVA	15 VA	16 VIA	17 VIIA	18 VIIIA
H																	He
Li	Be											B	C	N	O	F	Ne
Na	Mg											Al	Si	P	S	Cl	Ar
K	Ca	Sc	Ti	V	Cr	Mn	Fe	Co	Ni	Cu	Zn	Ga	Ge	As	Se	Br	Kr
Rb	Sr	Y	Zr	Nb	Mo	Tc	Ru	Rh	Pd	Ag	Cd	In	Sn	Sb	Te	I	Xe
Cs	Ba	57-71 Lanthanoids	Hf	Ta	W	Re	Os	Ir	Pt	Au	Hg	Tl	Pb	Bi	Po	At	Rn
Fr	Ra	89-103 Actinoids	Rf	Db	Sg	Bh	Hs	Mt	Ds	Rg	Cn	Nh	Fl	Mc	Lv	Ts	Og

La	Ce	Pr	Nd	Pm	Sm	Eu	Gd	Tb	Dy	Ho	Er	Tm	Yb	Lu
Ac	Th	Pa	U	Np	Pu	Am	Cm	Bk	Cf	Es	Fm	Md	No	Lr

FIGURE 1.2 The periodic table of the elements.

symbol. The number of protons in the nucleus determines what the element is; hence, every atom of any one element has the same number of protons as any other atom of that element. For example, carbon 14 is written as ${}^{14}_{6}C$ with the six protons that make the atom Carbon as a subscript to the left and the 14, the number of nucleons (protons+neutrons), as a superscript to the left.

It is important to note that different atoms of one particular element may have different numbers of neutrons. For example, ${}^{13}_{6}C$ refers to carbon that has 13 nucleons. Given that every atom of carbon has six protons (the subscript "6"), it means that the specified atom has seven neutrons (obtained by subtracting the number of protons, six, from the total number of nucleons, 13). Other ways of writing ${}^{13}_{6}C$ are ${}^{13}C$ and carbon-13. The relationship for the number of neutrons compared to the number of protons in the nucleus is not based on a balance, but a comparison to the most common arrangement found on Earth. If the nucleus has more or less neutrons than the most common combination of the element found on Earth, it is known as an isotope.

The other number associated with each element in the periodic table is the relative atomic mass of the element. Since it is relative, it was decided to compare the mass to that of (1/12th) the mass of a carbon-12 atom. A "weighted average" is used to represent the percentages of various isotopes that are expected for the specific element on Earth. For example, if a sample of borax, which contains the element boron, which has five protons in its nucleus, (${}_{5}B^{xx}$) was dug out of the ground, it would have more boron-11 (${}_{5}B^{11}$) than boron-10 (${}_{5}B^{10}$). In fact, if a sample contained 123 boron atoms, on average 100 would be boron-11 (${}^{11}B$) and only 23 would be boron-10 (${}^{10}B$).

Therefore, the total mass of the boron atoms would be $(23\times10)+(100\times11) = 1{,}330$.

The average mass of these 123 atoms would be $\frac{1{,}330}{123} = 18.81$.

Thus, on the periodic table, boron appears with a 5 in the upper left of the **B** and the number 10.81 below the name of the element.

Most atoms here on Earth have the same number of protons and electrons, so the positive charge of the protons cancels out the effect of the negative charge of the electrons, and vice versa. If the number of electrons in the shells of an atom is increased or decreased from their balanced levels, the atom still remains the same element, but it is considered an ion. An atom with excess electrons is a negative ion and if the atom has fewer electrons than protons it is a positive ion. To specify a particular ion of an element, a superscript is placed to the right of the name of the symbol for that element. The value of the superscript is the absolute value of the charge of the ion divided by the charge of a proton followed by a minus sign if the charge is negative and a plus sign if the charge is positive. As an example, ${}^{13}_{6}C^{2+}$ is a positive carbon ion having a charge that is +2 times the charge of a proton. ${}^{13}_{6}C^{2+}$ can be obtained from a carbon-13 atom by removing two electrons from the atom. A negatively charged ion has extra electrons relative to the number of electrons of the corresponding neutral atom.

1.2.3 Units

When we look at an object, we don't see the atoms and we normally don't even worry about the atomic scale of a material in most branches of physics. For example, when the forces on objects are studied, we usually don't think about the forces between the atoms making up an object and when we study the flow of electricity through a wire we don't compute the movement of each electron. To analyze all the different situation in physics a standard set of measurements must be agreed upon. It is amazing to realize that the results of all the different measurements that we can make in physics can be specified by means of units for only seven different base quantities: length, mass, time, electrical current, temperature, amount of substance, and luminous (light) intensity. In science, the International System of Units (SI Units, where the order of letters in the abbreviation is based on the French expression for "International System," namely, "Système International") is the standard system of units, and the seven base SI units are listed in Table 1.1.

These units are defined as follows:

For *time*, the *second* (s) is the time it takes for the electric field of the microwave radiation, emitted when an electron in a cesium 133 atom makes a transition from one specific orbital to another, to complete 9 192 631 770 oscillations.

For *length*, the *meter* (m) is the distance that light travels in a vacuum in $\frac{1}{299{,}792{,}458}$ s.

For *mass*, the *kilogram* (kg) for a long time was based upon a standard kilogram of a platinum-iridium cylinder, defined in 1889 and kept in France. Now, the kilogram is related to a precise measurement of Plank's constant of $6.626069934\times10^{-34}$ kg$\times$m^2/s.

For *electrical current*, the *ampere* (A) is the constant current in two long straight parallel wires 1 meter apart in vacuum that produce a force between the wires equal to 2×10^{-7}N per meter of length. (Note that a Newton (N) is a derived unit of force that combines mass, length, and time).

TABLE 1.1
The Seven Base Quantities and Their SI Units

Quantity	Name	Symbol
Length	Meter	m
Mass	Kilogram	kg
Time	Second	s
Electric current	Ampere	A
Temperature	Kelvin	K
Amount of substance	Mole	mol
Luminous intensity	Candela	cd

TABLE 1.2
Some Important Derived SI Units

Derived Quantity	Name	Expression in Terms of SI Base Units
Area (length)^2	Square meter (m^2)	m·m
Volume (length)^3	Cubic meter (m^3)	m·m·m
Speed, velocity (length/time)	Meter per second (m/s)	m/s
Acceleration (length/time^2)	(m/s^2)	m/s^2
Force	Newton (N)	$m \cdot kg \cdot s^{-2}$
Pressure, stress (force/area)	Pascal (Pa)	$m^{-1} \cdot kg \cdot s^{-2}$
Energy, work (force × distance)	Joule (J)	$m^2 \cdot kg \cdot s^{-2}$
Power, radiant flux (energy/time)	Watt (W)	$m^2 \cdot kg \cdot s^{-3}$
Electric charge (current × time)	Coulomb (C)	s·A
Voltage, electric potential (energy/charge)	Volt (V)	$m^2 \cdot kg \cdot s^{-3} \cdot A^{-1}$

For *temperature*, the unit is the *kelvin* (K), which is 1/273.16 of the temperature (relative to absolute zero) of the triple point of water, which is the temperature at which water can exist in all three states: ice, liquid water, and water vapor, at the same pressure.

For the *amount of a substance*, a *mole* (mol) is the number of objects (usually atoms or molecules) that is equal to the number of atoms in 0.012 kg (12 g) of carbon-12.

For *light intensity*, the *candela* (cd) is the luminous intensity of light having a radiant intensity of 1/683 watts per steradian (where the steradian is a unit of solid angle such that the solid angle of a sphere is 4π steradians) of a single color radiation with a frequency of 540×10^{12} Hz. (Luminous intensity is brightness and takes into account the response of the human eye, whereas radiant intensity is energy per time per solid angle.)

A list of some important derived units is given in Table 1.2.

Some of these derived units may be familiar, like area as square meters (m^2) and volume as cubic meters (m^3). One derived unit that may not be familiar, but will be used extensively in the class, is the unit for force known as the newton (N). A value having units of newtons is a measure of the strength of a force, which is a push or pull. The newton is the SI version of the unit of force in the English system called the pound (lb). The relationship between these units is that 4.448 N = 1 lb. Thus, you could give your weight in newtons.

1.2.4 Vectors and Scalars

Many physical properties can be expressed with only one number (with units), which represents the magnitude or value of the property. These quantities are called *scalars*. Time, mass, and charge are examples of scalars. A variable used to represent a scalar has the form of a single letter (with, in some cases, one or more subscripts); in printed text, the letter is in italics and is not bold. Some examples are t for time, m for mass, and q for charge. Some examples with subscripts are t_{01} for the duration of the time interval from instant 0 to instant 1, m_A for the mass of object A, and q_B for the charge of particle B.

Some physical quantities have both magnitude and direction. These quantities are called *vectors*. A variable used to represent a vector has the form of a single letter with an arrow over it. The letter may have one or more subscripts. In printed text, it is common for the letter to be in boldface and not in italics, and for the to represent a vector will be in boldface *and* have an arrow over it. The same letter without the arrow over it (in nonbold italics in printed text), with the same subscript or subscripts as applicable, represents the magnitude of the vector. So, for instance, if the symbol $\vec{\mathbf{F}}_g$ is used to represent a gravitational force vector, then the magnitude of that gravitational force vector

TABLE 1.3
Some Common Vectors Used in This Text

Vector	Magnitude	Direction
Force $\vec{\mathbf{F}}$	the strength of an ongoing push or a pull	direction of the ongoing push or pull
Displacement $\vec{\mathbf{d}}$	change in position from one place to another	direction in which the position changed
Velocity $\vec{\mathbf{v}}$	speed	direction of the movement
Acceleration $\vec{\mathbf{a}}$	how fast the velocity is changing	direction in which the velocity is changing

is represented by F_g. The magnitude of a vector can also be represented by the symbol for the vector inside a pair of vertical lines: $\left|\vec{\mathbf{F}}_g\right|$. Some vectors that we will use in this course are given in Table 1.3.

1.2.4.1 Instantaneous or Average

Vectors can be either instantaneous or average quantities. The instantaneous magnitude and direction of a vector is the magnitude and direction of that vector at an instant in time. For instance, the instantaneous velocity of an object is the speed of that object, and the direction in which the position of that object is changing, at one instant in time.

1.2.4.2 Graphical Representation of a Vector

Graphically, a vector is represented as an arrow. The direction of the arrow represents the direction of the vector and the length of the arrow can be used to represent the magnitude of the vector. If a two-dimensional Cartesian coordinate system has been established such that the vector lies in the x-y plane, the direction of the vector can be specified by the angle θ that the vector makes with the +x-axis, where the angle is measured counterclockwise from the +x direction. The vector $\vec{\mathbf{A}}$ has magnitude A and angle θ.

The vector $\vec{\mathbf{A}}$ depicted in Figure 1.3 can be considered to be partly in the x direction and partly in the y direction.

As such, it can be considered to be the sum of two vectors, one ($\vec{\mathbf{A}}_x$) pointing along the x-axis and one ($\vec{\mathbf{A}}_y$) pointing along the y-axis. This format is given in equation (1.1) as follows:

$$\vec{\mathbf{A}} = \vec{\mathbf{A}}_x + \vec{\mathbf{A}}_y \tag{1.1}$$

The symbol $\hat{\mathbf{i}}$ (which one reads as i-hat) is defined to mean "in the +x direction" and the symbol $\hat{\mathbf{j}}$ (which one reads as j-hat) is defined to mean "in the +y direction," then $\vec{\mathbf{A}}_x$ can be written as $A_x\hat{\mathbf{i}}$, and $\vec{\mathbf{A}}_y$ can be written as $A_y\hat{\mathbf{j}}$. Thus, $\vec{\mathbf{A}}$ can be written as shown in equation (1.2):

$$\vec{\mathbf{A}} = A_x\hat{\mathbf{i}} + A_y\hat{\mathbf{j}} \tag{1.2}$$

This notation is referred to as i j notation. The expressions A_x and A_y are referred to as the components of vector $\vec{\mathbf{A}}$. They are scalars. Thus, when comparing components of a vector, the absolute value of the components is what is considered. To emphasize that they are scalars, A_x and A_y can also be referred to as the *scalar components* of vector $\vec{\mathbf{A}}$.

Any vector $\vec{\mathbf{V}}$ that lies in the x-y plane can be written as a value-with-units times $\hat{\mathbf{i}}$ plus a value-with-units times $\hat{\mathbf{j}}$. The value-with-units that is multiplying $\hat{\mathbf{i}}$ is called the x-component of the vector $\vec{\mathbf{V}}$ and is represented by the symbol V_x. The value-with-units that is multiplying $\hat{\mathbf{j}}$ is called the y-component of the vector $\vec{\mathbf{V}}$ and is represented by the symbol V_y.

One can specify a vector by giving the magnitude and direction of the vector or by giving the x- and y-components of the vector. Plane geometry can be used to go back and forth. For instance, in the case of vector $\vec{\mathbf{A}}$ depicted in Figure 1.3, the values of A and θ are given and the task is to find the values of A_x and A_y. First, draw a component vector diagram as shown in Figure 1.4.

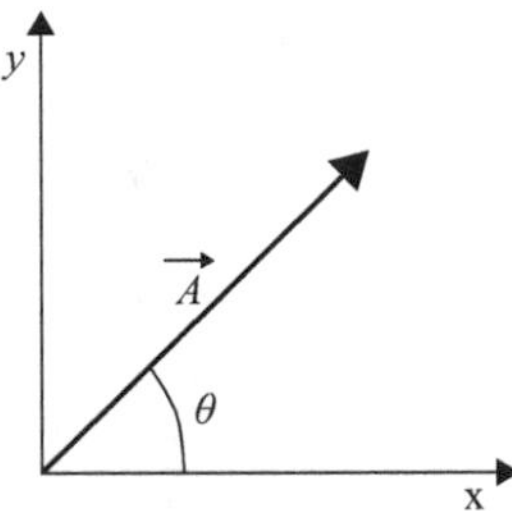

FIGURE 1.3 Graphical representation of a vector.

In Figure 1.4, the x-component of the vector $\vec{\mathbf{A}}_x$ and the y-component of the vector $\vec{\mathbf{A}}_y$ add up vectorially to vector $\vec{\mathbf{A}}$. The three vectors form a right triangle.

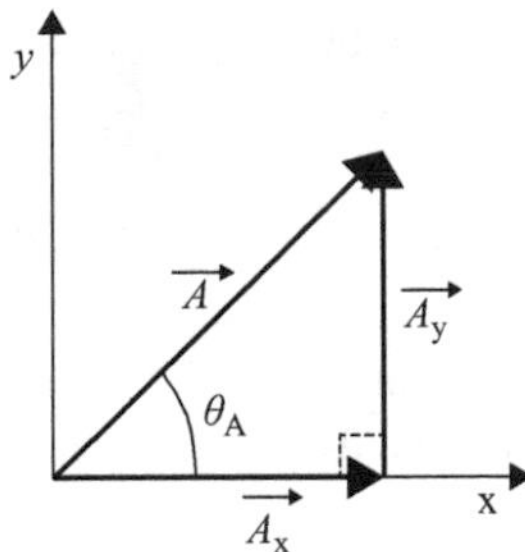

FIGURE 1.4 The components of a vector.

The Pythagorean Theorem and the trigonometric functions of sine, cosine, and tangent are the keys to understanding these two representations of a vector. The Pythagorean Theorem, which relates the length of the sides of a right triangle, and the sine, cosine, and tangent are defined with the right triangle, please see Figure 1.5.

For the triangle in Figure 1.5:

the Pythagorean Theorem is: $a^2+b^2=c^2$
the sine of the angle θ is: $\sin(\theta)=a/c$
the cosine of the angle θ is: $\cos(\theta)=b/c$
the tangent of the angle θ is: $\tan(\theta)=a/b$

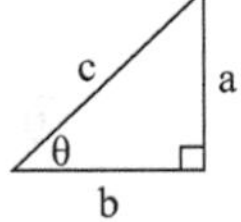

FIGURE 1.5 A right triangle.

From trigonometry, referring back to Figure 1.4, the length A_x of the side adjacent to the angle θ_A, divided by the hypotenuse (which has length A), is the cosine of θ_A.

$$\frac{A_x}{A} = \cos\theta_A$$

Solving this for A_x yields equation (1.3), which is the x-component of vector $\vec{\mathbf{A}}$ given as:

$$A_x = A\cos\theta_A \tag{1.3}$$

From trigonometry, we also know that, the length A_y of the side opposite the angle θ_A, divided by the hypotenuse (which has length A), is the sine of θ_A.

$$\frac{A_y}{A} = \sin\theta_A$$

Solving this for A_y yields equation (1.4), which is the y-component of vector $\vec{\mathbf{A}}$ which is:

$$A_y = A\sin\theta_A \tag{1.4}$$

Thus, given a vector of magnitude A and angle θ, we can determine the components of the vector using equations (1.3) and (1.4):

$$A_x = A\cos\theta_A \text{ and } A_y = A\sin\theta_A$$

An example of these two common representations of a vector ($\vec{\mathbf{A}}$) is best described with a simple example. The vector A, in Figure 1.6, can be expressed as a magnitude and an angle as follows:

$$A = 5\,[\angle 37°].$$

Another common representation is the vector as a combination of its components:

$$\vec{\mathbf{A}} = A_x\boldsymbol{i} + A_y\boldsymbol{j} = 4\boldsymbol{i} + 3\boldsymbol{j}.$$

As demonstrated, it is common to put an ***i*** next to the x-component and a ***j*** next to the y-component of the vector.

For the vector in Figure 1.6, the x- & y-components are:

$$A_y = 5\sin(37°) = 3 \quad A_x = 5\cos(37°) = 4$$

To compute the magnitude of a vector from its components, the Pythagorean Theorem is used as follows:

Given, $\vec{\mathbf{A}} = A_x\,\mathbf{i} + A_y\,\mathbf{j}$ then the magnitude of $|A| = \sqrt{{A_x}^2 + {A_y}^2}$.
For example: If $\vec{\mathbf{A}} = 4\,\mathbf{i} + 3\,\mathbf{j}$,
then the magnitude of $\vec{\mathbf{A}}$ is $|A| = \sqrt{4^2 + 3^2} = \sqrt{16+9} = \sqrt{25} = 5$.

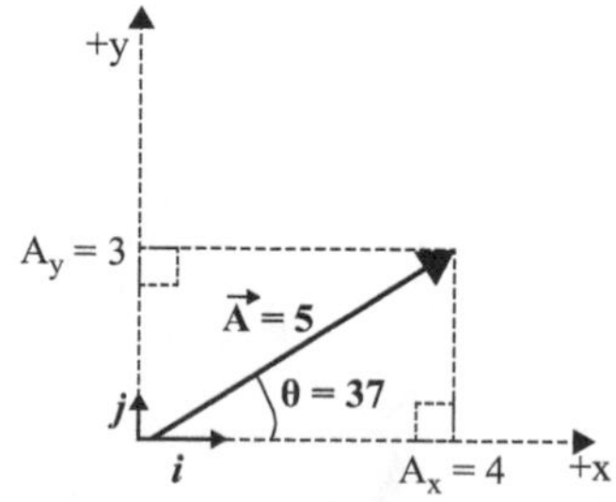

FIGURE 1.6 Vector $\vec{\mathbf{A}}$ represented in two different notations.

It is important to distinguish between the components of a vector and the magnitude of the vector. For example, as demonstrated in Figure 1.7, it is possible for two vectors to have different magnitudes and still have the same x-component.

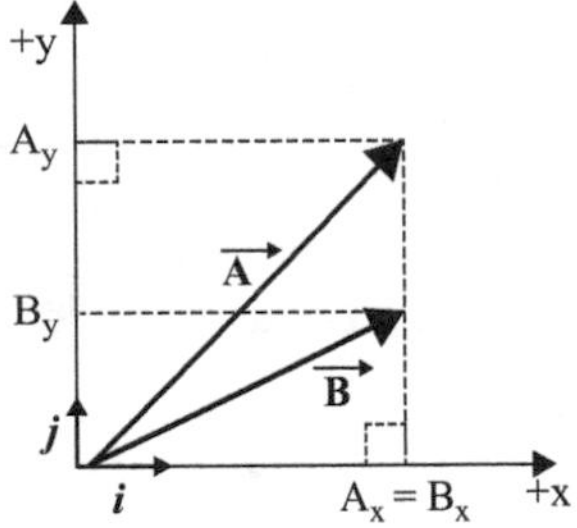

FIGURE 1.7 Two vectors with different magnitudes and the same x-component.

This is also the case for the y-component simply by changing the angle of the vector relative to the x-axis.

Also, as demonstrated in Figure 1.8, two vectors can have the same magnitude but different x- & y-components if the angles which they make with x-axis are different.

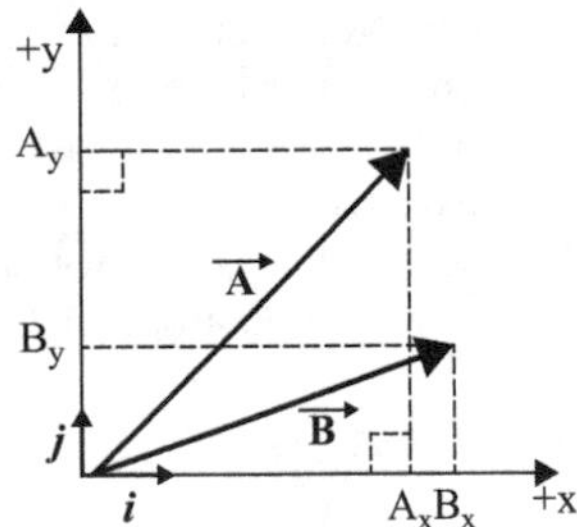

FIGURE 1.8 Graphical representation of two vectors.

It is important when working with vectors to be aware of the magnitude, angle, and components of the vectors.

1.2.4.3 Examples: Vectors

Example 1.1

Express the displacement vector $\vec{\mathbf{d}}$ = 5.00 m at 20.0° in i j notation.

Solution:

First, find the components d_x and d_y of the vector. After they are found, write: $\vec{\mathbf{d}} = d_x\ \hat{\mathbf{i}} + d_y\ \hat{\mathbf{j}}$ but with the values (with units) of the components in the places of the symbols d_x and d_y. First find the components of the vector:

$$d_x = d\cos\theta = 5\text{m}\cos 20° = 4.6985\ \text{m}$$

$$d_y = d\sin\theta = 5\text{m}\sin 20° = 1.7101\ \text{m}$$

Thus, the final answer is:

$$\vec{\mathbf{d}} = 4.6985\ \text{m}\ \hat{\mathbf{i}} + 1.7101\ \text{m}\ \hat{\mathbf{j}}$$

(Note: It is important to pay attention to the signs of the numbers in front of each component of a vector. The sign of the component signifies the direction of the vector.)

Example 1.2

Express the displacement vector $\vec{\mathbf{d}} = 5.00\,\text{m}$ at 160.0° in i j notation.

Solution:

Drawing diagrams for this example and the previous example makes it clear that the two problems are very similar (Figure 1.9).

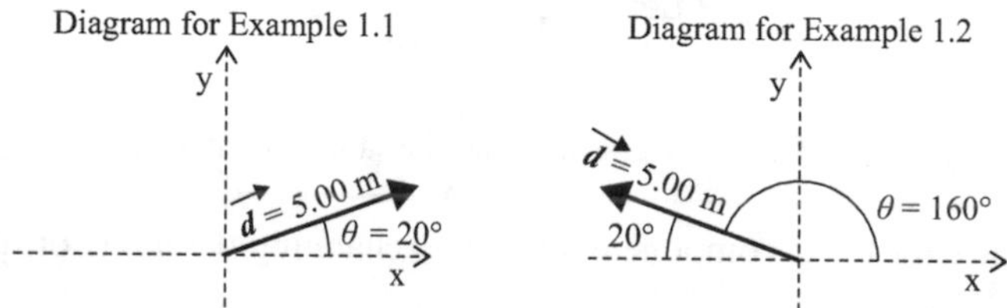

FIGURE 1.9 Diagrams of Examples 1.1 and 1.2.

The vector in Example 1.1 is partly in the +x direction and partly in the +y direction. The vector in Example 1.2 is partly in the −x direction and partly in the +y direction. Because the two vectors have the same magnitude, and the vector in Example 1.2 makes the same angle (20°) with the −x direction as the vector in Example 1.1 makes with the +x direction, the absolute values of the components will be the same for the two cases but the x-component of the vector in Example 1.2 will be negative. That is, based on the final answer to Example 1.1, the final answer to Example 1.2 is

$$\vec{\mathbf{d}} = -4.6985\text{ m}\,\hat{\mathbf{i}} + 1.7101\text{ m}\,\hat{\mathbf{j}}$$

Now, let's solve Example 1.2.

$$d_x = d\cos\theta = 5\text{m}\cos 160° = -4.6985\text{ m}$$

$$d_y = d\sin\theta = 5\text{m}\sin 160° = 1.7101\text{ m}$$

Thus, our final answer is:

$$\vec{\mathbf{d}} = -4.6985\text{ m}\,\hat{\mathbf{i}} + 1.7101\text{ m}\,\hat{\mathbf{j}}$$

just as we thought.

As another example, start with the components A_x and A_y of a vector and find the magnitude A and the angle θ of the vector. Given the vector $\vec{\mathbf{A}}$ in Figure 1.4, the magnitude of A can be determined using the Pythagorean Theorem given in equation (1.5) as:

$$A^2 = A_x{}^2 + A_y{}^2$$
$$A = \sqrt{A_x{}^2 + A_y{}^2} \tag{1.5}$$

The direction (the angle θ_A) can be determined by using the definition of the tangent of an angle from trigonometry; the tangent of an angle is the length of the side opposite the angle divided by the length of the side adjacent to the angle. This trigonometric expression can be employed such as equation (1.6):

$$\tan\theta_A = \frac{A_y}{A_x} \tag{1.6}$$

To solve for θ_A, apply the inverse tangent function to both sides. Just as the square root function "undoes" what the square function does, the inverse tangent function, also known as the arctangent function, "undoes" what the tangent function does. The inverse tangent function is written $\tan^{-1}$ or arctan. Since it is the inverse function to the tangent function, the inverse tangent of the tangent of θ_A is just θ_A. Applying the inverse tangent to both sides of the equation at hand thus yields:

$$\theta_A = \tan^{-1}\frac{A_y}{A_x}$$

This would be the final answer except for one complication. When taking the inverse tangent of a number in a calculator, the angle whose tangent is that number is found. The problem is that, for any given number, there are two angles between 0° and 360° whose tangents are the same number. When you take the inverse tangent of a number on your calculator, the calculator always gives you an angle between −90° and 90°. The other angle that has the same tangent is the angle your calculator gives you plus 180°. (For instance, both 20° and 200° have the same tangent; the value 0.3639702343 is the tangent of both angles. But if you take the inverse tangent of 0.3639702343 on your calculator, you only get the angle that is between −90° and 90°, namely 20°.) In finding the angle of a vector, you need to check if your calculator is giving you the correct answer or you have to add 180° to your calculator result to get the correct answer. An easy way to do that is to make a component vector sketch. This is illustrated in the following example.

Example 1.3

Find the magnitude and direction of the velocity vector.

$$\vec{v} = -3.00\ \text{m/s}\ \hat{\imath} + 4.00\ \text{m/s}\ \hat{\jmath}.$$

Solution:

By inspection, the components of $\vec{v}$ are $v_x = -3$ m/s and $v_y = 4$ m/s. Hence:

$$v = \sqrt{v_x^2 + v_y^2} = \sqrt{\left(-3\frac{\text{m}}{\text{s}}\right)^2 + \left(4\frac{\text{m}}{\text{s}}\right)^2} = 5\frac{\text{m}}{\text{s}}$$

To find the angle start by drawing a component vector diagram, as shown in Figure 1.10.

By inspection, θ is greater than 90°. So, subtract the angle calculated from 180° to find the angle of the vector:

$$\theta = 180 - \tan^{-1}\frac{v_y}{v_x}$$

$$\theta = 180° - \tan^{-1}\frac{4\ \text{m/s}}{3\ \text{m/s}}$$

$$\theta = 126.9°$$

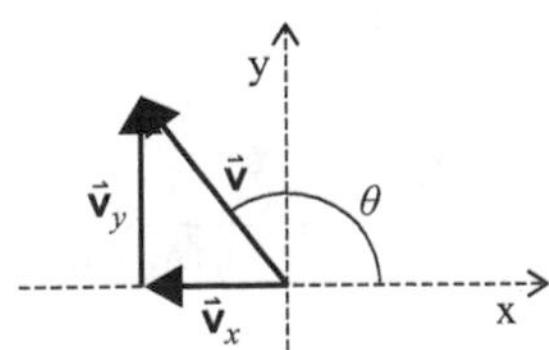

FIGURE 1.10 Vector example.

Remember that the angle of the vector is always measured counter clockwise from the +x-axis. A concept map of a vector is provided in Figure 1.11 to emphasize that a vector has a magnitude and direction and both the magnitude and the direction of the vector play a part in the x- and y-components a vector. That is, the magnitude of the x-component will increase as the y-component decreases, and vice versa.

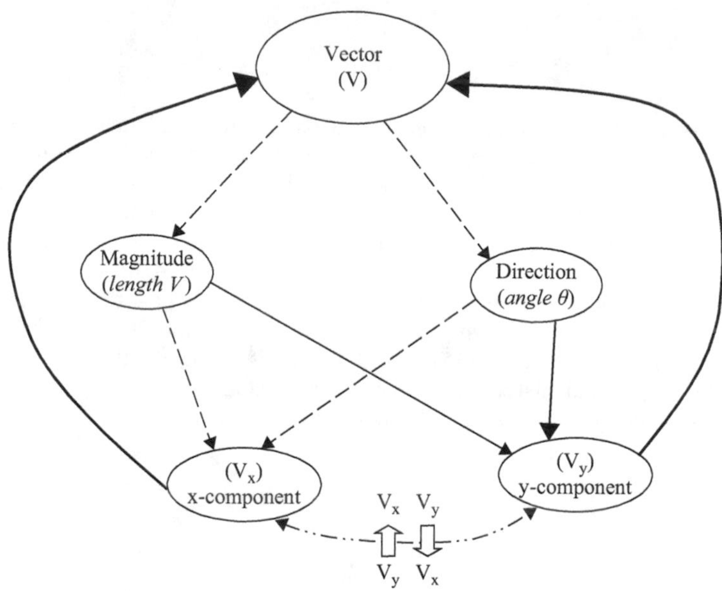

FIGURE 1.11 Conceptual map of a vector. A vector has a magnitude and direction. Both the magnitude and direction of the force contribute to both the x- and y-components of the vector. For a vector with a constant magnitude, if the x-component is increased the y-component decreases, as represented by the arrows at the bottom of the diagram.

1.2.4.4 Vector Addition

The operation of vector addition, $\vec{C} = \vec{A} + \vec{B}$, is performed by finding the components of each vector and then adding the like component (x-components to x-components and y-components to y-components) of each vector together to find the x- and y-components of the vector sum. An explanation of this process starts with the vectors in Figure 1.12.

Given the two vectors: $\vec{\mathbf{A}} = A_x\hat{\mathbf{i}} + A_y\hat{\mathbf{j}}$ and $\vec{\mathbf{B}} = B_x\hat{\mathbf{i}} + B_y\hat{\mathbf{j}}$, the sum $\vec{\mathbf{A}} + \vec{\mathbf{B}}$ is given by:

$$\vec{\mathbf{A}} + \vec{\mathbf{B}} = A_x\hat{\mathbf{i}} + A_y\hat{\mathbf{j}} + B_x\hat{\mathbf{i}} + B_y\hat{\mathbf{j}}$$

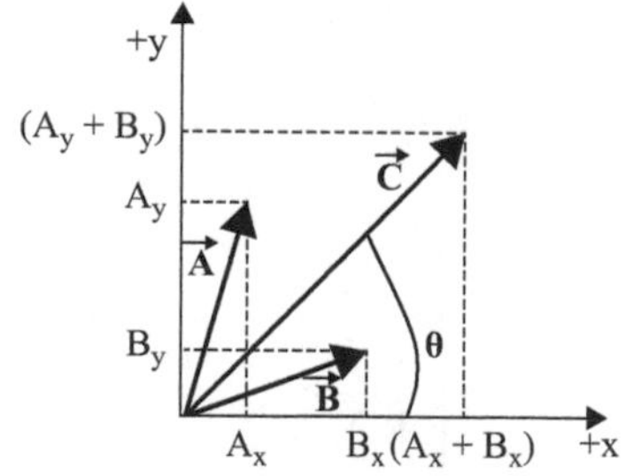

FIGURE 1.12 Vector addition.

Gathering the terms with $\hat{\mathbf{i}}$ in them and gathering the terms with $\hat{\mathbf{j}}$ in them yields:

$$\vec{\mathbf{A}}+\vec{\mathbf{B}}=\left(A_x\hat{\mathbf{i}}+B_x\hat{\mathbf{i}}\right)+\left(A_y\hat{\mathbf{j}}+B_y\hat{\mathbf{j}}\right)$$

Factoring out the $\hat{\mathbf{i}}$ in the first set of terms and factoring out the $\hat{\mathbf{j}}$ in the second set of terms gives:

$$\vec{\mathbf{A}}+\vec{\mathbf{B}}=\left(A_x+B_x\right)\hat{\mathbf{i}}+\left(A_y+B_y\right)\hat{\mathbf{j}} \text{ and since } \vec{C}=\vec{A}+\vec{B},$$

$$\vec{\mathbf{C}}=\left(A_x+B_x\right)\hat{\mathbf{i}}+\left(A_y+B_y\right)\hat{\mathbf{j}}$$

$$\vec{\mathbf{C}}=\left(C_x\right)\hat{\mathbf{i}}+\left(C_y\right)\hat{\mathbf{j}}$$

with, $C_x=A_x+B_x$ and $C_y=A_y+B_y$, respectively.

The magnitude of $\vec{\mathbf{C}}$ is $|C|=\sqrt{C_x^2+C_y^2}$ and the direction of $\vec{\mathbf{C}}$ is given by the angle $\theta=\tan^{-1}\left(\frac{C_y}{C_x}\right)$.

This is the final answer. The x-component of the sum of two vectors is the sum of the x-components of the two vectors, and the y-component of the sum of two vectors is the sum of the y-components of the two vectors. If the vectors are given in terms of magnitude and direction instead of i j notation, then we first have to calculate the components of the given vectors. If the answer is to be given in terms of magnitude and direction, then, after we find the sum of the two vectors, we have to calculate the magnitude and direction of the answer.

A special case of vector addition arises when the two vectors lie along the same line. The method above still works but the sum of the vectors can be arrived at more easily. If the two vectors that are being added together point in the same direction, the resultant vector (the sum of the two vectors) is in the same direction as the two vectors and the magnitude of the resultant vector is the sum of the magnitudes of the two vectors being added. If the two vectors being added point in exact opposite directions, the resultant vector points in the direction of the vector having the greater magnitude, and the magnitude of the resultant vector is the magnitude of the vector with the greater magnitude minus the magnitude of the vector with the smaller magnitude. A Conceptual Map of Vector Addition is provided in Figure 1.13 to emphasize that the components of each vector must be found and then the x-components are added together and the y-components are added together separately to find the components of the vector sum of $\vec{\mathbf{A}}+\vec{\mathbf{B}}=\vec{\mathbf{C}}$.

The conceptual map of vector addition, Figure 1.13, shows two vectors, but there can be any number of vectors. The first step is to find the components of the vectors individually, and add all the x-components together to get the x-component of the resultant vector. Then, separately, add all the y-components together to get the y-component of the resultant vector. Together, the x- and y-components of the resultant are the resultant vector.

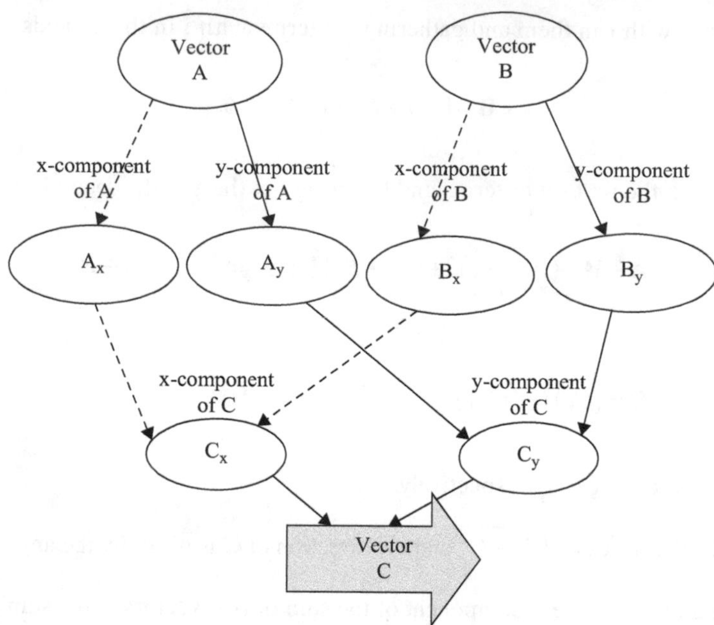

FIGURE 1.13 Conceptual map of vector addition.

1.2.4.5 Example of Vector Addition

Given that $\vec{A}$ has a magnitude of 5 and is at an angle of 70° and $\vec{B}$ has a magnitude of 4 and is at an angle of 20°.

Step 1: Make a sketch of the vectors as shown in Figure 1.14.

This is only a sketch for reference, and the location of vector $\vec{C}$, which is the vector sum of $\vec{A}$ and $\vec{B}$, is estimated by moving along the x-axis a amount of $A_x + B_x$ and moving up the y-axis an amount of $A_y + B_y$.

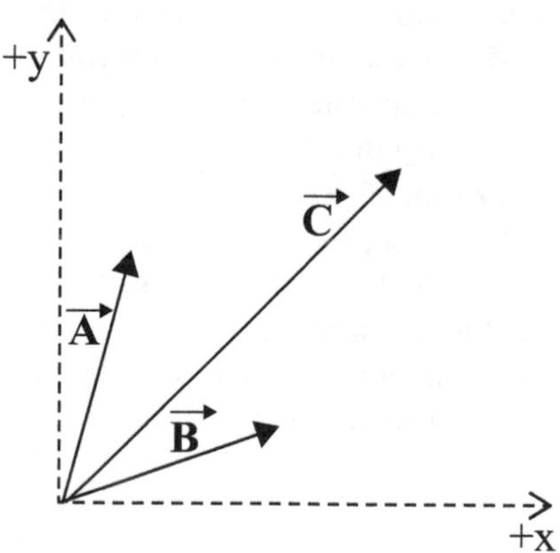

FIGURE 1.14 Vector addition.

Step 2: Finding the components of each vector

$$A_x = 5\cos(70°) = 1.71 \qquad A_y = 5\sin(70°) = 4.70$$
$$B_x = 4\cos(20°) = 3.76 \qquad B_y = 4\sin(20°) = 1.37$$

Step 3: Find the sum of the x-components and the sum of the y-components separately:

$$C_x = A_x + B_x \text{ and } C_y = A_y + B_y$$

$$C_x = A_x + B_x = 1.71 + 3.76 = 5.47$$

$$C_y = A_y + B_y = 4.70 + 1.37 = 6.07$$

Step 4: Generate an expression of the final answer:
One possible expression is: $\vec{\mathbf{C}} = (C_x)\hat{\mathbf{i}} + (C_y)\hat{\mathbf{j}} = = (5.47)\hat{\mathbf{i}} + (6.07)\hat{\mathbf{j}}$
Another format is with the magnitude and the angle.

The magnitude of $\vec{C}$ is $|C| = \sqrt{(C_x)^2 + (C_y)^2} = \sqrt{(5.47)^2 + (6.07)^2} = 8.17$ and the direction of $\vec{C}$ is given by the angle $\theta = \tan^{-1}(C_y/C_x) = \tan^{-1}(6.07/5.47) = 48°$.

1.3 CHAPTER QUESTIONS AND PROBLEMS

1.3.1 Multiple Choice Questions

1. What is the characteristic of an atomic nucleus that distinguishes it as the atom of a particular element?
   A. the types of protons. B. the types of neutrons.
   C. the number of protons. D. the number of protons and neutrons in the nucleus.

2. An atom-like entity with fewer electrons than protons is called an ion, isotope, or element?
   A. ion. B. isotope. C. element.

3. Which of the following quantities is not a base quantity?
   A. mass B. time C. speed D. electric current

4. Which of the following is the correct unit for speed?
   A. kg B. s C. m D. kg/s E. m/s

5. Vector $\vec{\mathbf{A}}$ has a magnitude of 5 and is at an angle of 50°. What is its x-component?
   A. 3.2 B. 3.8 C. 5 D. 6.5 E. 7.8

6. Vector $\vec{\mathbf{A}}$ has a magnitude of 5 and is at an angle of 50°. What is its y-component?
   A. 3.2 B. 3.8 C. 5 D. 6.5 E. 7.8

7. A vector, $\vec{\mathbf{F}}$, has a magnitude of 5 and is at an angle of 220°. What are the x- and y-components of this vector?
   A. $F_x = -3.21$, $F_y = -3.83$ B. $F_x = 3.21$, $F_y = 3.83$
   C. $F_x = 4.98$, $F_y = 0.442$ D. $F_x = -3.83$, $F_y = -3.21$

8. Vector $\vec{\mathbf{A}}$ is defined as: $\vec{\mathbf{A}} = -4\hat{\mathbf{i}} + 4\hat{\mathbf{j}}$ and vector $\vec{\mathbf{B}}$ is defined as $\vec{\mathbf{B}} = -3\hat{\mathbf{i}} - 2\hat{\mathbf{j}}$. What is the magnitude of the sum $\vec{\mathbf{A}} + \vec{\mathbf{B}}$ of these two vectors?
   A. 3.61 B. 5.66 C. 7.28 D. 9.26

9. Given that vector $\vec{\mathbf{A}}$ has a magnitude of 4 and an angle of 0° and vector $\vec{\mathbf{B}}$ has a magnitude of 3, if the sum of these two vectors has a magnitude of 1, the angle of vector $\vec{\mathbf{B}}$ is:
   A. 0° B. 90° C. 180° D. 270°

10. Figure 1.15 is the diagram for questions 10a–10e.
    Vectors $\vec{\mathbf{A}}$ and $\vec{\mathbf{B}}$ on the same axes.
    Both Vector $\vec{\mathbf{A}}$ and Vector $\vec{\mathbf{B}}$ have a magnitude of 4 and the angles as shown in Figure 1.15 are $\phi=20°$ and $\gamma=15°$

10a. The vector component with the greatest magnitude is:
A. A_x B. A_y C. B_x D. B_y

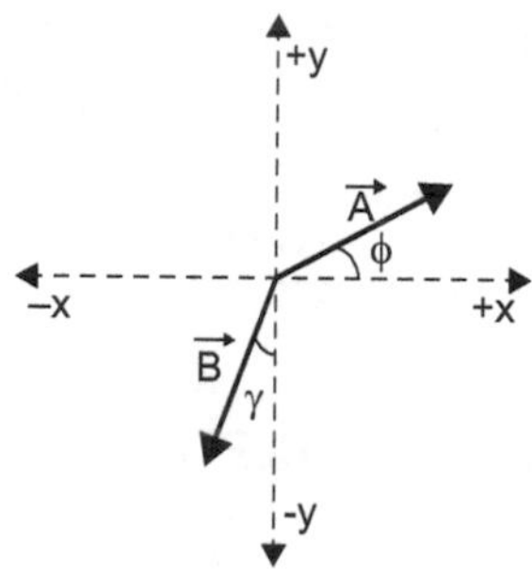

FIGURE 1.15 Multiple choice questions 10a–10e and problem 10.

10b. The vector component with the least magnitude is:
A. A_x B. A_y C. B_x D. B_y

10c. If the angle ϕ was set to 90°, the magnitude of the vector sum of A and B would
A. increase B. decrease C. remain unchanged

10d. If the angle γ was set to 0°, the magnitude of the vector sum of A and B would
A. increase B. decrease C. remain unchanged

10e. If both angles can be adjusted to be any angle relative to the x-axis, the maximum magnitude of the sum of vectors A and B is:
A. 0 B. 1.41 C. 2 D. 2.83 E. 4 F. 8

1.3.2 Problems

1. Using information provided in Chapter 1, calculate the temperature of the triple point of water.
2. In this chapter, it is stated that a *mole* (mol) is the number of objects (usually atoms or molecules) that is equal to the number of atoms in 0.012 kg of the isotope of carbon, carbon-12. Given that the mass of a single atom of carbon-12 is 1.992647×10^{-26} kg, calculate the number of atoms that is a mole of atoms.
3. A velocity vector, $\vec{\mathbf{v}}$, has an x-component of $v_x=-2$ m/s and a y-component of magnitude −3 m/s. Find the magnitude and direction of the velocity vector **v**.
4. A displacement vector, $\vec{d}$, has an x-component of $d_x=3$ m/s and a y-component of magnitude 4 m/s. Find the magnitude and direction of the velocity vector **v**.
5. A force vector, $\vec{F}$, has a magnitude of 10 N and is at an angle of 280°. Find the x- and y-components
6. A velocity vector, $\vec{\mathbf{v}}$, has a magnitude of 5 m/s and is at an angle of 30°. Find the x- and y-components.

7. Find the vector sum of the two displacement vectors:

$$\vec{d_1} = 3i + 2j \text{ and } \vec{d_2} = 4i - 3j$$

8. Find the vector sum of two velocity vectors: $\vec{v_1}$ has a magnitude of 2 m/s at an angle of 30° and $\vec{v_2}$ has a magnitude of 5 m/s at an angle of 55°.
9. Two force vectors are applied to a fastener, as depicted in Figure 1.16.

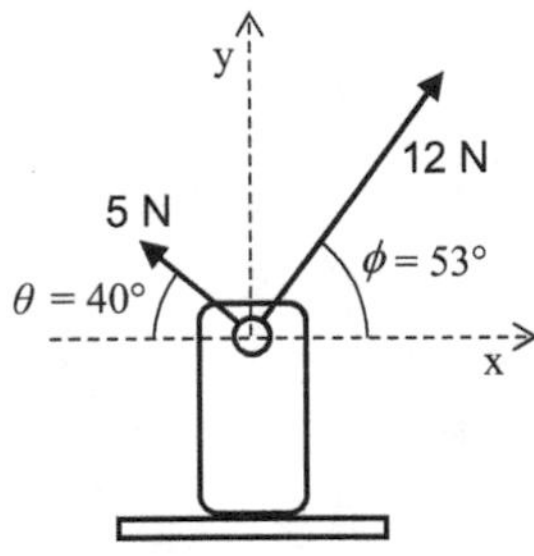

FIGURE 1.16 Problem 9.

a. Find the components of each force vector.
b. Compute the magnitude of the resultant (the sum) of the two force vectors.
c. Calculate the direction of the resultant force.
d. Sketch the resultant of the two force vectors.

10. Both Vector $\vec{\mathbf{A}}$ and Vector $\vec{\mathbf{B}}$ (in Figure 1.15) have a magnitude of 4 and the angles as shown in the diagram to the right are $\phi = 20°$ and $\gamma = 15°$. Compute the magnitude and direction of the sum of vectors $\vec{\mathbf{A}}$ and $\vec{\mathbf{B}}$.

2 Forces and Static Equilibrium

2.1 INTRODUCTION

Forces are simply a constantly applied push or a pull. One commonly experienced force is the weight of an object that we are holding. This will be the first force discussed in this chapter. The techniques developed in the previous chapter will be applied to force vectors in different static situations. It is important to mention, but not critical to the current understanding, that these static situations provide a simple starting point to develop the analysis techniques central to most of this volume. These situations allow us to focus on working with the vector components and becoming competent in analyzing situations that do not require the integration of trigonometry-based vector analysis and the equations associated with the motion of objects. Later in this volume, these concepts will be put together in the context of Newton's Laws and Kinematics, but it is important to first establish baseline knowledge of working with vectors before we try to put this all together.

Chapter Question: When a person weighing 100 lb. decides to perform a high wire act will a wire that is capable of sustaining a maximum tension equal to their weight be sufficient? This question will be answered at the end of this chapter, employing the concepts and methods developed throught this chapter (Figure 2.1).

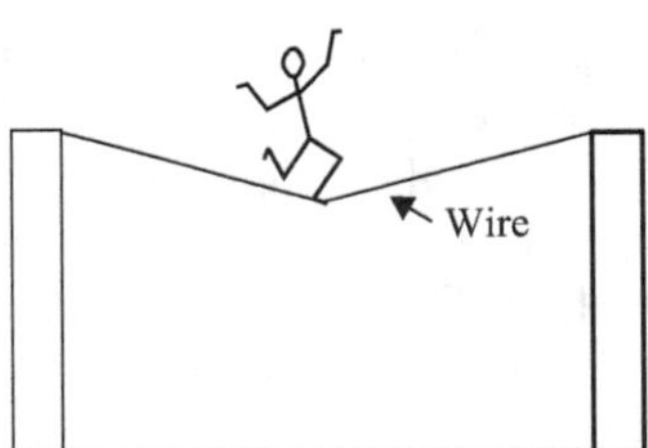

FIGURE 2.1 High wire act.

2.2 FORCES

As mentioned in the first chapter, a force is an ongoing push or pull on an object. Some examples of forces are the force of gravity, tension, and applied forces. The force of gravity is commonly referred to as the weight of an object, the pulling force of a string is known as tension, and the push on us by another person's hand is an example of an applied force. The magnitude of the force of gravity or weight is often measured in pounds (lb). In the SI units, the unit of force is the newton (N). A newton and a pound are both units of force, but a newton is a smaller unit of force than a pound. A newton is approximately one quarter of a pound or more precisely: 1 N=0.2248 lb.

2.3 WEIGHT AND MASS

As mentioned above, the magnitude of the force of gravity, $\vec{F}_g$, on an object is commonly called the weight of the object. Although this force depends on the mass of an object, weight and mass are not the same. The mass of an object is the inertia (inherent tendency to resist a change in its own motion) of that object; it is the same no matter where in the universe the object is located. On the other hand, weight is the pull of gravity on an object. Unfortunately, mass and weight are often confused

DOI: 10.1201/9781003308065-2

because most of the commodities that we purchase at the store (in the United States) are labeled in pounds when it is the amount of the contents in the box that we are most interested in, not how hard the Earth pulls on the contents. Since all the products will most likely be used on the planet in which they are purchased, the distinction between how hard the Earth pulls on the contents in the box and the amount of the contents in the box is inconsequential. The common labeling obscures the distinction between weight and the amount of material, but this distinction is important in physics.

The force of gravity on an object on the earth's surface is computed by multiplying the mass of the object in kg by how hard the Earth pulls down on each kilogram (1 kg) of mass that is near the surface of the earth. The average strength of the force with which the Earth pulls down on every kg of mass is the gravitational field strength, which is labeled as a g. Therefore, the magnitude of the force of gravity can be expressed in equation (2.1) as

$$F_g = mg \tag{2.1}$$

The average strength of the force with which the Earth pulls down on each kg of mass is $g = 9.81$ N/kg near the Earth's surface. Notice that the units of g are N/kg so that when you multiply this gravitational field strength by the mass, in kg, of an object, the kg will cancel and the force you calculate will have units of Newtons (N).

If you work out at a gym, you may have noticed that weights are labeled with both kg and lb. For example, the 9 kg plate on a machine has a weight of 20 lb. stamped on it. Since the plate has a mass of 9 kg, it weighs about 19.85 lb.

$$F_g = mg = (9\ \cancel{\text{kg}})(9.81\ \text{N}/\cancel{\text{kg}}) = 88.29\ \text{N} = (88.29\ \text{N})(0.2248\ \text{lb.}/\text{N}) = 19.85\ \text{lb.}$$

So, the force is rounded up and the plate is labeled as 20 lb. Although both labels are relevant to knowing if the weights are too light, too heavy, or just right to lift, only the value in pounds is actually the magnitude of the force of gravity on the weights. The value of g will be computed explicitly in Chapter 4, but until then it is important to note that this gravitational field strength of 9.81 N/kg is only accurate near the Earth's surface. Thus, an object with a given mass will have a different weight depending on its location. On the moon's surface, g is approximately 1.7 N/kg, so a 1 kg object that weighs 9.8 N on Earth weighs only 1.7 N on the moon, but it always has a mass of 1 kg. On another planet, there will be another value of g. In fact, the farther a point is from the center of the earth, the smaller the value of g.

2.4 FORCE VECTORS

In analyzing a physical situation, forces are represented as vectors. As stated in Chapter 1, vectors are mathematical entities that contain two pieces of information. The two pieces of information with regard to a force on an object are the magnitude of the force, how hard the ongoing pull or push on the object is, and the direction of the push or pull on the object.

In Figure 2.2, the person is pushing on a crate with a force directed to the right. The force vector is represented by an arrow pointing in the direction of the force. The length of the arrow can be used to indicate the magnitude of the force. The longer the vector arrow, the stronger the force.

2.4.1 Vector Components

Consider a case in which an xy coordinate system has been defined and a force that is directed neither along the x-axis nor along the y-axis is acting on an object. This is the case when the coordinate system has been defined so that the x-axis is horizontal and the y-axis is vertical, when the force on an object is neither completely horizontal nor completely vertical. A force which is aligned

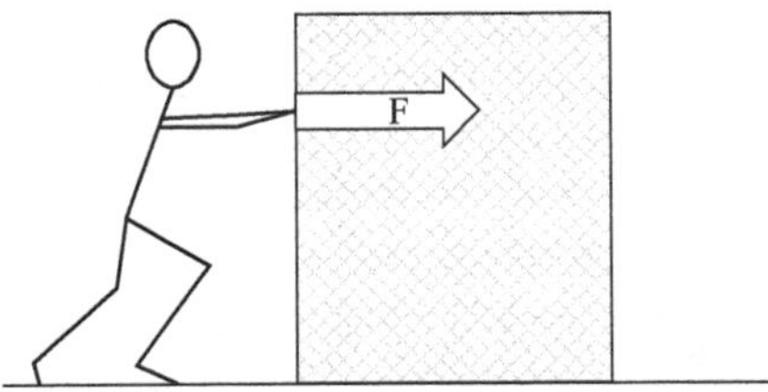

FIGURE 2.2 Force vector.

with neither axis will have some part of the force in each direction (+x and +y). The amount of force in the +x-direction and the amount of force in the +y-direction is known as the components of the force vector.

For example, the person in Figure 2.3 is pulling on a rope attached to a wagon, so that the force of the tension in the rope and applied to the wagon has a magnitude F and is directed at an angle of θ from the +x-axis. This force of tension has two components, an x-component and, a y-component.

The x- and y-components of the force vector are found in exactly the way that the x and ycomponents of any vector are computed with the method outlined in Chapter 1. Once the strength and direction of the force are established as the magnitude and angle of a vector, the method used to calculate the x- and y-components is the same for all vector quantities.

Vector Example 2.1

Find the components of the tension in the rope in Figure 2.3, assuming that the tension force has a magnitude of 20 N and an angle of $\theta = 38°$.

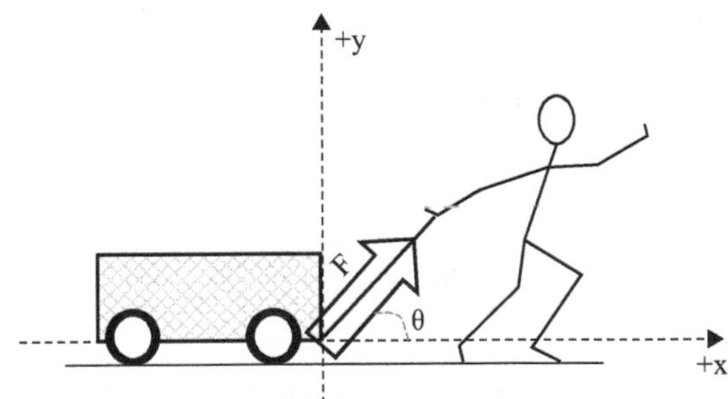

FIGURE 2.3 A person pulling on a rope that is attached to a wagon, which they are pulling across the floor.

Step 1: Since the diagram is already given in Figure 2.3, it does not need to be redrawn here. As specified in Chapter 1, the angle θ must always be measured counterclockwise from the +x-axis. In this case, the x- and y-components can be computed with the relationships specified by the trigonometric functions of sine and cosine as follows:
The x-component of the force of tension is: $F_{Tx} = F_T \cos\theta = (20\,\text{N})\cos(38°) = 15.76\,\text{N}$
The y-component of the force of tension is: $F_{Ty} = F_T \sin\theta = (20\,\text{N})\sin(38°) = 12.31\,\text{N}$
Step 2: Find the components of the vectors.

Finding the components of the forces is a critical step in analyzing many situations, since the analysis of the forces is done in each direction (+x and +y), separately. A concept map of a force vector is given in Figure 2.4, which shows that the force vector has two parts, the strength of the force applied and a direction in which the force is applied, the angle. Both of these qualities, magnitude and direction, play a part in computing the x- and y-components of the force. Together the two components are another way to define the force vector.

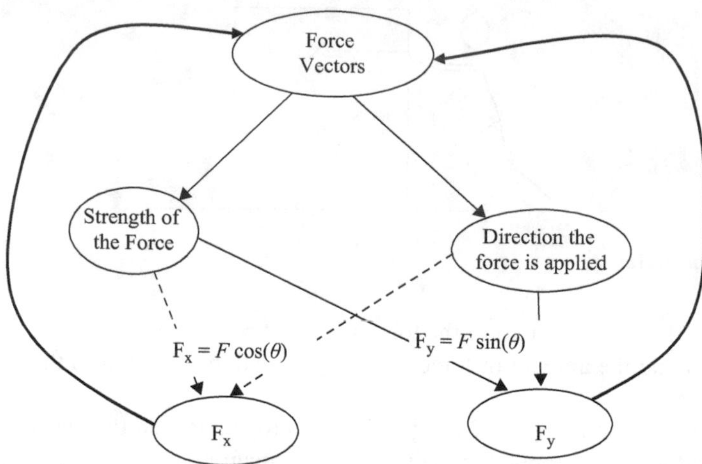

FIGURE 2.4 A force vector has a magnitude and direction. The magnitude is the strength of the force and the direction is determined by the direction in which the force is applied. Both the magnitude and direction of the force contribute to both the x- and y-components of the force.

2.5 NET FORCE

Another important definition is that of the *net force* acting on an object. Like the net income from your paycheck, which is the gross amount you earned minus the taxes and other deduction taken from your paycheck, the net force is the result of adding all the forces both positive and negative on an object. In equation form, the Greek letter Σ (sigma) is used to indicate the summation. So, the statement that the net force acting on an object is equal to the sum of all the forces acting on the object can be written in equation form in equation (2.2) as:

$$\vec{\mathbf{F}}_{\text{net}} = \sum_{i=1}^{n} \vec{\mathbf{F}}_{\text{i}} \tag{2.2}$$

where i takes on the integer values 1, 2, 3, … up to the total number of forces, with each value of i corresponding to a unique one of the forces in the problem. It is important to remember that this is not the $\boldsymbol{i}$ which indicates the x-component of the vector in the i, j, k notation, but just the number of forces involved in the situation which is to be analyzed. In the analysis of physical situations, the net force is computed in each coordinate direction separately. The components of the net force are computed by separately adding all the x-components of the forces together, all the y-components of the forces together, and all the z-components of the force together:

$$F_{\text{net,x}} = SF_{\text{ix}} = F_{1\text{x}} + F_{2\text{x}} + F_{3\text{x}} + \ldots..,$$

$$F_{\text{net,y}} = SF_{\text{iy}} = F_{1\text{y}} + F_{2\text{y}} + F_{3\text{y}} + \ldots..,$$

$$F_{\text{net,z}} = SF_{\text{iz}} = F_{1\text{z}} + F_{2\text{z}} + F_{3\text{z}} + \ldots...$$

If the analysis can be done in two dimensions (2-D) the plane created by the x and y axes are used and the value of force in the z-direction is set to zero. Thus, the introductory analysis in this chapter will take place with the x- and y-components.

2.6 TRANSLATIONAL EQUILIBRIUM

The simplest system to analyze is one involving a single object that is at rest and remains at rest. In such a case, the net force on the object must be equal to zero. This condition, in which the sum of all the forces on an object is zero, is called *translational equilibrium*. This can be expressed as shown in equation (2.3) as:

$$\vec{\mathbf{F}}_{\text{net}} = \sum_{i=1}^{n} \vec{\mathbf{F}}_{\text{i}} = 0 \tag{2.3}$$

Thus, the fundamental scalar equations for two-dimensional translational equilibrium are:

$$F_{\text{net,x}} = 0 \qquad F_{\text{net,y}} = 0$$

Statics is the name of the process of analyzing a system in translational equilibrium.

2.6.1 ONE-DIMENSIONAL TRANSLATIONAL EQUILIBRIUM

The process of the static analysis is introduced with the simple example of a box suspended by a string, as depicted in Figure 2.5.

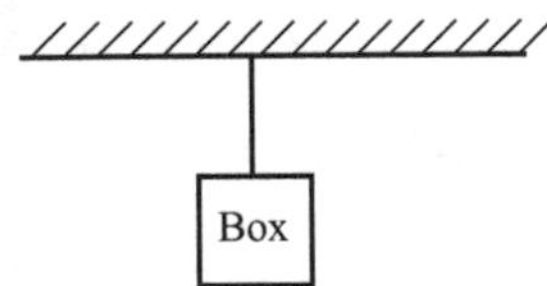

FIGURE 2.5 Box hanging from a string.

Given that the mass of the box is known as m (kg) and the task is to find the tension in the string, the two forces on the box in equilibrium are the gravitational force of the earth on the box and the tension force of the string.

Step 1: The starting point of the analysis of an object in equilibrium is making a drawing known as a free-body diagram. The first step in drawing a free-body diagram is to draw and label the coordinate system.

It is common to choose the +x-direction to the right and the –x-direction to the left and to choose the +y-direction to be upward and the –y-direction to be downward. With the coordinate system defined, the object is drawn at the origin of the coordinate system. Arrows representing the forces on the object are drawn and labeled. A perfect reproduction of the object is not required, and it is acceptable to substitute a square or a circle for the box, or any other object, in a free-body diagram. Please see Figure 2.6 for an example of a free-body diagram of the box hanging from a string. Notice that the string is not included in the free-body diagram, the object to be analyzed is the box.

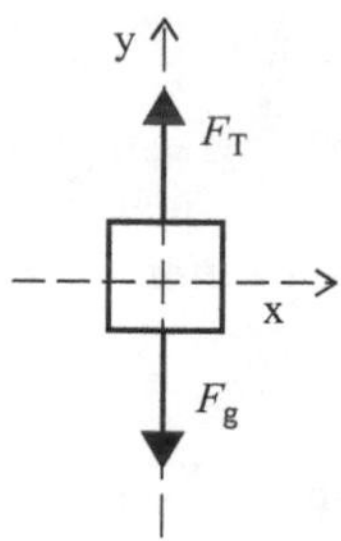

FIGURE 2.6 Free-body diagram of the forces on the box.

Step 2: Write the standard expressions for the components of the force vectors:

$$F_{Tx} = F_T \cos(\theta_{FT}) \qquad F_{Ty} = F_T \sin(\theta_{FT})$$

$$F_{gx} = F_g \cos(\theta_{Fg}) \qquad F_{gy} = F_g \sin(\theta_{Fg})$$

where θ_{FT} and θ_{Fg} are the angles of the force vectors as measure from the +x-direction. For this situation: $\theta_{FT}=90°$ and $\theta_{Fg}=270°$. Substituting these angles into the expressions for the components of the force vectors, we find:

$$F_{Tx} = F_T \cos(\theta_{FT}) = F_T \cos(90°) = 0 \quad F_{Ty} = F_T \sin(\theta_{FT}) = F_T \sin(90°) = F_T$$

$$F_{gx} = F_g \cos(\theta_{Fg}) = F\theta_{Fg} \cos(270°) = 0 \quad F_{gy} = F_g \sin(\theta_{Fg}) = F_g \sin(270°) = -F_g$$

Note that the general procedure for finding the components has been provided, but the example chosen is so simple that you can write the components by inspection. The tension force is straight upward, in the +y-direction. Thus, the x-component of the tension force is 0 by inspection, and the y-component of the tension force is equal to the magnitude F_T of the tension force, also by inspection. The gravitational force is straight downward, in the –y-direction. By inspection, the x-component of the gravitational force is 0, and the ycomponent of the gravitational force is the negative of the magnitude of the gravitational force; $F_{gy}=-F_{g.}$

Step 3: Apply the translational equilibrium equations $F_{net,\,x}=0$ and $F_{net,\,y}=0$.

$F_{Tx+}\,F_{gx}=0$ (Write the equilibrium equation for the x-direction.)

$0+0=0$ (Substitute the x-components.)

$F_{Ty}+F_{gy}=0$ (Write the equilibrium equation for the y-direction.)

$F_T+(-F_g)=0$ (Substitute the y-components.)

Step 4: The two final expressions in Step 3 must be solved to find the unknown quantity.

The first final expression above ($0+0=0$) yields no information.

The second final expression ($F_T+-\,F_g=0$) tells us that $F_T=F_g$.

This is the final answer.

2.6.2 Two-Dimensional Translational Equilibrium

The addition of forces in both the horizontal and vertical direction does not change the method outlined in the above section. A box pulled on in four different directions: Up, Down, East, and West, but still at rest, is analyzed with the exact same method as the hanging box.

Step 1: As described in Figure 2.7, the analysis begins by selecting an orientation for the axes for the free-body diagram.

In this case, the +x-direction is defined to be eastward and the +y-direction is defined to be upward. The free-body diagram is completed by sketching the box and the four forces on the box.

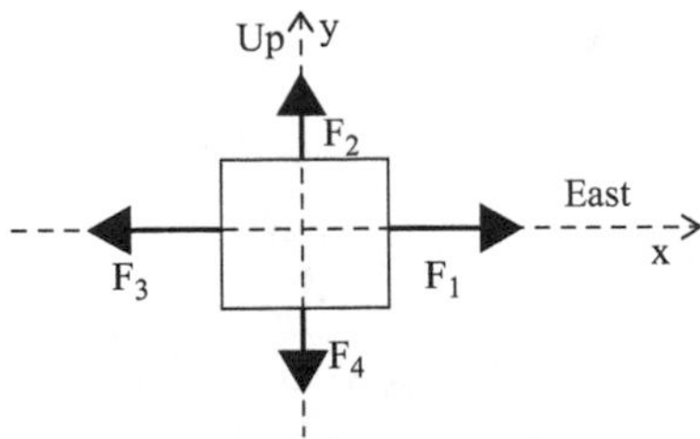

FIGURE 2.7 Two-dimensional free-body diagram.

Step 2: The components of the forces are found by means of the expressions:

$$F_{1x} = F_1 \cos(\theta_{F1}) \qquad F_{1y} = F_1 \sin(\theta_{F1})$$

$$F_{2x} = F_2 \cos(\theta_{F2}) \qquad F_{2y} = F_2 \sin(\theta_{F2})$$

$$F_{3x} = F_3 \cos(\theta_{F3}) \qquad F_{3y} = F_3 \sin(\theta_{F3})$$

$$F_{4x} = F_4 \cos(\theta_{F4}) \qquad F_{4y} = F_4 \sin(\theta_{F4})$$

For the forces in Figure 2.7, the angles to the forces from the +x-axis are: $\theta_{F1}=0°$, $\theta_{F2}=90°$, $\theta_{F3}=180°$, and $\theta_{F4}=270°$. Plugging these angles into the standard expressions for the components of the forces in this situation gives:

$$F_{1x} = F_1 \cos(0°) = F_1 \qquad F_{1y} = F_1 \sin(0°) = 0$$

$$F_{2x} = F_2 \cos(90°) = 0 \qquad F_{2y} = F_2 \sin(90°) = F_2$$

$$F_{3x} = F_3 \cos(180°) = -F_3 \qquad F_{3y} = F_3 \sin(180°) = 0$$

$$F_{4x} = F_4 \cos(270°) = 0 \qquad F_{4y} = F_4 \sin(270°) = -F_4$$

Because the vectors are along the axes, all eight of the results above can be obtained by inspection, for instance, $F_{1x}=F_1$, $F_{2x}=0$, without doing the calculations.

Step 3: Summing the forces in each direction and setting the sum equal to 0 to satisfy the condition for translational equilibrium gives, in the x-direction:

$$F_{1x} + F_{2x} + F_{3x} + F_{4x} = 0 \quad \text{so,} \quad F_1 + 0 + (-F_3) + 0 = 0,$$

and in the y-direction:

$$F_{1y} + F_{2y} + F_{3y} + F_{4y} = 0 \quad \text{so,} \quad 0 + F_2 + 0 + (-F_4) = 0.$$

Solving these two expressions yields the obvious result that the magnitude of F_1 must equal the magnitude of F_3 and the magnitude of F_2 must equal the magnitude of F_4. This technique can be used to find the forces in any arrangement, as demonstrated in the following examples.

2.7 EXAMPLES OF 2-D STATICS

Example 2.1

Find the magnitude and direction of the force exerted by the father in the stalemated tug-of-war, depicted in Figure 2.8.

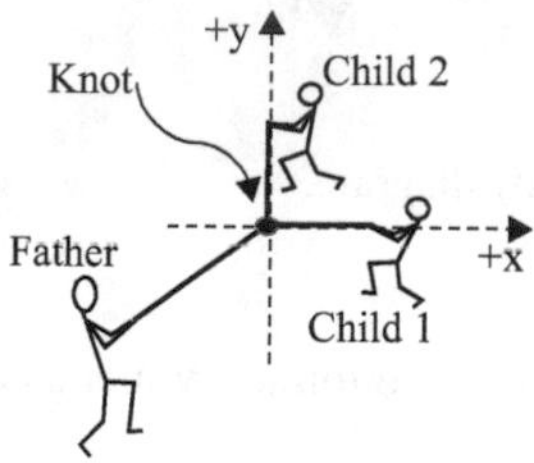

FIGURE 2.8 Diagram for Example 2.1.

Notice that Figure 2.8 is a view from directly above the tug-of-war and the father pulls in a direction in the South-West quadrant of the image, child-1 pulls to the East with a magnitude of 20 N and child-2 pulls North in the +y-direction with a magnitude 10 N. The gravitational force on the knot, where the three rope segments meet, is perpendicular to the forces we are studying, so it is not part of this analysis.

Step 1: Draw the free-body diagram of the forces on the knot, as shown in Figure 2.9.

This free-body diagram is drawn as if the analysis is done from a view directly above the tug-of-war. The additional right triangle and the angle θ_{F3} is not needed in the free-body diagram but will be used to compute the direction of F_3.

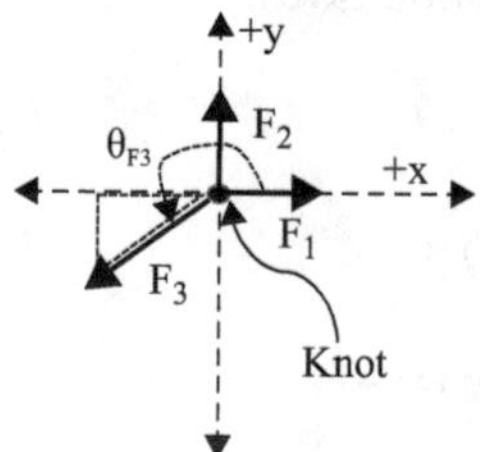

FIGURE 2.9 Free-body diagram with the angle and right triangle needed of Example 2.1.

Step 2: Find the components of each force:

$$F_{1x} = F_1 \cos(0°) = F_1 = 20 \text{ N}$$

$$F_{1y} = F_1 \sin(0°) = 0$$

$$F_{2x} = F_2 \cos(90°) = 0$$

$$F_{2y} = F_2 \sin(90°) = F_2 = 10 \text{ N}$$

$$F_{3x} = ?$$

$$F_{3y} = ?$$

Step 3: Set the sum of the force components along each axis to zero to satisfy the condition for translational equilibrium. In the x-direction:

$$F_{1x} + F_{2x} + F_{3x} = 0$$

$$F_{3x} = -(F_{1x} + F_{2x})$$

$$F_{3x} = -(20\text{ N} + 0)$$

$$F_{3x} = -20\text{ N}$$

and in the y-direction:

$$F_{1y} + F_{2y} + F_{3y} = 0$$

$$F_{3y} = -(F_{1y} + F_{2y})$$

$$F_{3y} = -(0 + 10\text{ N})$$

$$F_{3y} = -10\text{ N}$$

Step 4: Solve for the unknowns.

The magnitude of the force F_3 is:

$$F_3 = \sqrt{F_{3x}^2 + F_{3y}^2} = \sqrt{(-20\text{N})^2 + (-10\text{N})^2} = 22.36\text{N}$$

Thus, the father does not pull with a force that is equal to the algebraic sum (20 N+10 N=30 N) of the forces applied by the two children to stay in equilibrium. The force applied by the father is much smaller than the algebraic sum of the force magnitudes of the children.

To find the direction of the force applied by the father, look back at Figure 2.9. Since the x- and y-components of the father's force are negative, the angle of the force that the father applies is greater than 180° and less than 270°. So, starting with 180°, the angle between the –x-axis and F_3 is found with the right triangle in the diagram using the inverse tangent. Thus, the angle of the force that the father applies is found with the following expression:

$$\theta_{F3} = \tan^{-1}\frac{F_{3y}}{F_{3x}} + 180°.$$

Substituting values with units yields:

$$\theta_{F3} = \tan^{-1}[(-10\text{N})/(-20\text{N})] + 180°$$

$$\theta_{F3} = 206.6°$$

Example 2.2

For the configuration depicted in Figure 2.10, with m=1.00 kg, and ropes A and B making angles of 30° and 50° with the horizontal direction, respectively, find the tensions in the ropes.

Solution:

First: Draw the free-body diagram, Figure 2.11, and label the forces with reasonable variable names.

Note that from Figure 2.10 and the fact that the x-axis and the ceiling are both horizontal, the angles of the force vectors are:

$$\theta_A = 180° - 30° = 150°$$

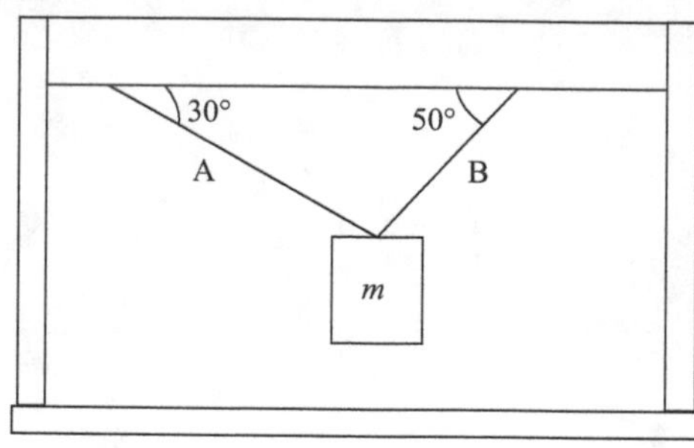

FIGURE 2.10 Block suspended from the ceiling by two ropes, A & B.

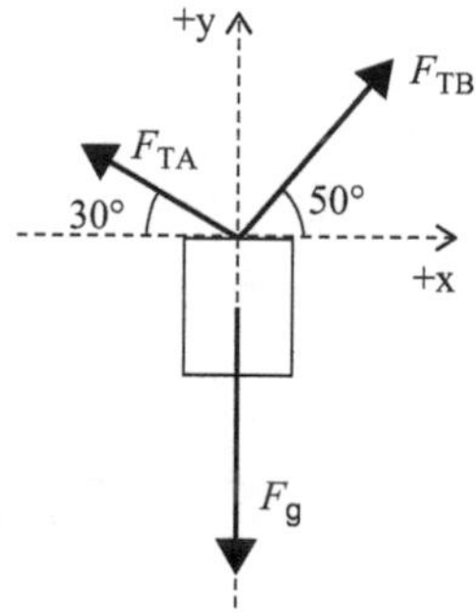

FIGURE 2.11 Free-body diagram of Example 2.2.

$$\theta_B = 50°$$

$$\theta_g = 270°$$

and the magnitude of the gravitational force being exerted by the earth on the block is:

$$F_g = mg = (1\ \text{kg})(9.8\ \text{N/kg}) = 9.8\ \text{N}$$

Second: Find expressions for the components of the vectors.

$$F_{gx} = F_g \cos(\theta_g) = 9.8\ \text{N} \cos(270°) = 0 \qquad F_{gy} = F_g \sin(\theta_g) = 9.8\ \text{N} \sin(270°) = -9.8\ \text{N}$$

$$F_{TAx} = F_{TA} \cos\theta_A \qquad F_{TAy} = F_{TA} \sin\theta_A$$

$$F_{TBx} = F_{TA} \cos\theta_B \qquad F_{TBy} = F_{TB} \sin\theta_B$$

Third: Set the sum of the components equal to zero.

$$F_{TAx} + F_{TBx} + F_{gx} = 0$$
$$F_{TA} \cos\theta_A + F_{TB} \cos\theta_B + 0 = 0 \tag{E.1}$$

$$F_{TAy} + F_{TBy} + F_{gy} = 0$$
$$F_{TA} \sin\theta_A + F_{TB} \sin\theta_B + F_{gy} = 0 \tag{E.2}$$

Fourth: Solve equations (E.1) and (E.2) to find the magnitudes of the tension forces.

For this example, this step is more complicated, but it is important to note that the following solution is still only one step in the process. Sometimes, the algebra is simple and other times it is a bit more complicated, but it is still only one step in the process.

Equation (E.1) can be rearranged to yield an expression for F_{TA} in terms of F_{TB}:

$$F_{TA} = -\frac{\cos\theta_B}{\cos\theta_A} F_{TB} \tag{E.3}$$

Substituting this expression for F_{TA} into equation (E.2) yields:

$$-\frac{\cos\theta_B}{\cos\theta_A} F_{TB} \sin\theta_A + F_{TB} \sin\theta_B + F_{gy} = 0$$

$$F_{TB} = \frac{F_{gy}}{\frac{\cos\theta_B}{\cos\theta_A}\sin\theta_A - \sin\theta_B}$$

$$F_{TB} = \frac{-9.80\text{ N}}{\frac{\cos 50^\circ}{\cos 150^\circ}\sin 150^\circ - \sin 50^\circ}$$

$$F_{TB} = 8.618\text{ N}$$

Substituting this result (and the values of the angles) into equation (E.3) yields:

$$F_{TA} = -\frac{\cos 50^\circ}{\cos 150^\circ} 8.618\text{ N}$$

$$F_{TA} = 6.3965\text{ N}$$

The process of solving a static problem described in this section is summarized in a concept map in Figure 2.12.

Starting with all the forces, Figure 2.12 shows three forces, but there can be any number of forces. Find the components of the forces individually, add all the x-components together and set the sum equal to zero and separately add all the y-components together and set the sum equal to zero.

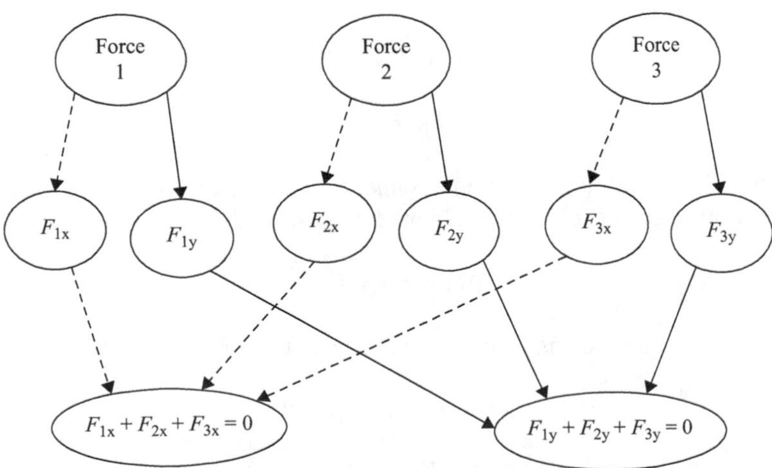

FIGURE 2.12 Concept map of the process of solving a static problem.

2.8 ANSWER TO CHAPTER QUESTION

When a 100 lb. person decides to perform a high wire act, as depicted in Figure 2.1, will a wire that is capable of sustaining a maximum tension equal to their weight be sufficient? If not, is a wire with a strength of twice their weight safe?

The answer is: *Don't count on it!* Let's do the analysis.

First: Draw a free-body diagram of the person on the wire (Figure 2.13) for analysis.

In this free-body diagram, the circle on the origin represents the person, F_g is the weight of the person, which is the gravitational force, and F_{TL} and F_{TR} are the left and right tensions, respectively. The angles to the vectors are from the +x-direction:

$$\theta_L = \theta, \theta_R = 180° - \theta, \theta_g = 270°.$$

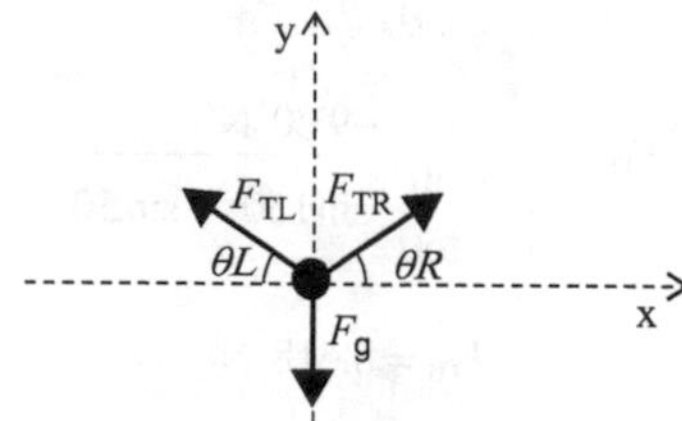

FIGURE 2.13 Free-body diagram of the person on the wire.

Second: Find the components of $\vec{\mathbf{F}}_{TL}$, $\vec{\mathbf{F}}_{TR}$, and $\vec{\mathbf{F}}_g$

$$F_{gx} = F_g \cos(\theta_g) = (10 \text{ lbs.})\cos(180°) = 0 \qquad F_{gy} = F_g \sin(\theta_g) = -F_g \sin(270°) = -F_g$$

$$F_{TLx} = F_{TL} \cos(\theta) \qquad F_{TLy} = F_{TL} \sin(\theta)$$

$$F_{TRx} = F_{TR} \cos(180° - \theta) = -F_{TR} \cos(\theta) \qquad F_{TRy} = F_{TR} \sin(180° - \theta) = F_{TR} \sin(\theta)$$

Third: Set the sum of the components of the forces equal to zero and solve.

Start with the x-components:

$$F_{TLx} + F_{TRx} + F_{gx} = 0$$

$$F_{TL} \cos(\theta) + [-F_{TR} \cos(\theta)] + 0 = 0$$

$$F_{TL} = F_{TR}$$

Having found that both tensions are the same, we will use the symbol F_T to represent the tension in both segments of the wire. In other words, we define F_T such that:

$$F_T = F_{TL} = F_{TR}$$

Fourth: Solve for the unknowns. The y-components of the forces:

$$F_{TLy} + F_{TRy} + F_{gy} = 0$$

$$F_{TL} \sin(\theta) + F_{TR} \sin(\theta) + (-F_g) = 0$$

Substituting F_T for both F_{TL} and F_{TR} yields:

$$F_T \sin(\theta) + F_T \sin(\theta) + (-F_g) = 0$$

$$F_T = \frac{F_g}{2\sin\theta} = \frac{100 \text{ lbs.}}{2\sin\theta}$$

Thus, if sin $\theta<0.5$, then $F_T>100$ lb. This occurs when $\theta<30°$. This seems likely for a tightly stretched wire with a person walking across it. So, make sure the wire can support more than the weight of the person walking across it or the angle of the wire is $\theta\geq30°$.

2.9 CHAPTER QUESTIONS AND PROBLEMS

2.9.1 Multiple-Choice Questions

1. A man is pulling on one end of a rope tied to a donkey with a force of magnitude 50 N and the donkey is pulling on the other end of the rope with a force of magnitude 50 N, so the rope is in static equilibrium. Thus, the tension in the rope is:
   A. 0 N B. 50 N C. 100 N

2. Which cable (A or B) is under greater tension in Figure 2.14?
   A. A B. B C. Neither. The tensions are equal.

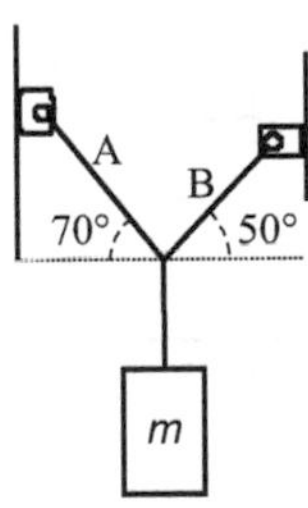

FIGURE 2.14 Multiple-choice question 2.

3. What is the tension in the cord tied between the boxes m_1 and m_2 that is strung over the pulley, in Figure 2.15, given that each box weighs 10 N?
   A. 0 N B. 10 N C. 20 N D. None of the other answers is correct.

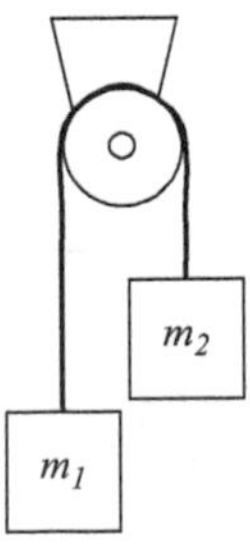

FIGURE 2.15 Multiple-choice question 3.

4. In Figure 2.16, block 1 has a weight of 100 N and block 2 has a weight of 40 N. Consider the pulley to be frictionless and the cable to be massless. Block 1 rests on the ground and block 2 is suspended by the cable that passes over the pulley and is connected to the top of each mass. What is the tension in the cord tied between the boxes?
   A. 0 N B. 40 N C. 60 N D. 100 N E. 140

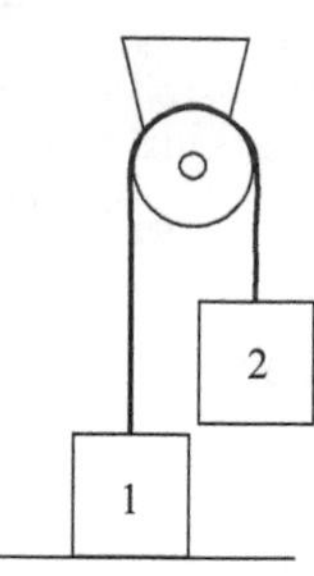

FIGURE 2.16 Multiple-choice question 4.

5. In Figure 2.17, which cable has the greatest tension: 1, 2, or 3?
   A. 1 B. 2 C. 3

6. In Figure 2.17, which cable has the lowest tension: 1, 2, or 3?
   A. 1 B. 2 C. 3

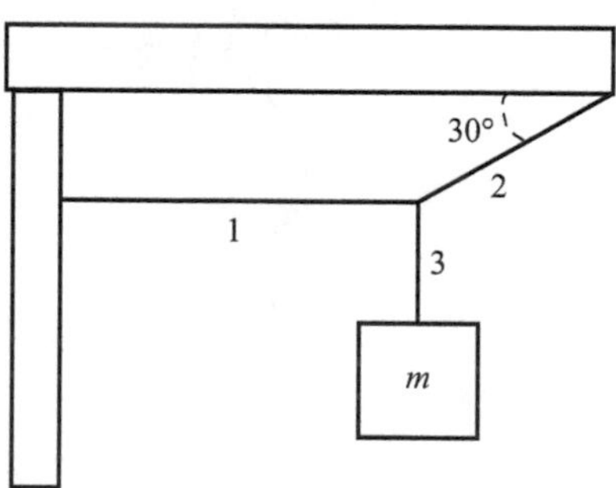

FIGURE 2.17 Multiple-choice questions 5 & 6.

7. Which of the cables in Figure 2.18 has the least tension (A, B, or C)?
   A. A B. B C. C

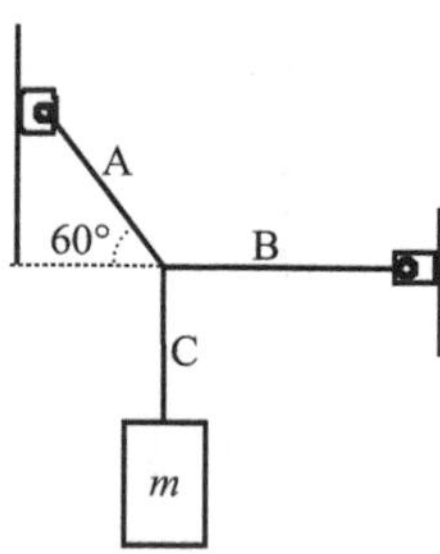

FIGURE 2.18 Multiple-choice question 18.

Questions 8–11 refer to Figure 2.19. The angles of the ropes to the horizontal direction are given as 50° & 40° and the weight of the hanging cube is 50 N.

8. Which cable is under a greater tension (A or B)?
   A. A B. B C. Equal Tension

9. Which cable is under a smaller tension (A or B)?
   A. A B. B C. Equal Tension

10. If the angle of 50° is adjusted to 40°, both cables will have the same tension.
    A. True B. False

11. As compared to the original arrangement, if the right end of cable B is moved so that cable B is perfectly horizontal, will the tension in cable A increase, decrease, or remain the same?
    A. increase B. decrease C. remain the same

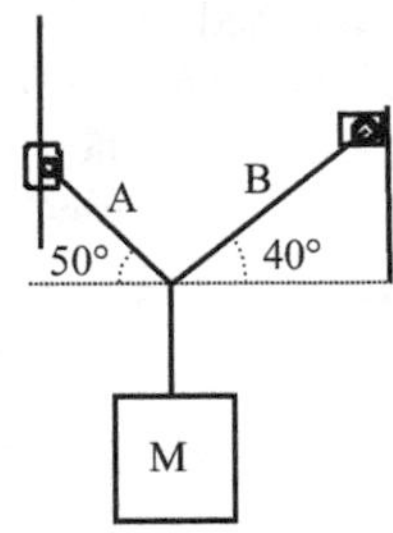

FIGURE 2.19 Multiple-choice questions 8–11 and problem 7.

2.9.2 Problems

1. Find the magnitude and direction of $\vec{\mathbf{F}} = 3.00\ \text{N}\ \hat{\mathbf{i}} + 4.00\ \text{N}\ \hat{\mathbf{j}}$
2. Find the magnitude and direction of $\vec{\mathbf{F}} = 3.00\ \text{N}\ \hat{\mathbf{i}} - 4.00\ \text{N}\ \hat{\mathbf{j}}$
3. A force vector, $\vec{\mathbf{F}}$, has x- and y-components of $F_x = 1.74\,\text{N}$ and $F_y = -9.84\,\text{N}$. What is its approximate magnitude and angle?
4. Two forces act on an object. An x-y coordinate system has been established within which two of the forces can be expressed as:

   $\vec{\mathbf{F}}_A = 5.0$ N at 140 and $\vec{\mathbf{F}}_B = 12.0$ N at 53. Find the net force acting on the object.

5. Three forces act on an object. The object is not necessarily in equilibrium (meaning we cannot assume that the net force is zero). An x-y coordinate system has been established and in that system, the three forces are: $\vec{\mathbf{F}}_1 = 3\ \text{N}\ \hat{\imath}$, $\vec{\mathbf{F}}_2 = 5\ \text{N}\ \hat{\jmath}$, and $\vec{\mathbf{F}}_3 = 6\ \text{N}\ \hat{\imath}$. Compute the magnitude of the net force on the object.
6. Three forces act on an object. An x-y coordinate system has been established within which two of the forces can be expressed as:

   $$\vec{\mathbf{F}}_1 = 3.00\ \text{N}\ \hat{\mathbf{i}} + 5.00\ \text{N}\ \hat{\mathbf{j}} \text{ and } \vec{\mathbf{F}}_2 = 4.00\text{N}\ \hat{\mathbf{i}} - 7.00\ \text{N}\ \hat{\mathbf{j}}$$

   The net force acting on the object is zero. Find the third force.

7. In Figure 2.19, the angles of the ropes relative to the horizontal direction are given as 50° & 40° and the mass of the hanging cube is $M = 5\,\text{kg}$. Find the tension in cable A.
8. Find the mass of the object hanging in Figure 2.20 if the tension in rope B is 2 N.

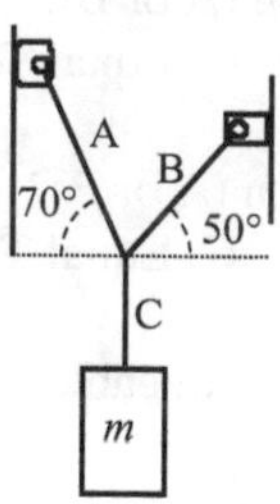

FIGURE 2.20 Problem 8.

9. The system depicted in Figure 2.21 is in equilibrium. All angles are given in degrees, and the support structure is made up of a horizontal top bar and a vertical side. The string at $\theta = 60°$ to the horizontal top bar is string-1, the string connected to the right vertical bar is string-2, and the vertical string attached to the mass is string-3. The mass (m) of the hanging mass is 4 kg and the angles θ and ϕ are 60° and 70°, respectively, as depicted in the diagram to the right. Neglect the mass of the ropes. For the situation depicted in the diagram to the right, compute the tension in string-1.

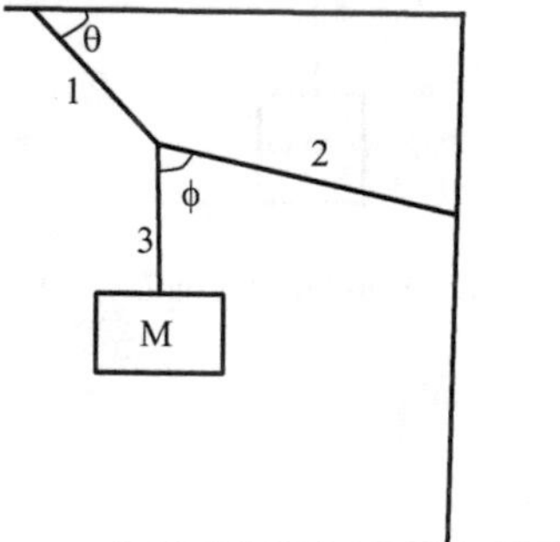

FIGURE 2.21 Problem 9.

10. The system depicted in Figure 2.22 is in equilibrium. All angles in the problem are given in degrees and the ceiling is horizontal. As demonstrated in the diagram to the right, the right end of rope-1 makes an angle of $\theta_1 = 55°$ while the ceiling and the left end of rope-2 makes an angle of $\theta_2 = 25°$ with the ceiling. The weight of the mass M is 500 N. For the situation depicted in the diagram to the right, find the tension in rope-1.

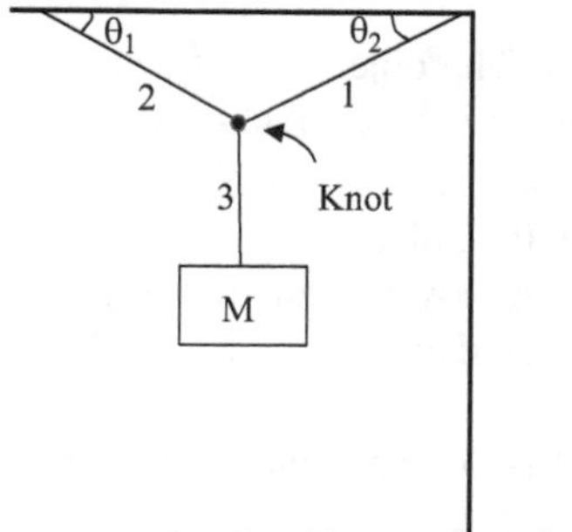

FIGURE 2.22 Problem 10.

3 Torque and Rotational Equilibrium

3.1 INTRODUCTION

In some situations, a force is applied using a tool like a wrench or a screwdriver and a twisting force, called a torque, is produced. These twisting forces are central to the understanding and analysis of some objects in equilibrium. In this chapter, the concepts of torque will be studied and then applied to the analysis of systems in rotational equilibrium.

Chapter question: A person suspends a block of mass m from his or her hand, as depicted in Figure 3.1.

Assuming their forearm has a mass m_A, is the force of tension in their bicep muscle equal to, greater than, or less than the sum of the weight of their arm and the weight of the object? To answer this question, the concepts of torque and rotational equilibrium will be developed and they will be applied at the end of this chapter to answer this question.

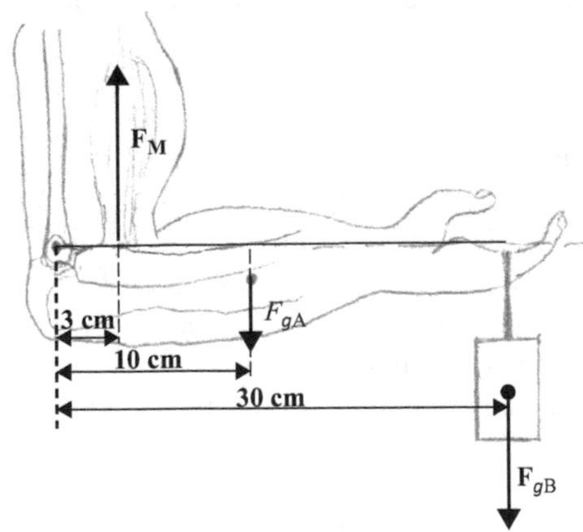

FIGURE 3.1 Diagram of an arm.

3.2 TORQUE

Torque is a twisting force, which is denoted by the Greek letter tau (τ). When opening the lid of a jar, the more force you apply the more twisting force (torque) you apply to open the jar. As you may also know from experience, if you are trying to loosen or tighten a bolt with a wrench, the harder you pull on the wrench, the force (F), the more torque there is on the bolt. But sometimes the bolt will not loosen no matter how hard you pull on the wrench. To loosen the bolt you can get a longer wrench or even slip a pipe on the end of the wrench. So, it is clear that the torque depends not only on the force applied but also on the distance away from the rotation point at which the force is applied.

In the previous example, the angle between the force and the distance away from the rotation point is assumed to be perpendicular as shown in Figure 3.2, but that is not always the case. So, the

DOI: 10.1201/9781003308065-3

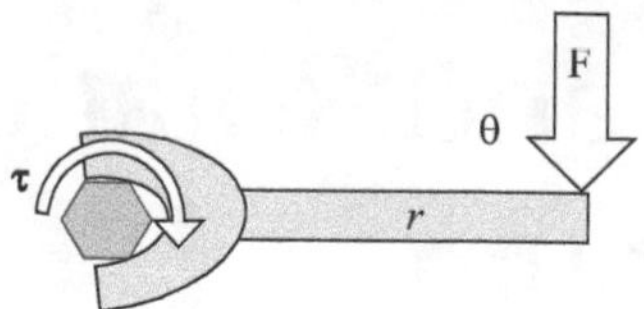

FIGURE 3.2 A wrench used to tighten a bolt.

torque (τ) is defined as the product of the applied force (F) and the perpendicular distance ($r_\perp$) from the location of the applied force and the rotation point, and given in equation (3.1) as:

$$\tau = r_\perp F \tag{3.1}$$

The perpendicular distance $r_\perp$ is known as the moment arm of the force $\vec{\mathbf{F}}$ that produces the torque, $\vec{\boldsymbol{\tau}}$.

For a more general definition of torque that takes the angle of the force into consideration, start with Figure 3.3, in which the rotation point is the center of the bolt and the force is applied to the right end of the wrench in the direction indicated by the force vector ($\vec{\mathbf{F}}$).

The distance r is the magnitude of the position vector, relative to the rotation point, of the point of application of the force. From the right triangle in the diagram we see that $r_\perp = r\,|\sin(\theta)|$ so the magnitude of the torque can be expressed as shown in equation (3.2) as:

$$\tau = r\ F\,|\sin(\theta)| \tag{3.2}$$

where θ is the angle between $\vec{\mathbf{r}}$, the position vector of the point at which the force is applied, relative to the point about which you are calculating the torque, and, $\vec{\mathbf{F}}$, the force vector, when the two vectors are placed tail to tail. Remember it is the direction that the vector is pointing that determines its angle, so the angle of the Force vector, $\vec{\mathbf{F}}$, in Figure 3.3 is θ. The angle θ is measured counterclockwise, as viewed from a point on the +z-axis, from the direction of $\vec{\mathbf{r}}$ to the direction of $\vec{\mathbf{F}}$. In a case in which an x-y-z coordinate system has been defined, θ_r is the angle the position vector makes with the +x-direction, θ_F is the angle the force vector makes with the +x-direction, and τ_z can be expressed as equation (3.3) as:

$$\tau_z = r\ F \sin(\theta_F - \theta_r) \tag{3.3}$$

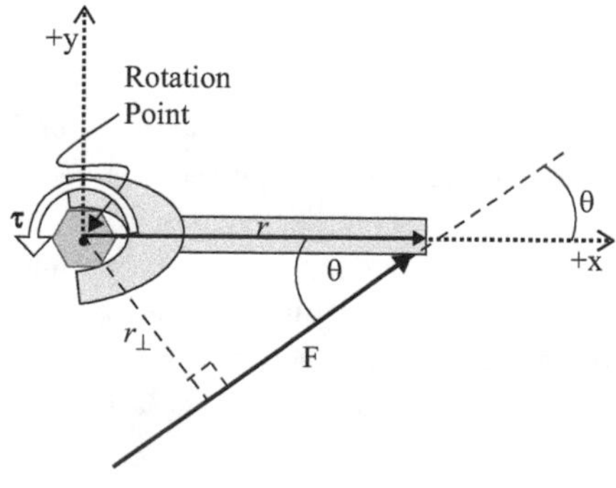

FIGURE 3.3 Definition of torque.

3.2.1 Direction of the Torque

For the two-dimensional systems that will be analyzed in this text, the direction of the torque is clockwise or counterclockwise as viewed from above the plane created by the two vectors of force $\vec{\mathbf{F}}$ and the distance $\vec{\mathbf{r}}$. The torque in Figure 3.2 is counterclockwise as viewed from a position along the +z-axis. (The +z-axis is not shown but is directed out of the page in Figure 3.2.) To make it easier to communicate the direction of the torque, it is said that if observed from a position along the +z-axis a counter clockwise torque is in the +z-direction and a clockwise torque is in the −z-direction. The right-hand rule (RHR) provides a way to remember this labeling of the sign of the z-component of the torque. If you point the thumb of your right hand at your nose, your fingers can rotate in a counter clockwise direction and if you point the thumb of your right hand directly away from your nose your fingers can rotate in a clockwise direction. See Figure 3.4 for an example of a +z torque, due to a rotation that is counter clockwise as seen by someone looking down from a point above the rotation on the +z-axis.

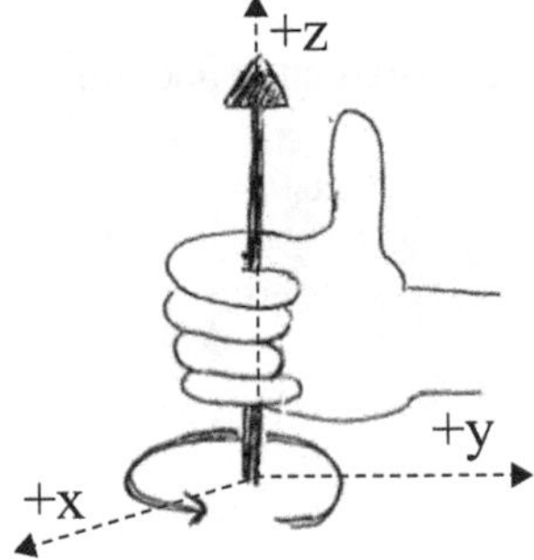

FIGURE 3.4 Right-hand rule.

Thus, the torque vector due to the force in Figure 2.3 is in the +z-direction (out of the page). This sign of direction (+z or −z) will be in agreement with the z-component of the torque, as calculated with equation (3.3).

3.2.2 Examples of Calculating Torques

A wrench is used to loosen or tighten a bolt. For each of the following examples, the distance r from the center of the bolt to the point of application of the force is 25 cm (which is 0.25 m) and the magnitude of the force applied at the end of the wrench is 10 N. In each example, the quantity to be computed is the z-component of the torque.

Example 3.1

See Figure 3.5, the force is in the +y-direction and the vector $\vec{\mathbf{r}}$ is in the +x-direction.

So,

$$\theta_F = 90° \text{ and } \theta_r = 0°, \text{ thus:}$$

$$\tau_z = r\,F\sin(\theta_F - \theta_r)$$

$$\tau_z = (0.25\text{ m})(10\text{ N})\sin(90° - 0°)$$

$$\tau_z = +2.50\text{ Nm}.$$

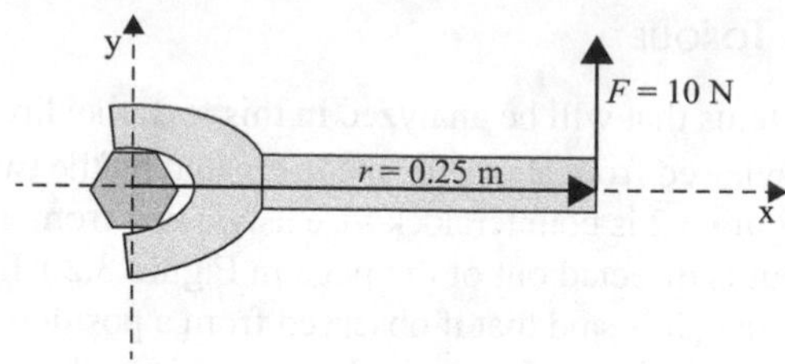

FIGURE 3.5 Torque Example 3.1.

RHR: To rotate the fingers of your right hand in the direction of the twisting, your thumb must be pointing directly away from the page.

Example 3.2

See Figure 3.6, the force is in the +y- & –x-direction and the vector $\vec{\mathbf{r}}$ is in the +x-direction, so

$$\theta_F = 90° \text{ and } \theta_r = 0°, \text{ thus:}$$

$$\tau_z = r\ F \sin(\theta_F - \theta_r)$$

$$\tau_z = (0.25 \text{ m})(10 \text{ N})\sin(110° - 0°)$$

$$\tau_z = +2.35 \text{ Nm.}$$

RHR: Same RHR explanation as Example 1.

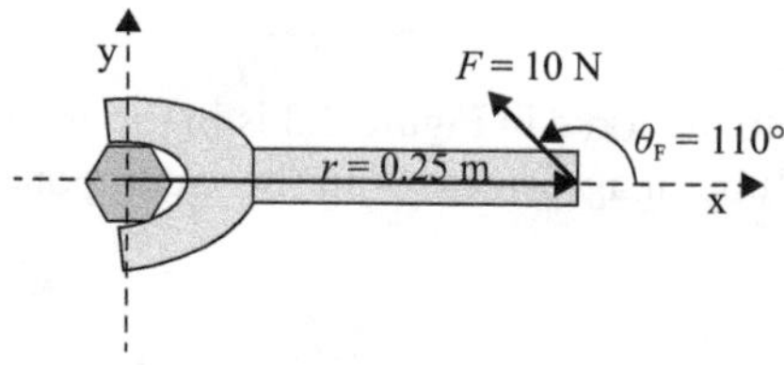

FIGURE 3.6 Torque Example 3.2.

Example 3.3

See Figure 3.7, the force is in the –x and +y directions and the vector $\vec{\mathbf{r}}$ is in the +x and +y directions at 40° from the +x-axis, so

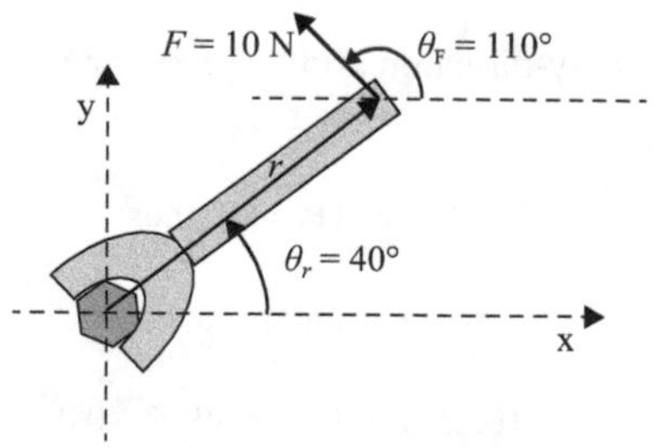

FIGURE 3.7 A wrench used to loosen a bolt.

$$\theta_F = 110° \text{ and } \theta_r = 40°, \text{thus:}$$

$$\tau_z = r\ F \sin(\theta_F - \theta_r)$$

$$\tau_z = (0.25\ \text{m})(10\ \text{N})\sin(110° - 40°)$$

$$\tau_z = +2.35\ \text{Nm}.$$

RHR: Same RHR explanation as Example 1.

Example 3.4

See Figure 3.8, the force is in the –y-direction and the vector $\vec{\mathbf{r}}$ is in the +x and the +y directions at 40° from the +x-axis, so

$$\theta_F = 270° \text{ and } \theta_r = 40°, \text{thus:}$$

$$\tau_z = r\ F \sin(\theta_F - \theta_r)$$

$$\tau_z = (0.25\ \text{m})(10\ \text{N})\sin(270° - 40°)$$

$$\tau_z = -1.92\ \text{Nm}.$$

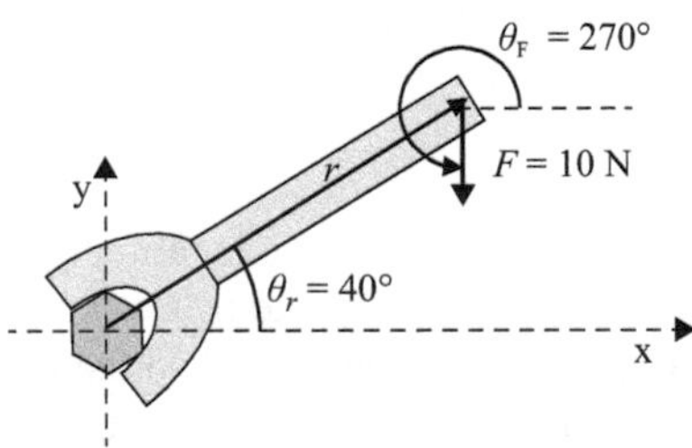

FIGURE 3.8 Torque Example 3.4.

RHR: To rotate the fingers of your right hand in the direction of the twisting, your thumb must be pointing directly into the page. This negative value for the z-component of the torque vector means that the torque vector is directed into the page.

Example 3.5

See Figure 3.9, the force is in the –y-direction and the vector $\vec{\mathbf{r}}$ is in the –x-direction, so, $\theta_F = 270°$ and $\theta_r = 180°$, thus:

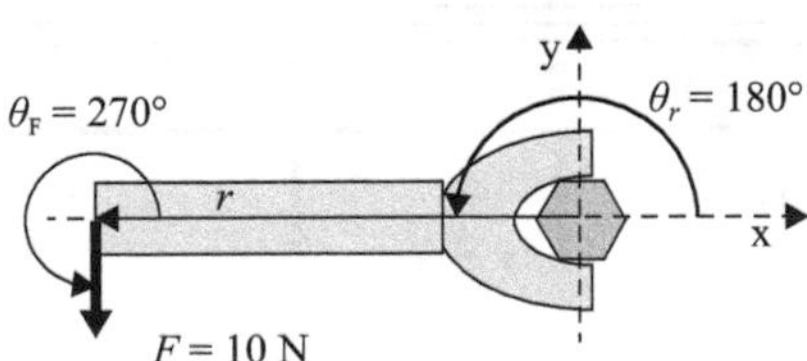

FIGURE 3.9 Torque Example 3.5.

$$\tau_z = r\ F \sin(\theta_F - \theta_r)$$

$$\tau_z = (0.25\ \text{m})(10\ \text{N})\sin(270° - 40°)$$

$$\tau_z = +2.50\ \text{Nm}.$$

RHR: To rotate your fingers of your right hand in the direction of the twisting your thumb must be pointing directly out of the page. This positive value for the z-component of the torque vector means that the torque vector is directed out of the page.

3.3 NET TORQUE

Like the net force, the *net torque* on an object is the sum of the torques on an object and is given in equation (3.4) as:

$$\vec{\tau}_{\text{Net}} = \sum \vec{\tau} \tag{3.4}$$

If the forces and relative displacement all lie in the xy plane, as will be the case for the problems you will be asked to solve, the torques will all be along the z-axis. Thus, the net torque will be along the z-axis and of the three scalar equations that the vector equation $\vec{\tau}_{\text{Net}} = \sum \vec{\tau}$ implies, namely, $\tau_{\text{Net, x}} = \Sigma\ \tau_x$, $\tau_{\text{Net, y}} = \Sigma\ \tau_y$, and $\tau_{\text{Net, z}} = \Sigma\ \tau_z$, the one that will matter is the z-component given in equation (3.5) as

$$\tau_{\text{Net, z}} = \sum \tau_z \tag{3.5}$$

3.4 CENTER OF MASS

All extended objects have a center of mass. This is the location at which the object can be balanced and it is the location about which the object will rotate if it was tossed in the air. It is also the location at which all the mass of the object can be considered so the torque due the weight of the object can be computed. For a uniform bar, its center of mass is at the center of the bar. It should not be surprising that you would put your finger at the 50 cm mark on a meter stick if you were trying to balance that meter stick.

The reason for this can be understood by dividing a bar of length (L) into equal parts. In Figure 3.10, the bar is divided into four equal segments, so that each segment has a weight of one-quarter of the total weight (F_g). So the sum of the torques, using equation (3.5), for this bar due to the weight of each segment is:

$$\tau_{\text{Net, z}} = (F_g/4)r_1 + (F_g/4)r_2 + (F_g/4)r_3 + (F_g/4)r_4$$

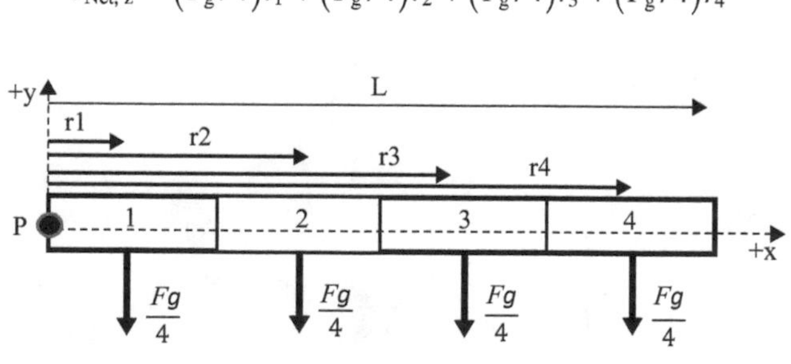

FIGURE 3.10 Center of mass of a uniform bar.

$$\tau_{\text{Net, z}} = (F_g/4)[(L/8)+(3L/8)+(5L/8)+(7L/8)]$$

$$\tau_{\text{Net, z}} = (F_g/4)\ [(L/8)+(3L/8)+(5L/8)+(7L/8)]$$

$$\tau_{\text{Net, z}} = (F_g/4)\ [(16L/8)]$$

$$\tau_{\text{Net, z}} = [F_g(L/2)]$$

$$\tau_{\text{Net, z}} = [\text{mg}(L/2)]$$

Thus, as shown in Figure 3.11, if an object with a mass, m, is placed at the end of a light weight rod a distance (L/2) away from the rotation point, P, the torque about the left end of the rod due to the weight of the mass will be the same as the torque about the left end of a uniform bar of mass m and length L, due to the weight of the bar.

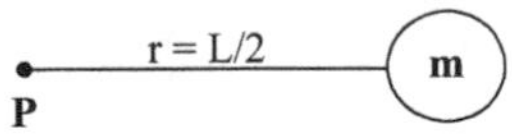

r = L/2
m
P

FIGURE 3.11 An object of mass *m* at the end of the low-mass rod of length (L/2).

No matter how many equal parts a uniform bar is divided up into the center of mass will always be at the center. For objects that are not uniform, the center of mass will need to be measured or if the function that described the object is known an integral needs to be used in a process like the one described for the uniform bar, but with different values for each part of the object. It cannot be assumed that the center of mass is at the center of the object that is not uniform.

3.5 ROTATIONAL EQUILIBRIUM

In Chapter 2, translational equilibrium was defined as the condition in which all the forces in each coordinate direction add to zero. In some situations, translational equilibrium is not enough to explain what is happening. For example, if the bar in Figure 3.12 is 10 m long and pinned at its center, it is free to rotate in the x-y plane about its center, and 10 N forces are applied to each end as shown in the figure, the bar would be in translational equilibrium, but it would not be at rest.

It is obvious that the bar would rotate counterclockwise about the center. Thus, to understand this situation, another condition is needed. This additional condition for this system is rotational equilibrium.

Similar to translational equilibrium of forces, rotational equilibrium is the sum of all torques adding up to zero, this can be written mathematically in equation (3.6) as

$$\vec{\tau}_{\text{Net}} = 0, \text{which means that } \Sigma\vec{\tau} = 0 \tag{3.6}$$

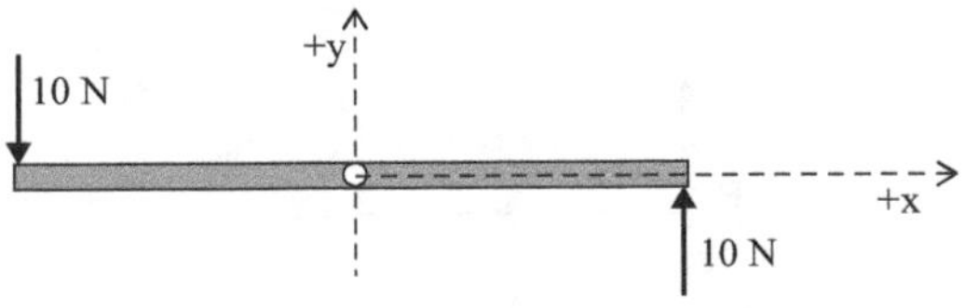

FIGURE 3.12 Bar in translational equilibrium but not in rotational equilibrium.

which states that the sum of the torques acting on an object in equilibrium must equal zero. For the problems in this text, all the forces will lie in the x-y plane, so all the torques will be parallel to the z-axis so the relevant torque component equation for rotational equilibrium will be given by equation (3.7) as:

$$\tau_{\text{Net, z}} = 0 \text{ which means that} \sum \tau_z = 0 \tag{3.7}$$

which states that the sum of the z-components of the torques acting on an object in equilibrium must equal zero.

3.5.1 Rotational Equilibrium Concept Map

In Figure 3.13, a concept map of the process of solving a problem in rotational equilibrium is presented. In this concept map, it is described that the magnitude and sign of z-components of each of the torques must be found and then the sum of these z-components of torques are added and the unknown is found so that the sum of the z-components of the torques add to zero.

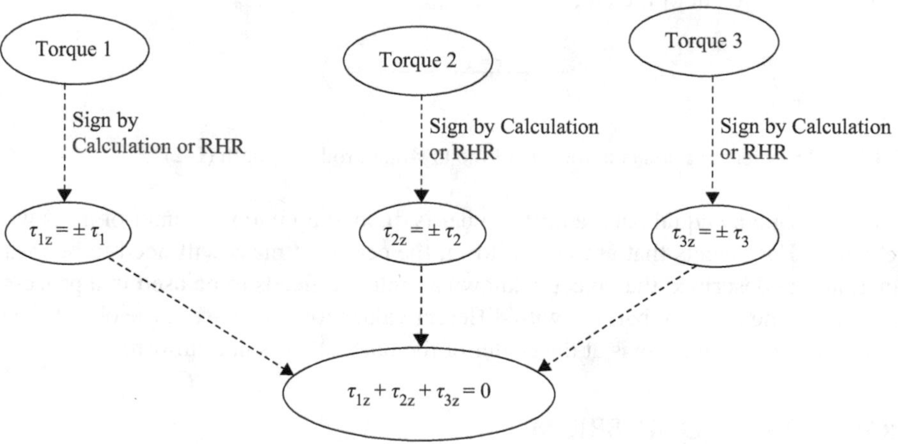

FIGURE 3.13 Concept map of rotational equilibrium.

3.5.2 Rotational Equilibrium Examples

Example 3.1

A uniform bar that is 4 m long and weighs 10 N (as depicted in Figure 3.14) is supported horizontally by two vertical ropes, one at each end of the bar. What is the tension in the ropes? Note that for a uniform bar, the beam weight is considered to act at the center of the bar. This location is the *center of mass* of the bar. The forces F_{T1} and F_{T2} can be found with the conditions of translational and rotational equilibrium:

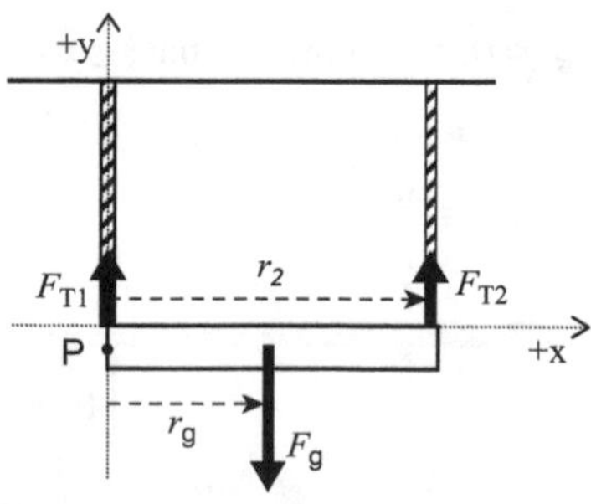

FIGURE 3.14 Free-body diagram for Example 3.1.

Step 1: Draw the free-body diagram for this example. Note that the forces and relative locations (r's) are sketched on the actual diagram in Figure 3.14 for this example.
Step 2: By inspection, the y-components of the forces are:

$$F_{\text{T1y}} = F_{\text{T1}},\ F_{\text{gy}} = -F_{\text{g}},\ \text{and}\ F_{\text{T2y}} = F_{\text{T2}}$$

Step 3.A: Employ the translational equilibrium equation for the y-components:

$$F_{\text{Net,y}} = 0$$

$$F_{\text{T1y}} + F_{\text{gy}} + F_{\text{T2y}} = 0$$

$$F_{\text{T1}} + (-F_{\text{g}}) + F_{\text{T2}} = 0$$

$$F_{\text{T1}} + F_{\text{T2}} = F_{\text{g}}$$

F_g is known so we have two unknowns and so far, just one equation. Our other equation comes from rotational equilibrium.
Step 3.B: Apply rotational equilibrium.
A rotation point must be chosen. For this example, there is no clear rotation point, so point P on the far left-hand side, where rope 1 is attached, is chosen as the rotation point. About P=$\Sigma\tau_z$=0. Notice that $\vec{\mathbf{F}}_{\text{T1}}$ acts on the bar at the rotation point P so it creates no torque about P.

$$\tau_{2z} + \tau_{gz} = 0$$

$$F_{\text{T2}}\ r_2 \sin(\theta_{\text{F2}} - \theta_{\text{r2}}) + F_{\text{g}} r_{\text{g}} \sin(\theta_{\text{Fg}} - \theta_{\text{rg}}) = 0$$

$$F_{\text{T2}} = \frac{-F_{\text{g}} r_{\text{g}} \sin\left(\theta_{\text{Fg}} - \theta_{\text{rg}}\right)}{r_2 \sin\left(\theta_{\text{F2}} - \theta_{\text{r2}}\right)}$$

$$F_{\text{T2}} = \frac{-10\text{N}(2\text{m})\sin(270° - 0°)}{4\text{m}\ \sin(90° - 0°)}$$

$$F_{\text{T2}} = 5\ \text{N}.$$

Step 4: Plugging this back into the expression for translational equilibrium gives:

$$F_{\text{T1}} = 5\ \text{N}.$$

Example 3.2

A uniform bar that is 4 m long and weighs 10 N (as depicted in Figure 3.15) is supported horizontally by two vertical ropes, one at each end of the bar. This time, a block weighing 5 N is placed so that its center is located 1 m to the left of the right end of the bar. What is the tension in each rope?

Again, the beam weight is considered to act at the center of the bar and the weight of the 5 N block is at the center of mass of the block, 1 m from the end of the bar.
Step 1: Generate a free-body diagram of the combination object consisting of the bar and the block is drawn. Please see Figure 3.15.
Step 2: Find the components of the forces:
By inspection, F_{T1y}=F_{T1}, F_{gy}=$-F_g$, F_{gBy}=$-F_{gB}$, and F_{T2y}=F_{T2}.

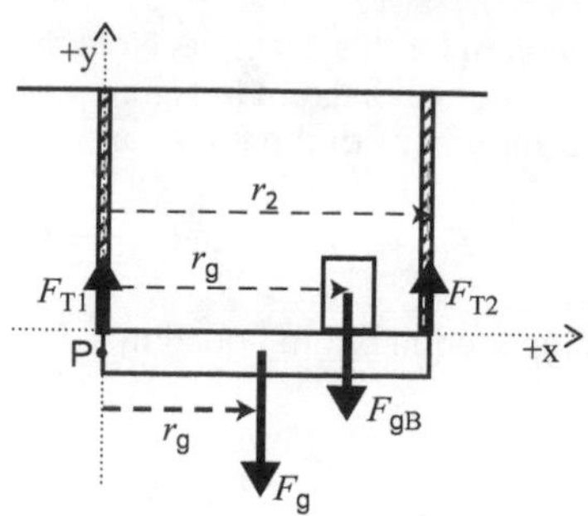

FIGURE 3.15 Free-body diagram for Example 3.2.

Step 3A: Employ the translational equilibrium equation:

$$F_{\text{Net, y}} = 0$$

$$F_{\text{T1y}} + F_{\text{gy}} + F_{\text{gBy}} + F_{\text{T2y}} = 0$$

$$F_{\text{T1}} + (-F_{\text{g}}) + (-F_{\text{gB}}) + F_{\text{T2}} = 0$$

$$F_{\text{T1}} + F_{\text{T2}} = F_{\text{g}} + F_{\text{gB}}$$

F_g and F_{gB} are known, so we have two unknowns (F_{T1} and F_{T2}) and so far, just one equation.
Step 3B: Apply rotational equilibrium.

First, a rotation point must be chosen. For this example, point P on the far left-hand side, where rope 1 is attached, is chosen as the rotation point. About P : Σ_z=0. Notice that $\vec{\mathbf{F}}_{T1}$ is at the rotation point P, so it creates no torque.

$$\tau_{2z} + \tau_{gz} + \tau_{gBz} = 0$$

$$F_{\text{T2}}\ r_2\ \sin(\theta_{\text{FT2}} - \theta_{\text{r2}}) + F_{\text{g}} r_{\text{g}} \sin(\theta_{\text{Fg}} - \theta_{\text{rg}}) + F_{\text{gB}} r_{\text{gB}} \sin(\theta_{\text{FgB}} - \theta_{\text{rgB}}) = 0$$

$$F_{\text{T2}} = -\frac{F_{\text{g}} r_{\text{g}} \sin\left(\theta_{\text{Fg}} - \theta_{\text{rg}}\right) + F_{\text{gB}} r_{\text{gB}} \sin\left(\theta_{\text{FgB}} - \theta_{\text{rgB}}\right)}{r_2 \sin\left(\theta_{\text{F2}} - \theta_{\text{r2}}\right)}$$

$$T_2 = -\frac{10\text{N (2m) } \sin(270° - 0°) + 5\text{N (3m) } \sin(270° - 0°)}{4\text{m} \sin(90° - 0°)} \lim_{x \to \infty}$$

$$F_{\text{T2}} = -\frac{10\text{N}(2\text{m})\sin(270° - 0°) + 5\text{N}(3\text{m})\sin(270° - 0°)}{4\text{m}\sin(90° - 0°)}$$

$$F_{\text{T2}} = 8.75 \text{ N.}$$

Step 4: Plugging this back into the expression for translational equilibrium gives F_{T1}=6.25 N.

Example 3.3

The 10 kg sign of a restaurant is supported by a uniform horizontal beam of length 6 m and mass 2 kg, as depicted in Figure 3.16.

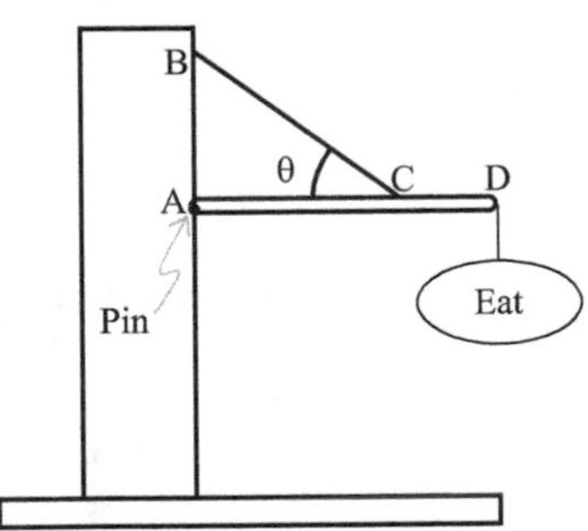

FIGURE 3.16 Example 3.3.

The distance from A to B is 3 m and the horizontal distance from A to C is 4 m. Calculate the tension in the cable that extends from B to C and the components of the force being exerted on the beam by the pin at A. Neglect the thickness of the beam.

Solution:

Given that the mass of sign is m_S=10 kg and the mass of beam: m=2 kg

Note: The right triangle (ABC) [$\tan^{-1}(\theta)=(3/4)$], $\theta=36.87°$.

Step 1: Draw the free-body diagram of the beam: Figure 3.17 is a separate free-body diagram for this example.

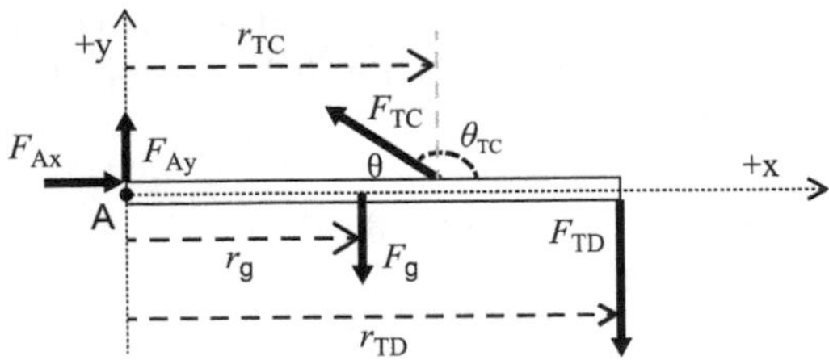

FIGURE 3.17 Free-body diagram for Example 3.3.

Step 2: Write expressions for the components of the forces:

$$F_{TD} = F_{g(sign)} = m_S g = (10\text{ kg})\ (9.8\text{ N/kg}) = 98\text{ N}$$

$$F_{TCx} = F_{TC}\cos\theta_{TC}, F_{gx} = 0, F_{TDx} = 0$$

$$F_{TCy} = F_{TC}\sin\theta_{TC}, F_{gy} = -F_g, F_{TDy} = -F_{TD}$$

Step 3A. Translational equilibrium is applied.

Set the sum of the x-components of the forces equal to zero.

$$F_{Ax} + F_{TCx} + F_{gx} + F_{TDx} = 0$$

$$F_{Ax} + F_{TC}\cos\theta_{TC} + 0 + 0 = 0$$

Set the sum of the y-components of the forces equal to zero.

$$F_{Ay} + F_{TCy} + F_{gy} + F_{TDy} = 0$$

$$F_{Ay} + F_{TC}\sin\theta_{TC} + (-F_g) + (-F_{TD}) = 0$$

The two equations that we get from translational equilibrium have a total of three unknowns (F_{Ax}, F_{Ay}, and F_{TC}), so we can't solve for the unknowns yet. We must apply the concept of rotational equilibrium at this point.

Step 3B: Apply Rotational Equilibrium:

Set the sum of the z-components of torques about point A equal to zero (recognizing that the force of the pin at A creates no torque about A because that force acts on the beam at point A).

$$\tau_{TCz} + \tau_{gz} + \tau_{Dz} = 0 \text{ Nm}$$

$$F_{TC} r_{TC} \sin(\theta_{FTC} - \theta_{rTC}) + F_{TD} r_{TD} \sin(\theta_{FTD} - \theta_{rTD}) + F_g r_g \sin(\theta_{Fg} - \theta_{rg}) = 0 \text{ Nm}$$

With the three equations from the static in the x-direction, the y-direction, and the rotational equilibrium and the information from the question, the tension in the cable and the components of the force of the pin on the beam can be found.

Step 4: Solve for unknowns:

From rotational equilibrium: F_{TC}=269.75 N
From translational equilibrium y: F_{AY}=−45.05 N
From translational equilibrium x: F_{Ax}=215.8 N

3.6 CHAPTER QUESTION: ANSWER

Enter in some reasonable numbers to answer this question. A person suspends a block of mass m=5 kg from his or her hand, as depicted in Figure 3.1. Their forearm, which has a mass of 2 kg with a center of mass at 10 cm from the elbow joint, is held horizontally and the upper arm is held vertically. Given these masses, the gravitational forces on the forearm and the block are:

$$F_{gA} = m_A g = (2\text{kg})(9.8 \text{ N/kg}) = 19.6 \text{ N}$$

$$F_{gB} = m_B g = (5 \text{ kg})(9.8 \text{ N/kg}) = 49 \text{ N}$$

So, the total weight of the forearm plus block is 68.6 N.

To find F_M, the tension in the bicep muscle, we need to draw a free-body diagram of the arm. One of the forces on the arm is the force of the tension in the string with which the block is suspended. Before we draw the free-body diagram of the arm, let's figure out what that force is. Start with a free-body diagram of the block in Figure 3.18.

Because the block is in equilibrium, that is, it is at rest and it remains at rest, we know that the net force on it must be zero. By inspection, the x-components of both forces are zero so we can neglect those. Also, by inspection:

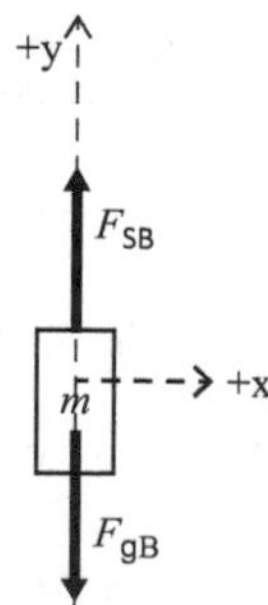

FIGURE 3.18 Free-body diagram of the block.

$$F_{SBy} = F_{SB} \quad \text{and} \quad F_{gBy} = -F_{gBy}$$

Setting the sum of the y-components of the forces equal to zero (and substituting) yields:

$$F_{SBy} + F_{gBy} = 0$$

$$F_{SB} + (-F_{gB}) = 0$$

$$F_{SB} = F_{gB}$$

$$F_{SB} = 49 \text{ N}$$

This is the force of the string on the block. The analysis of the arm requires the force of the string on the hand. The tension in a string is the same throughout, so however hard the string is pulling upward on the block, that is how hard the string is pulling downward on the hand, so $F_{SH}=F_{SB}$. The magnitude of the force of the string on the hand is equal to the magnitude of the force of the string on the block so

$$F_{SH} = 49 \text{ N}$$

The force of the string on the hand, F_{SH}, is included in the free-body diagram of the arm in Figure 3.19.

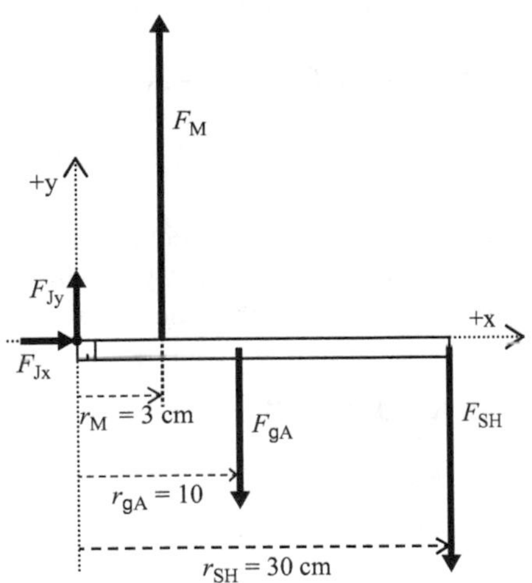

FIGURE 3.19 Free-body diagram of the arm.

The arm is at rest and remains at rest, so it is in linear and rotational equilibrium. The fact that it is in rotational equilibrium implies that the net torque on the arm is zero. This is true for the torque about any axis, so in the solution here the net torque about the elbow joint is set equal to zero. The only component of the torque due to any of the forces is the z-component, so we will set the sum of the z-components of the torque about the elbow joint (point J in the diagram) equal to zero (recognizing that the force $\vec{\mathbf{F}}_J$ of the elbow joint creates no torque about point J because force $\vec{\mathbf{F}}_J$ acts on point J; its moment arm is 0):

$$\tau_{Mz} + \tau_{gAz} + \tau_{SHz} = 0$$

$$r_M F_M - r_{gA} F_{gA} - r_{SH} F_{SH} = 0$$

$$F_M = (r_{gA}F_{gA} + r_{SH}F_{SH}) / r_M$$

$$F_M = [(0.1\text{ m})(19.6\text{ N}) + (0.3\text{ m})(49\text{ N})] / (0.03\text{ m})$$

$$F_M = 555.33\text{ N}$$

So, the tension in the bicep is much larger than the sum of the weight of the arm and the block. Are the forces still balanced? Yes, the elbow joint applies the force $\vec{\mathbf{F}}_J$ to the forearm. We can find a value for F_{Jx} and a value for F_{Jy} by applying the principle of translational equilibrium to the forearm.

For the x-components of the forces:

$$F_{\text{Net, x}} = 0, F_{Jx} = 0$$

The y-components of the forces are given by:

$$F_{Jy} = F_{Jy}, F_{My} = F_M, F_{gAy} = -F_{gA}, F_{SHy} = -F_{SH}$$

Setting the sum of the y-components equal to zero yields:

$$F_{Jy} + F_{My} + F_{gAy} + F_{SHy} = 0$$

$$F_{Jy} + F_M + (-F_{gA}) + (-F_{SH}) = 0$$

$$F_{Jy} = F_{gA} + F_{SH} - F_M$$

$$F_{Jy} = 19.6\text{ N} + 49\text{ N} - 555.33\text{ N} = -486.7\text{ N}$$

Remember that a newton is about ¼ lb so holding an object that weighs about 12 lb (holding it in the manner given in the problem) requires a tension of roughly 140 lb in the bicep muscle and a downward force whose magnitude is roughly 120 lb by the elbow joint. That is impressive!

3.7 CHAPTER QUESTIONS AND PROBLEMS

3.7.1 Multiple-Choice Questions

1. The direction of the torque on the nut in Figure 3.20 is:
   A. +x B. –x C. +y D. –y E. +z F. –z

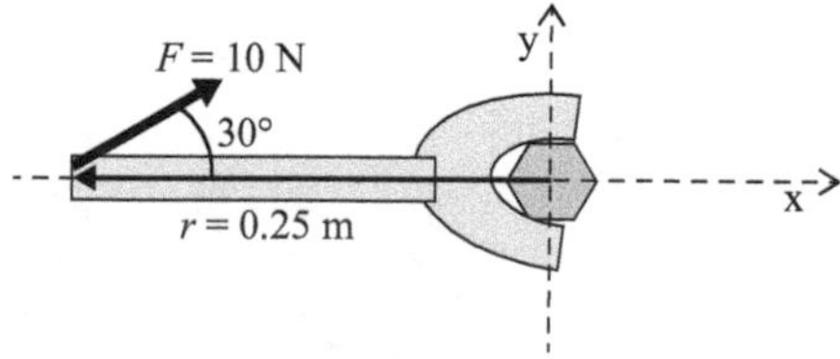

FIGURE 3.20 Multiple-choice question 1 and problem 1.

2. The z-component of the torque on the nut in Figure 3.21 is:
   A. positive B. negative C. zero

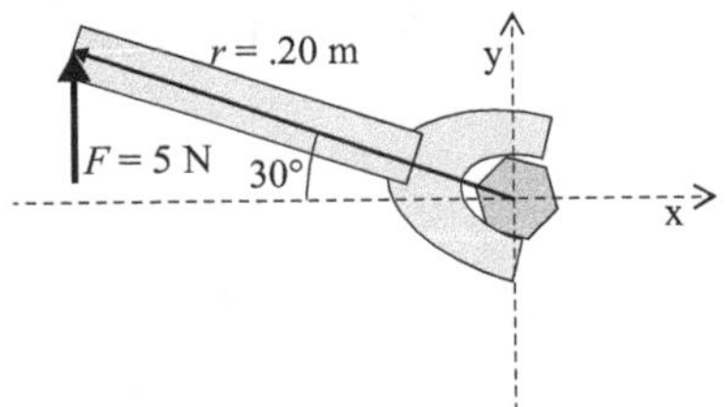

FIGURE 3.21 Multiple-choice question 2 and problem 2.

3. Depicted in Figure 3.22 is a door as viewed from above the door. A person is applying a horizontal force of magnitude *F*, on the doorknob of the door.
   What is the direction of the force on the door knob?
   A. Northward B. Eastward C. Southward
   D. Westward E. Upward F. Downward

4. Depicted in Figure 3.22 is a door as viewed from above the door. A person is applying a horizontal force of magnitude *F*, on the doorknob.
   What is the direction of the torque, on the door, about the hinge, generated by that force F?
   A. Northward B. Eastward C. Southward
   D. Westward E. Upward F. Downward

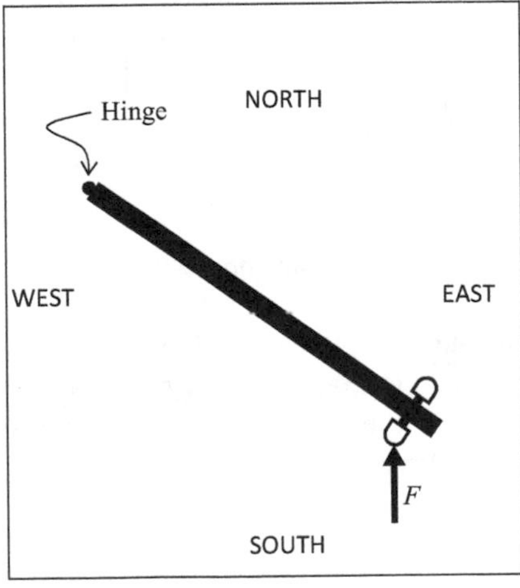

FIGURE 3.22 Multiple-choice questions 3 and 4 and problem 5.

5. Referring to Figure 3.23, the uniform beam is secured to the wall on its left side and the force *F* is producing a clockwise torque about the left end of the beam. The person applying the force doubles the magnitude of the force (without changing the direction or location of the force). Thus, will the torque about the left end of the beam, increase, decrease, or remain the same?
   A. Increase B. Decrease C. Remain the same.

6. Referring to Figure 3.23, the uniform beam is secured to the wall on its left side and the force *F* is producing a clockwise torque about the left end of the beam. The location of the force is changed so r is ½ its original value (without changing the magnitude or direction

of the force). Thus, will the torque about the left end of the beam, increase, decrease, or remain the same?

A. Increase B. Decrease C. Remain the same.

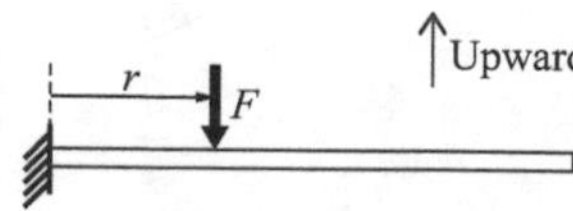

FIGURE 3.23 Multiple-choice questions 5 and 6 and problem 4.

7. A uniform horizontal beam of mass 2 kg is supported by a pin at its left end and a vertical string at its right end, as depicted in Figure 3.24. How does the tension F_T in the string compare with the magnitude F_g of the gravitational force being exerted on the beam by the earth?

   A. $F_T < F_g$ B. $FT = F_g$ C. $F_T > F_g$

8. What is the direction of the force being exerted by the pin on the uniform horizontal beam of mass 2 kg supported by the pin and a vertical string, as depicted in Figure 3.24?

   A. Straight upward.
   B. Straight downward.
   C. Up and to the right.
   D. Up and to the left.
   E. Down and to the right.
   F. Down and to the left.

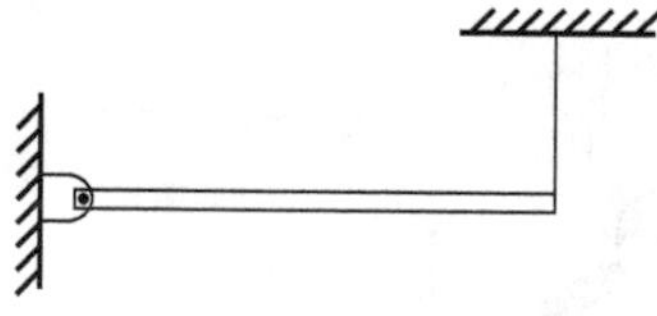

FIGURE 3.24 Multiple-choice questions 7 and 8 and problem 6.

9. Figure 3.25 depicts a uniform horizontal beam of mass 2 kg that is supported by a pin at its left end and a vertical string that is attached to the beam at a point that is less than half the length of the beam to the right of the pin. How does the tension F_T in the string compare with the magnitude F_g of the gravitational force being exerted on the beam by the earth?

   A. $F_T < F_g$ B. $F_T = F_g$ C. $F_T > F_g$

10. What is the direction of the force being exerted by the pin on the uniform horizontal beam of mass 2 kg supported by the pin and a vertical string, as depicted in Figure 3.25?

   A. Straight upward.
   B. Straight downward.

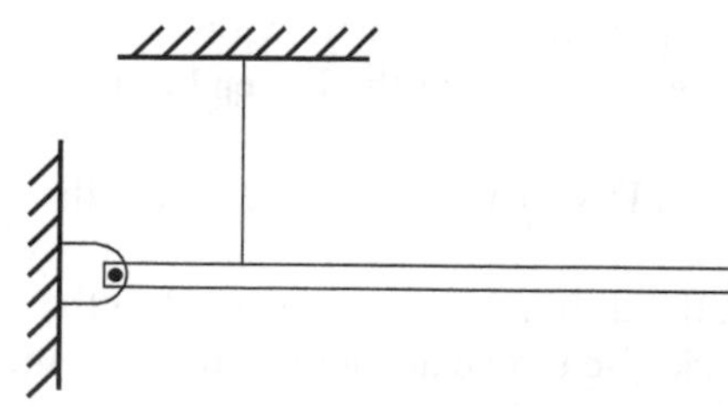

FIGURE 3.25 Multiple-choice questions 9 and 10 and problem 7.

C. Up and to the right.
D. Up and to the left.
E. Down and to the right.
F. Down and to the left.

11. In Figure 3.26, is the magnitude of the tension force in *cable-1* greater than, less than, or equal to the magnitude of the tension force in *cable-2*?
    A. greater than B. less than C. equal to

12. In Figure 3.26, is the *horizontal-component of the force* on the bar by the pin to keep the bar in equilibrium to the right, to the left, or zero?
    A. right B. left C. zero

13. In Figure 3.26, if the location of point C, on the vertical wall, is lowered so that the bar now makes an angle of 0° with the horizontal and cable-1 still makes a 35° angle with the horizontal, will the tension in *cable-1* for this new situation be greater than, less than, or equal to the tension in *cable-1* in the original configuration in Figure 3.25?
    A. greater than B. less than C. equal to

14. In Figure 3.26, if the location of point C, on the vertical wall, is lowered so that the bar now makes an angle of 0° with the horizontal and cable-1 still makes a 35° angle with the horizontal, will the tension in *cable-2* for this new situation be greater than, less than, or equal to the tension in *cable-2* in the original configuration in Figure 3.26?
    A. greater than B. less than C. equal to

15. Is the magnitude of the *y-component of the force on the bar due to the pin* for this situation greater than, less than, or equal to magnitude of the y-component of the force on the bar due to the pin in the original configuration in Figure 3.26?
    A. greater than B. less than C. equal to

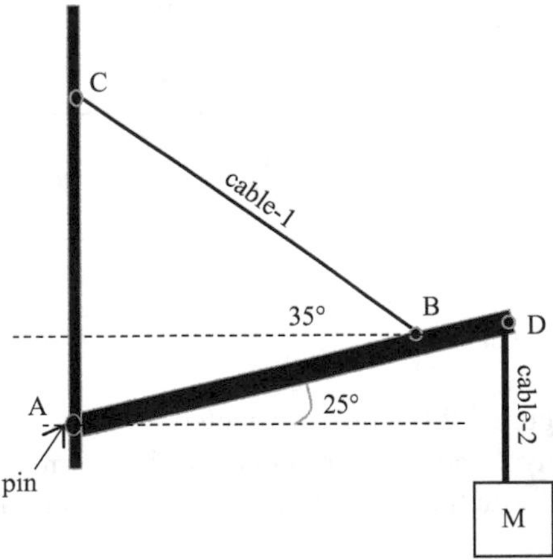

FIGURE 3.26 Multiple-choice questions 11–15 and problem 10.

3.7.2 Problems

1. Calculate the magnitude and direction of the torque on the nut in Figure 3.20.
2. Calculate the magnitude and direction of the torque on the nut in Figure 3.21.

3. Assuming the direction of the force on the wrench in Figure 3.20 can be adjusted, what is the magnitude of the maximum possible torque that can be generated with the given magnitude of the force and the given wrench?
4. Referring to Figure 3.23, the uniform beam that is 8 m long and has a mass of 10 kg is secured to the wall on its left side. The force F is downward with a magnitude of 50 N applied to a location of $r = 2$ m on the beam. Calculate the net torque on the left end of the beam due to the weight of the beam and the downward force (F) on the beam.
5. For the door in Figure 3.22, the position vector of the point of application of the force relative to the hinge is horizontal and is in the compass direction 40.5° south of east. The magnitude of the force being applied by the person, on the doorknob, is 2.42 N. The point on the doorknob at which the person is applying the force is 1.10 m, measured horizontally, from the hinge. As depicted, the force is directed horizontally northward. Find the magnitude of the torque on the door, about the hinge, due to the force being applied by the person, on the doorknob.
6. The uniform beam in Figure 3.24 is attached to the wall on its left side with a pin, so that it is free to rotate about that point. The beam has a mass of 4 kg and a length of 8 m. Compute the tension in the cable of the right side of the bar if this cable provides the upward force that holds the beam in its horizontal position.
7. The uniform beam in Figure 3.25 is attached to the wall on its left side with a pin, so that it is free to rotate about that point. The beam has a mass of 4 kg and a length of 8 m. Compute the tension in the cable that is attached 2 m of the left end of bar if this cable provides the upward force that holds the beam in its horizontal position.
8. A small child with a mass of 20 kg is positioned at the very left end of a 4-m long, uniform, seesaw, which pivots at the center (Figure 3.27). The child's big brother has a mass of 40 kg and he is positioned at a distance d from the center of the seesaw, which is balanced in a perfectly horizontal orientation. What must the distance d be for the seesaw to be in rotational equilibrium?

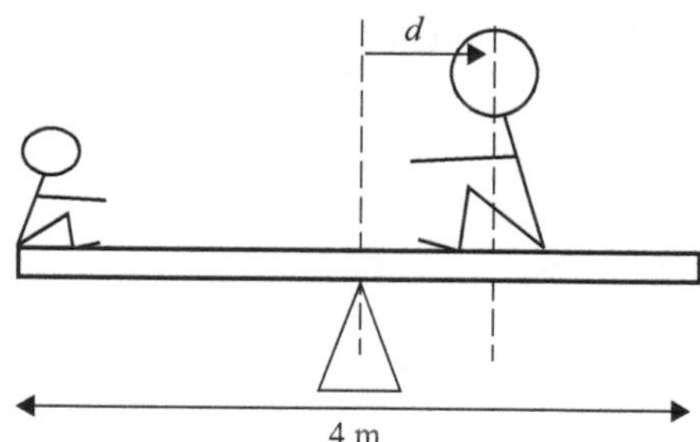

FIGURE 3.27 Diagram for problem 8.

9. As demonstrated in Figure 3.28, a uniform meter stick, which weighs 1 N, is supported on the left end with a pivot and on the right end with a string. A block that weighs 2 N is attached to the meter stick at the location depicted in the diagram at right. The strings are vertical.

   The meter stick is in static equilibrium at an angle of 20° to the horizontal. For this arrangement compute:

   a. the tension in the string attached to the right end of the meter stick.
   b. the magnitude and direction of the force that the pivot is exerting on the meter stick.

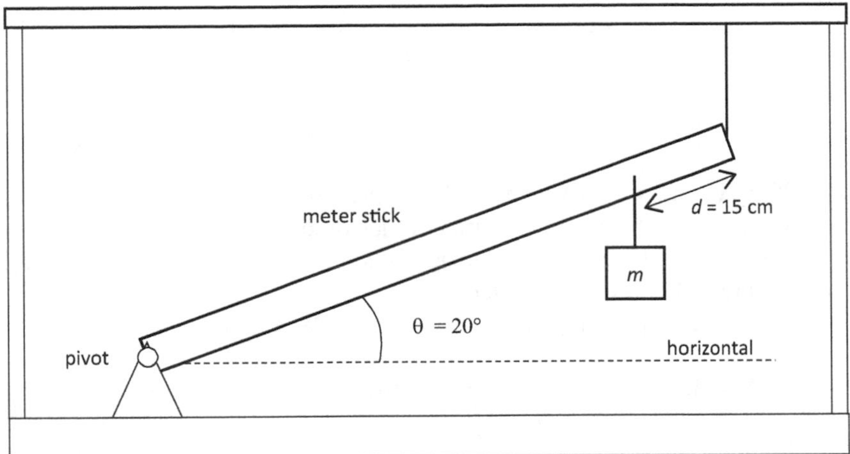

FIGURE 3.28 Diagram for problem 9.

10. A uniform bar, 10 m long with a mass of 20 kg, is in an equilibrium situation at an angle of 25° relative to the horizontal, as described in Figure 3.26. A mass, (*M*) which weighs 400 N, hangs from cable-2 that is attached to the right end of the bar. The left end of the bar is attached to a vertical wall with a pin that allows the bar to rotate freely about the point A. In this equilibrium situation, cable-1 is attached to a point 2 m from the end of the right side of the bar and cable-1 makes an angle of 35° to the horizontal, as described in the diagram. Compute the tension in cable-1.

APPENDIX 1: CROSS-PRODUCT FORM OF THE TORQUE EQUATION

The mathematically formal representation of the equation for torque and the RHR is the cross-product:

$$\vec{\tau} = \vec{\mathbf{r}} \times \vec{\mathbf{F}}$$

The cross-product is a way of multiplying two vectors together that results in a third vector, which is perpendicular to the plane in which the original two vectors lie. That is, for the cross-product of two vectors, $\vec{\mathbf{A}}$ and $\vec{\mathbf{B}}$, as demonstrated in Figure 3.29, the tails of vectors $\vec{\mathbf{A}}$ and $\vec{\mathbf{B}}$ are arranged so that their tails are at a common point.

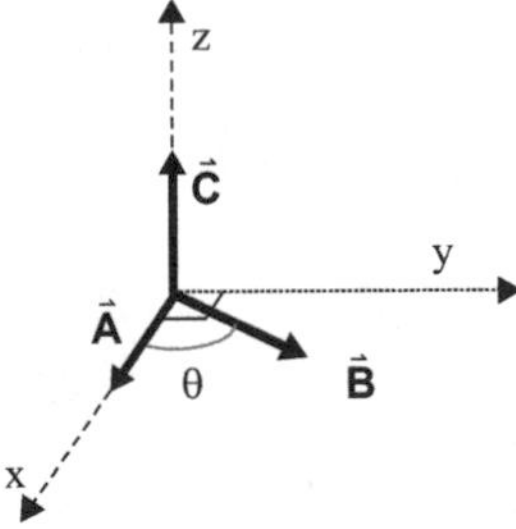

FIGURE 3.29 Cross-product.

Then, their cross-product, $\vec{\mathbf{A}} \times \vec{\mathbf{B}}$, gives a third vector, $\vec{\mathbf{C}}$, that points in a direction perpendicular to both $\vec{\mathbf{A}}$ and $\vec{\mathbf{B}}$. In Figure 2.4, vectors $\vec{\mathbf{A}}$ and $\vec{\mathbf{B}}$ are in the x and y plane so the vector $\vec{\mathbf{C}}$ is in the z-direction.

Given two arbitrary vectors $\vec{\mathbf{A}} = A_x\hat{\imath} + A_y\hat{\mathbf{j}} + A_z\hat{\mathbf{k}}$ and $\vec{\mathbf{B}} = B_x\hat{\mathbf{\imath}} + B_y\hat{\mathbf{j}} + B_z\hat{\mathbf{k}}$ (where $\hat{\mathbf{k}}$ is a unit vector in the +z-direction), the cross-product is defined as:

$$\vec{\mathbf{A}} \times \vec{\mathbf{B}} = \left(A_y B_z - A_z B_y\right)\hat{\mathbf{\imath}} + \left(A_z B_x - A_x B_z\right)\hat{\mathbf{j}} + \left(A_x B_y - A_y B_x\right)\hat{\mathbf{k}}$$

where $\hat{\mathbf{\imath}}$, $\hat{\mathbf{j}}$, and $\hat{\mathbf{k}}$ represent the +x, +y, and, +z directions, respectively.

The cross-product is a mathematical operation, which produces a result that matches the physical relation between a torque and the force producing that torque and the position vector of the point of application of the force relative to the rotation point.

Using the cross-product formula with the vectors $\vec{\mathbf{A}}$ and $\vec{\mathbf{B}}$ in Figure 2.4 and noting that:

$$A_x = A, A_y = 0, A_z = 0; B_x = B\cos(\theta), B_y = B\sin(\theta), \text{ and } B_z = 0$$

$$\vec{\mathbf{A}} \times \vec{\mathbf{B}} = \left[(0)(0) - (0)B\sin\theta\right]\hat{\mathbf{\imath}} + \left[(0)B\cos\theta - A\,(0)\right]\hat{\mathbf{j}} + \left[AB\sin\theta - (0)B\cos\theta\right]\hat{\mathbf{k}}$$

$$\vec{\mathbf{A}} \times \vec{\mathbf{B}} = AB\sin\theta\;\hat{\mathbf{k}}$$

$$\vec{\mathbf{C}} = AB\sin\theta\;\hat{\mathbf{k}}$$

The cross-product gives the same directional answer as the RHR. For this example, if vector $\vec{\mathbf{A}}$ is placed along the +x-axis and $\vec{\mathbf{B}}$ is in the +x & +y quadrant, the cross-product is in the $\hat{\mathbf{k}}$ direction and the RHR gives the same direction, as shown in Figure 3.30.

Since we can usually choose the orientation of the axis to best suit the problem, we commonly put the vectors to be crossed in the x-y plane. With this arrangement, we can express the z-component of the torque as:

$$\tau_z = r\,F\sin(\theta)$$

where θ is the angle between $\vec{\mathbf{r}}$, the position vector of the point at which the force is applied, relative to the point about which you are calculating the torque, and, $\vec{\mathbf{F}}$, the force vector, when the two vectors are placed tail to tail. The angle θ is measured counterclockwise, as viewed from a point on the +z-axis, from the direction of $\vec{\mathbf{r}}$ to the direction of $\vec{\mathbf{F}}$. In a case in which an x-y-z coordinate system has been defined, θ_r is the angle the position vector makes with the +x-direction, θ_F is the angle the force vector makes with the +x-direction, and θ_z can be expressed as:

$$\tau_z = r\,F\sin(\theta_F - \theta_r)$$

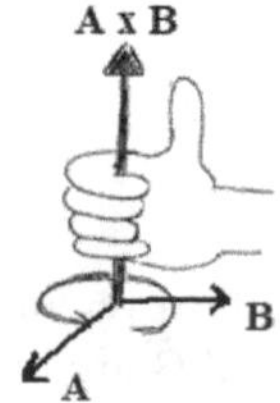

FIGURE 3.30 The cross-product and the right-hand rule.

MNEMONIC DEVICE TO USE FOR THE CROSS-PRODUCT

There is a mnemonic device often used to remember the cross-product equation. Start with a matrix (a series of rows and columns) with the unit vector ($\hat{i}$, $\hat{j}$, $\hat{k}$) in the top row and the vectors to be crossed in the next two rows. The order in which the vectors are entered into the matrix is important so the matrix below is used to compute the cross-product of $\vec{\mathbf{A}} \times \vec{\mathbf{B}}$:

$$\begin{matrix} \hat{\mathbf{i}} & \hat{\mathbf{j}} & \hat{\mathbf{k}} \\ A_x & A_y & A_z \\ B_x & B_y & B_z \end{matrix}$$

where $\hat{\mathbf{i}}$, $\hat{\mathbf{j}}$, and $\hat{\mathbf{k}}$ represent the +x, +y, and +z directions, respectively. Draw a line down through the first column to find the $\hat{i}$-component of the cross-product. To do this, we multiply the components on the diagonal and subtract the other components on the other diagonal. So, in this case: $A_yB_z - A_zB_y$

$$\begin{matrix} \hat{i} & \hat{j} & \hat{k} \\ A_x & A_y & A_z \\ B_x & B_y & B_z \end{matrix}$$

$(A_yB_z - A_zB_y)\hat{i}$

Then, draw a line down the middle column and cross the x and z columns to get the middle term. To do this, we multiply the components on the diagonal and subtract the other components on the other diagonal. So in this case: $A_xB_z - A_zB_x$. Wait, notice that the terms are the opposite way in the cross-product equation so we need to remember that the signs of the terms switch from + for $\hat{i}$, – for $\hat{j}$, and + again for $\hat{k}$. So, the second term in the cross-product is:

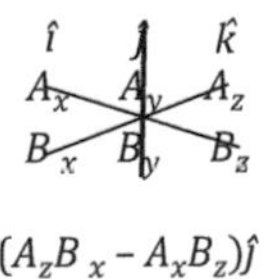

$(A_zB_x - A_xB_z)\hat{j}$

This switching of the signs from + to – to + gives the correct arrangement of the terms in the cross-product. Notice that the A_zB_x term comes first.

Then, draw a line down the right column and cross the x and y columns to get the last term.

$$\begin{matrix} \hat{i} & \hat{j} & \hat{k} \\ A_x & A_y & A_z \\ B_x & B_y & B_z \end{matrix}$$

$(A_xB_y - A_yB_x)\hat{j}$

Remember that the middle term is negative and the cross-product equation is produced.

4 Gravity and the Forces of Nature

4.1 INTRODUCTION

Gravity is one of the four forces of nature, and it is the force that we often call the weight of an object. In this chapter, the Newton's Law of Universal Gravitation is presented and examples are given in which the force of gravity between objects are calculated. As the first example of a field in the text, the gravitational field is defined and examples are given in which some of them result in calculations of values that will hopefully seem familiar.

4.2 THE FOUR FORCES OF NATURE

In the previous chapters, forces were discussed in a phenomenological sense, that is, in the way we experience them as a push or a pull. If a package is suspended by a string, as depicted in Figure 2.5, it is not difficult to explain that the weight of the object is opposed by the force of tension in the string. It is obvious that the weight of the object is due to the force of gravity, but it may not be obvious what the fundamental forces are that result in the tension in the rope. At this time, it is understood that all the forces that exist on Earth and in the Universe can be reduced down to four forces of nature, which are: *gravity, electromagnetic, nuclear weak, and nuclear strong.* In this and the following two chapters, the gravitational, electrical, and magnetic forces are presented separately. In the late 1600s, Isaac Newton formalized the study of gravity that Einstein reformulated in the early 1900s. Since the theoretical work of James Maxwell in the late 1800s, the electrical and magnetic forces have been grouped together as the electromagnetic force. What joins these two forces together is that both of these forces depend on electrical charge, either at rest or moving in a current. Along with gravity as the weight of an object, we experience electromagnetic forces first hand in trivial events such as when: a balloon that you rub on your hair sticks to the wall or a paperclip is pulled off the table with a magnet. The electromagnetic force also accounts for the contact forces between objects, including us. At the atomic and molecular level, it is electromagnetic forces that hold matter together. Most bonds can be understood by interactions between the outer electrons of the atoms that make up an object. In fact, when you push your hand against a table, the electromagnetic forces between the atoms in your hand keep it from going through the table. Together gravity and electromagnetic forces allow you to stand on the Earth. Gravity pulls objects down toward the center of the Earth and the electromagnetic forces hold the atoms and molecules of the string together so that a tension is possible. Of the four, only the gravitational and the electromagnetic forces are perceivable on the human scale.

The nuclear weak and nuclear strong forces are needed to explain the models of the nucleus but are not directly observed by humans. These forces are significant only at extremely short distances, of approximately the radius of an atomic nucleus, 1×10^{-15}m, and decreases rapidly as the distance between particles increases so that they are insignificant at the distance from the nucleus of an atom to the electron shells, about 1×10^{-10}m. These nuclear forces are not studied extensively in this text but are discussed conceptually in the appendix of this chapter and the topics associated with these forces are studied in the second volume of the text in the context of energy.

DOI: 10.1201/9781003308065-4

Chapter question: Is the force of gravity on the International Space Station (ISS) in its orbit around the Earth (405 km above the surface of the Earth) closer to zero or closer to the same value as it was when it was on the surface of the Earth? The theories presented and the examples completed in this chapter will provide the background needed to make the necessary calculations to answer this question at the end of the chapter.

4.3 UNIVERSAL LAW OF GRAVITY

In the late 17th century, Isaac Newton studied the motion of the planets and the Moon in an effort to produce better star charts for naval navigation. During his studies, he proposed that the same force that causes objects to have a weight here on Earth also keeps the Moon in its orbit around the Earth and the Earth and the other planets in their orbits around the Sun. As described in Figure 4.1, Newton proposed that gravity is an attractive force between two objects with a magnitude that is proportional to the product of the masses of the objects involved and inversely proportional to the square of the distance between the objects, as measured from the center of one object to the center of the other object.

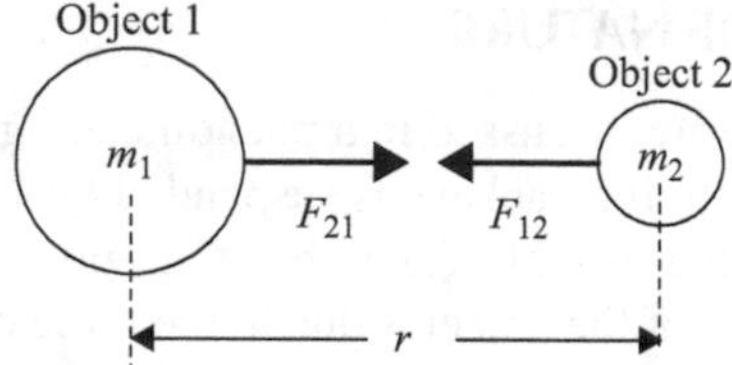

FIGURE 4.1 Force of gravity between two objects.

Newton's Universal Law of Gravity gives the magnitude of the gravitational force, F_g, as shown in equation (4.1) as:

$$F_g = \frac{Gm_1m_2}{r^2} \tag{4.1}$$

In this expression, G is the universal gravitation constant, m_1 is the mass of one of the objects, m_2 is the mass of the other object, and r is the separation of the centers of mass of the two objects. Equation (4.1) gives the magnitude of the force on each object. The direction of the force on either object is along the line connecting the two centers, and pointing toward the center of the other object.

4.3.1 FINDING THE VALUE OF G

The value of the universal gravitation constant, G, was not part of Newton's original work. Approximately 100 years after Newton's work, Henry Cavendish built a torsion balance similar to the one pictured in Figure 4.2.

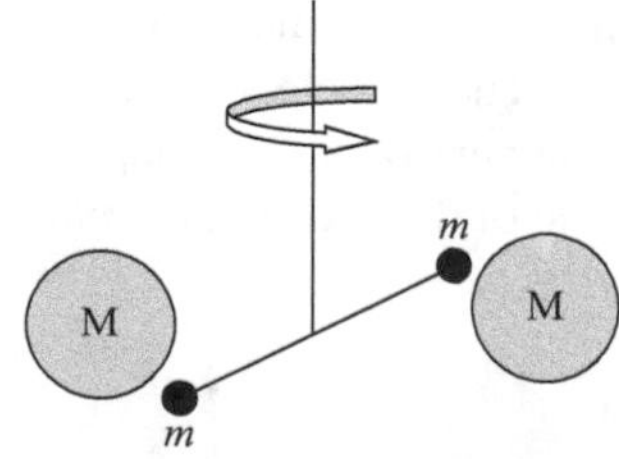

FIGURE 4.2 Cavendish's torsion balance.

His torsion balance consisted of a long thin wooden rod with a small solid lead sphere mounted on each end. The rod was suspended horizontally by means of a vertical metal fiber attached to the center of the rod. Cavendish carefully measured the torque needed to twist the metal fiber through very small angles. Large solid lead spheres were brought close to, but not touching, the smaller lead spheres, as shown in the diagram. The gravitational attraction between the small and large lead spheres caused the torsion balance to twist to an angle that corresponded to a known torque. From this torque, the magnitude of the force of gravity being exerted by each large sphere on the small sphere nearer it was calculated. Since the masses of the spheres and the distance between their centers were known, G was found directly from Newton's equation for the Universal Law of Gravitation. Cavendish published his result in 1798 and since then the measurement of G has been refined and the currently accepted value of G is 6.67×10^{-11}N m²/kg².

Some important notes about Newton's Universal Law of Gravity:

1. Consider two objects, object 1 of mass m_1 and and object 2 of mass m_2, as depicted in Figure 4.1. The gravitational force exerted by object 2 on object 1 has the same magnitude, but opposite direction to, the gravitational force exerted by object 1 on object $F_{21}=F_{12}$ but $\vec{\mathbf{F}}_{21}=-\vec{\mathbf{F}}_{12}$. Even if the objects have very different masses, the magnitudes of the two forces are equal.
2. For small objects, like a person, in contact with a large object, like the Earth, the center-to-center distance, r, is approximated as the radius of the larger object. Since the radius of the Earth is usually given as 6,370,000 m and the center of mass of a person is about 1 m from their feet, the uncertainty in the given value of the Earth's radius is much larger than the 1 m.

Example

Calculate the gravitational force that the Moon exerts on the Earth.

Step 1: Draw the free-body diagram. In this case, we are free to put the Earth and the Moon at any location on the page, so in this example the Earth is put at the origin and the Moon is put to the right of the Earth on the +x axis.

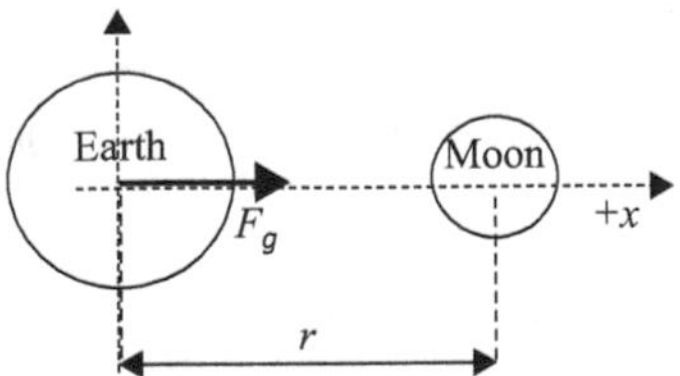

FIGURE 4.3 The Earth and the Moon.

Step 2: Compute the magnitude of the gravitational force with Newton's Universal Law of Gravity:

$$F_g = \frac{Gm_{moon}m_{earth}}{r^2}$$

by plugging in the values for G, the mass of the Moon (7.34×10^{22} kg), the mass of the Earth (5.97×10^{24} kg), and the mean center-to-center distance between the Earth and the Moon (384 399 km). The expression with all the necessary values plugged in is:

$$F_G = \frac{\left(6.67\times10^{-11}\,\frac{\text{Nm}^2}{\text{kg}}\right)\left(7.35\times10^{22}\,\text{kg}\right)\left(5.97\times10^{24}\,\text{kg}\right)}{\left(3.844\times10^{8}\,\text{m}\right)^2}$$

Which results in a value of:

$$F_G = 1.98 \times 10^{20}\,\text{N}$$

Step 3: Find the direction of the force from the diagram.

Note that the value found is the magnitude of the force of gravity that the Moon exerts on the Earth and it is also the magnitude of the force of gravity that the Earth exerts on the Moon. In Figure 4.3, the Earth is at the origin and the Moon is at a distance of r to the right in the positive x-direction. Thus, the x-component of the force of gravity that the Moon exerts on the Earth is:

$$F_{G_x} = \left(1.98 \times 10^{20}\,\text{N}\right)\cos(0°) = 1.98 \times 10^{20}\,\text{N}$$

and the y-component of the force of gravity that the Moon exerts on the Earth is:

$$F_{G_y} = \left(1.98 \times 10^{20}\,\text{N}\right)\sin(0°) = 0\ \text{N}$$

Since the Moon pulls the Earth toward its center, which is to the right of the origin, the force is in the +x-direction and the y-component of the force is zero.

If the Moon was at a different location in its orbit around the Earth, as shown in Figure 4.4, such that the distance between the Earth and Moon was still the same but the Moon was at an angle $\theta = 30°$ relative to the +x axis, steps 1 and 2 in the process would be performed the same way and step 3 would change to find the x- and y-components of the force.

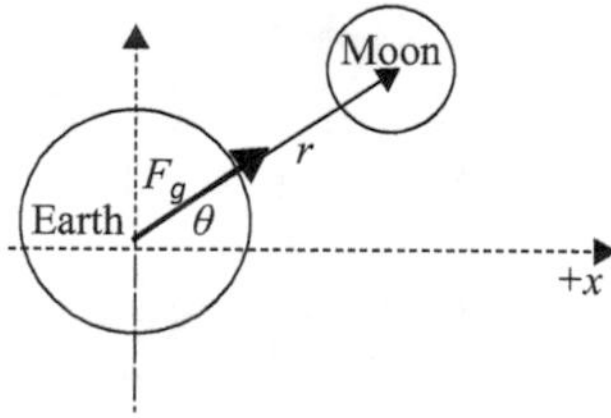

FIGURE 4.4 The Earth and the Moon.

Thus, for the arrangement presented in Figure 4.4, the x-component of the force of gravity that the Moon exerts on the Earth is:

$$F_{G_x} = \left(1.98 \times 10^{20}\,\text{N}\right)\cos(30°) = 1.72 \times 10^{20}\,\text{N}$$

and the y-component of the force of gravity that the Moon exerts on the Earth is:

$$F_{G_y} = \left(1.98 \times 10^{20}\,\text{N}\right)\sin(30°) = 9.9 \times 10^{19}\,\text{N}$$

4.4 GRAVITATIONAL FIELD

The concept of a field is widely used in physics and is important for the understanding of many different phenomena including electricity, magnetism, and light. A force field, in general, is the force per quantity that the magnitude of force depends upon. Therefore, the gravitational field ($\vec{g}$) is the gravitational force per mass in a region of space. Thus, the magnitude of this field (g) is given in equation (4.2) as:

$$g = \frac{F_g}{m} = \frac{GM\ m/_{r^2}}{m} = \frac{GM}{r^2} \quad (4.2)$$

The object with a mass m that was used in equation 4.2 to arrive at the expression for g was completely arbitrary, so for any object near the surface of the Earth multiply the mass of the object by the gravitational field strength g to get the magnitude of the gravitational force on that object:

$$F_g = mg$$

The gravitational field of an object like the Earth is depicted with field lines similar to those shown in Figure 4.5.

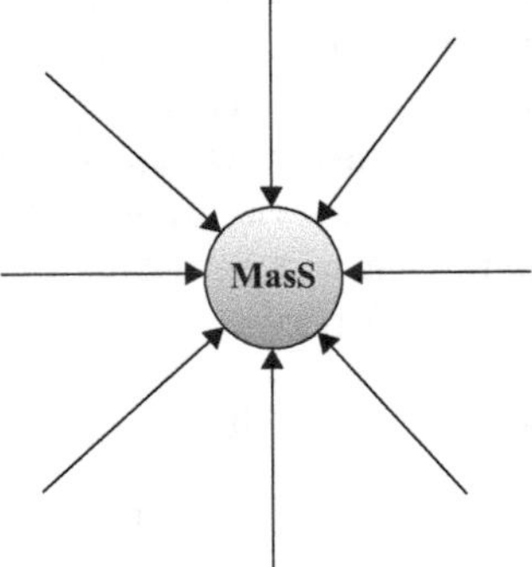

FIGURE 4.5 Gravitational field lines.

These gravitational field lines depict the direction of the force-per-mass in the vicinity of the object, in two dimensions, 2-D. Since gravity is an attractive force, field lines always point toward the center of the object that creates the field. The lines are radial, like the spokes on a wheel in 2-D or like toothpicks in an orange in 3-D. These lines are never parallel and never cross. The gravitational field strength (g) decreases as the distance away from the object increases, but the field will always be pointing inward and never actually reaches zero even when it gets too small to measure. Thus, the gravitational field of an object reaches infinitely far away from an object.

As shown in Figure 4.6, the field lines (represented as the dashed lines) due to the mass M demonstrate the direction the gravitational force will be applied to an object with mass, m, if it was placed in the region of space surrounding the mass, M.

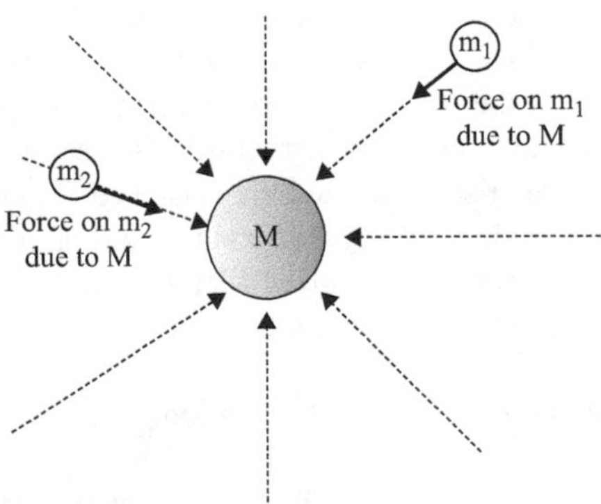

FIGURE 4.6 The gravitational field is the gravitational force per mass around an object.

If two similar masses, m_1 and m_2, are placed at different distances away from the center of M, the force (represented as the bold solid lines) on the two masses will be different magnitudes and directions, but both will point toward the center of M.

From the Universal Law of Gravity, the gravitational field strength near the surface of the Earth is:

$$g = \frac{GM_{\text{earth}}}{r^2} = \frac{\left(6.67\times\frac{10^{-11}\,\text{Nm}^2}{\text{kg}^2}\right)\left(5.97\times 10^{24}\,\text{kg}\right)}{\left(6.378\times10^6\,\text{m}\right)^2} = 9.78\,\frac{\text{N}}{\text{kg}} = 9.8\,\frac{\text{N}}{\text{kg}}$$

This is the *9.8 N/kg* that has already been used to compute the force of gravity or the weight of an object in previous chapters. The value of the gravitational field strength on the surface of the Earth (g=9.8 N/kg) is a rounded average value for the gravitational field strength near the surface of the Earth. There are minor variations of g over the surface of the Earth, but 9.8 N/kg is an acceptable approximation. This is lower-case g, which is different for different planets, unlike the upper-case G, which is the same for the entire universe. For example, the gravitational field strength on the surface of the Moon is much smaller than that on Earth, but it is calculated the same way using the same gravitational constant G:

$$g = \frac{GM_{\text{moon}}}{r^2} = \frac{\left(6.67\times10^{-11}\,\text{Nm}^2/\text{kg}^2\right)\left(7.35\times 10^{22}\,\text{kg}\right)}{\left(1.737\times10^6\,\text{m}\right)^2} = 1.62\,\frac{\text{N}}{\text{kg}}$$

This is approximately 1/6th of the gravitational field strength on the Earth's surface. So, please don't confuse G and g. They are related but they are not the same thing.

4.5 GRAVITY IS UNIVERSAL

We live on a planet that orbits around a star that is part of the galaxy called the Milky Way, which is a pinwheel-shaped collection of billions and billions of stars approximately 100,000 light-years across. The Milky Way is just one of billions and billions of galaxies that make up that part of the universe that we can see, a part that is billions of light-years wide.

4.5.1 General Relativity

Today, the theory of *General Relativity* is considered the best theory of gravity, since it can explain all the phenomena that Newton's Law of Universal Gravitation can and some that it cannot. It is a theory that predicts the effects of gravity on the geometry of space and time. One example of the unexplained phenomena is the precession of the perihelion of Mercury's orbit. The perihelion is a part of the elliptical orbit with the longest radius. The precession of Mercury's perihelion is a small rotation of approximately 1.58° per century of the entire orbit's orientation. This precession cannot be explained with Newton's Law of Gravitation and the known structure of the solar system. In fact, before Einstein proposed his explanation, some scientists suggested the existence of another planet called Vulcan that pulled on Mercury from the other side of the Sun. It was not until Einstein published his work on the general theory of relativity that the perihelion shift of Mercury was truly understood.

The basic concepts behind General Relativity are:

1. Mass changes the geometry of space in its location, like a bowling ball dropped in the middle of a pillow.
2. That space dictates the motion of objects, by forcing them to travel along the lines of the fabric of space. This fabric of space can be conceptualized as a grid drawn onto the couch cushion, before the bowling ball is dropped onto it.

The equations of General Relativity are used to compute the curvature of space in a region with a mass distribution. The rules of geometry of curved space are known as Non-Euclidean Geometry, which are used in general relativity.

If a large mass curves the fabric of space, as shown in Figure 4.7, a smaller object will follow this curved path, like a planet in orbit around the Sun.

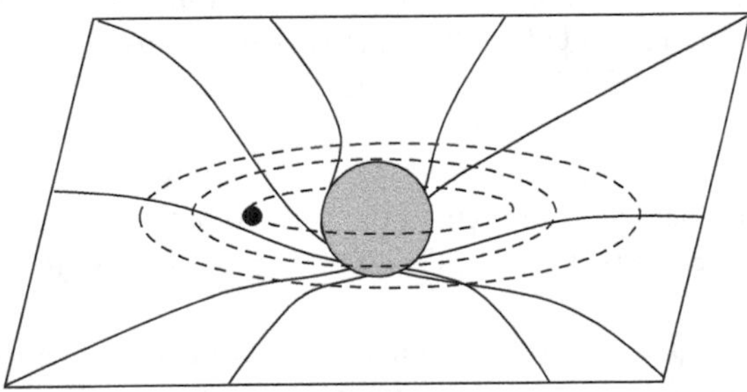

FIGURE 4.7 Warping of space due to a massive object.

One of the greatest proofs of general relativity was the prediction and subsequent observations of bending of light rays by the curvature of space around the Sun during an eclipse. Since light has no mass, Newton's Law of Universal Gravitation predicts that light will not experience gravitational forces, but Einstein's General Relativity explains that light just follows along the lines of space, which are warped by the mass of the Sun. It sounds like science fiction, but so do Black Holes which were hypothesized based on this theory and are now considered part of the accepted picture of the universe.

4.6 CHAPTER QUESTION ANSWER

It is interesting to note, that g for the Earth does not decrease as quickly as you may think. The International Space Station orbits the Earth in a fairly circular orbit that has an average altitude of h=405 km above the Earth's surface. So, the gravitational field strength in the orbit of the space station is:

$$g_o = \frac{G\, m_{\text{earth}}}{r_{\text{orbit}}^2} = \frac{G\, m_{\text{earth}}}{(r_{\text{earth}} + h)^2} = \frac{(6.67\times10^{-11}\,\text{Nm}^2/\text{kg}^2)(5.97\times10^{24}\,\text{kg})}{(6.37\times10^6\text{m} + 405\times10^3\text{m})^2} = 8.7\frac{\text{N}}{\text{kg}}$$

g_o=8.7 N/kg, which is only 11% less than g=9.8 N/kg on the surface of the Earth.

$$\frac{(9.8-8.7)}{9.8}\times100\% = 11.22\%$$

Thus the actual gravitational force of the Earth on the International Space Station in the space station's orbit is closer to what it would be if the space station were on the surface of the Earth as compared to a value of zero.

4.7 CHAPTER QUESTIONS AND PROBLEMS

4.7.1 Multiple-Choice Questions

1. An object weighs 100 N on the Earth's surface. When it is moved to a point in an orbit, which is one-hundredth of an Earth radius above Earth's surface, the gravitational force on it will be closest to approximately 25 N, 50 N, 100 N, or 400 N?
   A. 25 N B. 50 N C. 100 N D. 400 N

2. An object weighs 100 N on the Earth's surface. When it is moved to a point in an orbit, which is one Earth radius above Earth's surface, the gravitational force on it will be closest to approximately 25 N, 50 N, 100 N, or 400 N?
A. 25 N B. 50 N C. 100 N D. 400 N

3. A planet that has a mass which is ≈ 0.1 M_{Earth} and a radius which is ≈ ½$R_{Earth.}$ What is the approximate gravitational field strength at the surface of this planet?
A. 0.98 N/kg B. 1.96 N/kg C. 3.92 N/kg D. 9.8 N/kg E. 24.5 N/kg

4. A planet that has a mass which is ≈ 10 M_{Earth} and a radius which is ≈ 2 R_{Earth}. What is the approximate gravitational field strength at the surface of this planet?
A. 0.98 N/kg B. 1.96 N/kg C. 3.92 N/kg D. 9.8 N/kg E. 24.5 N/kg

5. Given the current weight of a block on Earth as W, if the radius of the Earth was reduced to half its original value and all masses remained the same, the weight of the block on the smaller and denser Earth will be:
A. (¼ W) B. (½ W) C. the same D. (2 W) E. (4 W)

6. Given the current weight of a block on Earth as W, if the radius of the Earth was doubled to twice its original value and all masses remained the same, the weight of the block on the larger and less dense Earth will be:
A. (¼ W) B. (½ W) C. the same D. (2 W) E. (4 W)

7. Given the current weight of a block on Earth as W, if the mass of the Earth was doubled to twice its original value and the mass of the block and the radius of the Earth remained the same, the weight of the block on the more massive Earth will be:
A. (¼ W) B. (½ W) C. the same D. (2 W) E. (4 W)

8. Given the current weight of a block on Earth as W, if the mass of the Earth was doubled to twice its original value and the mass of the block was cut in half and the radius of the Earth remained the same, the weight of the block on the more massive Earth will be:
A. (¼ W) B. (½ W) C. the same D. (2 W) E. (4 W)

9. The mass of Saturn is roughly 100 times that of the Earth and Saturn's orbit has a radius, which is roughly 10 times that of the Earth's orbit around the Sun. Thus, the gravitational force of the Sun on Saturn is approximately 1/100, 1/10, equal to, 10×, or 100× the gravitational force of the Sun on the Earth.
A. 1/100th B. 1/10th C. equal to D. 10× E. 100×

10. Geosynchronous weather satellites are in a fixed orbit above the Earth at a constant distance of approximately six Earth radii ($6R_E$) from the surface of the Earth. Thus, the gravitational field strength at the location of these satellites is approximately:
A. 0.2 N/kg B. 0.27 N/kg C. 1.4 N/kg D. 1.63 N/kg E. 9.8 N/kg

4.7.2 Problems

Use the following values to solve the problems in this section.

Celestial Object	Equatorial Diameter (km)	Mean Orbital Radius (km)	Mass ($\times 10^{24}$kg)
Sun	1,391,400	–	1,988,500
Mercury	4,879	57,900,000	0.330
Venus	12,104	108,200,000	4.87
Earth	12,756	149,600,000	5.97
Mars	6,792	227,900,000	0.642
Jupiter	142,984	778,600,000	1,898
Saturn	120,536	1,433,500,000	568
Uranus	51,118	2,872,500,000	86.8
Neptune	49,528	4,495,100,000	102
Moon (Earth)	3,475	0.384	0.0735

1. Calculate the magnitude of the force of gravity of the Sun on Mars.
2. Calculate the magnitude of the force of gravity of the Sun on the Earth.
3. Calculate the magnitude of the force of gravity of the Sun on Venus.
4. Calculate the magnitude of the force of gravity of the Sun on Jupiter.
5. Calculate the magnitude of the force of gravity of the Sun on Saturn.
6. Calculate the gravitational field strength at the surface of Mercury.
7. Calculate the gravitational field strength at the surface of Mars.
8. Calculate the gravitational field strength at the surface of Venus.
9. Calculate the gravitational field strength at a distance of 1 Earth Radii away from the surface of Earth.
10. Calculate the gravitational field strength at a distance of 6 Earth Radii away from the surface of Earth.

5 Electric Forces and Fields

5.1 INTRODUCTION

The electric force is another force of nature that is associated with electric charge, which is another property of matter. This chapter begins with a discussion of electric charge and then the format of the electric force is presented and examples are provided. Similar to the previous chapter, the concept of the electric field is discussed along with several examples.

Chapter question: Gel electrophoresis is a lab technique employed to separate and identify biological molecules, based on size. It is used to study proteins and DNA for medical and forensic investigations. The question is, what is the role of the electric field in the lab technique known as gel electrophoresis? This question will be answered at the end of this chapter after the concept of the electric field is developed throughout this chapter.

5.2 CHARGE

Like mass, charge is a property of matter and it is the starting place for the study of all of electricity and magnetism. Unlike mass that has only one type under normal conditions here on Earth (antimatter is possible but not common), there are two types of charge in common matter that balance each other. These two types are called positive and negative and, in most matter, there are equal amounts of both, so in most objects the net amount of charge is zero. So, it is the imbalance of charge that is measured and referred to in our analysis as charge. It is important to remember that the term charge on an object is not the total charge, but just the imbalance of excess positive or negative charge. The symbol for charge is either a capital or lower case (Q or q). Both symbols are used to represent the imbalance of charge of an object, and it is common to use the upper-case Q for larger charges in a problem and the lower-case q for the smaller charges in a problem. It is also acceptable to use only upper- or lower-case q in all problems. It is important to note that in this text both will be used to represent the quantity of charge.

As mentioned in Chapter 1, the two kinds of charge, positive and negative, are associated with protons and electrons, respectively. A macroscopic neutral object consists of an astronomically large number of protons and an equal number of electrons (as well as an astronomically large number of neutrons, which have no charge). Each proton has the same amount of positive charge that every other proton does and each electron has the same amount of negative charge that every other electron does. In addition, each electron has the same amount of negative charge as the positive charge of a proton. It is important to note that positive charge is in no way stronger or more important than negative charge, it is just the other kind of charge. The idea that there are two kinds of charge comes from simple experiments in which the two different types of charge can be separated from each other, and it can be observed that like types of charge repel and unlike types of charge attract.

A charged object has an imbalance in the number of protons and electrons. A negatively charged object has more electrons than protons and a positively charged object has more protons then electrons. Given that the protons are part of the nucleus and electrons exist in the orbital around the nucleus, electrons tend to move from atom to atom more freely then protons. For instance, when you charge an object made of one material by rubbing it with an object made of a different material, it is electrons that move from one to the other. In doing so, both objects become charged. If you rub a balloon on your head, you are removing electrons from your hair and putting them on the balloon.

DOI: 10.1201/9781003308065-5

Electrons are negatively charged. Hence, you are giving the balloon a negative charge. At the same time, you leave your hair with an equal amount of positive charge.

As depicted in Figure 5.1, if one negatively charged balloon is suspended from a string and another negatively charged balloon is brought close to the suspended balloon, it will be repelled. Each balloon will exert a force on the other balloon. Each force is directed away from the balloon that is exerting that force. If a glass rod is rubbed with a piece of silk, the silk will acquire electrons from the glass, leaving the glass with more positive charge than negative charge, meaning it will have a net positive charge. (The silk will have a net negative charge of the same amount.) The glass rod will then attract the suspended negatively charged balloon.

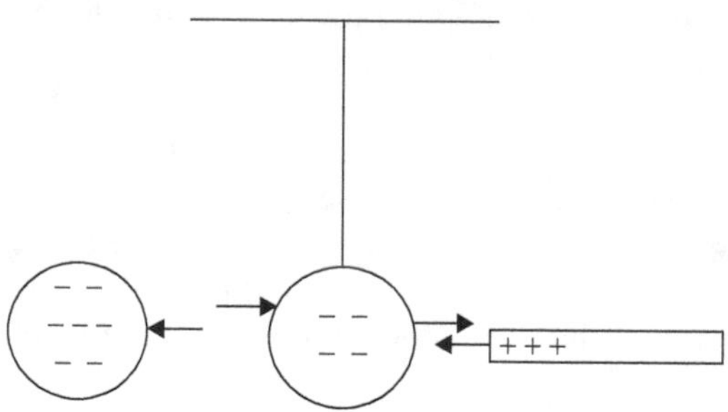

FIGURE 5.1 Charge on a balloon.

5.2.1 Units of Charge

The SI unit of charge is the Coulomb denoted by the symbol C. Like the kg for mass, it denotes an amount of a quantity and, in the case of the Coulomb, 1 C is roughly the amount of charge of 6.24×10^{18} protons. Thus, a Coulomb is a large amount of charge when working on a problem when the amount of charge is in the atomic or molecular levels, such as in many chemistry or biochemistry interactions. Therefore, another unit, the elementary charge, *e*, is often used. This unit is the charge of a single proton, where $1\ e = 1.602 \times 10^{-19}$ C. (In units of e, the charge of an electron is –1 *e*.) The charge associated with each of the elementary particles and their masses are given in Table 5.1.

5.2.2 Types of Materials

Materials are divided into three electrical categories: conductors, insulators, and semiconductors, based on how easily electrons can move along the surface of, or through, the material. Metals, like copper, are good conductors of electric charge, while plastics, wood, and rubber are very poor conductors referred to as insulators. Charge flows easily through conductors but does not flow easily through insulators; this is why wires are wrapped in a rubber coating. The difference between conductors and insulators is how tightly or loosely their electrons are bound to the atoms that make up the material. In fact, it isn't even all the electrons that matter; it is just the outermost electrons in the atoms that matter. In a conductor, the outermost electrons are so weakly bound to the atoms that they

TABLE 5.1
Charge and Mass of Elementary Particles

Particle	Charge (C)	Charge (e)	Mass (kg)
Electron	-1.602×10^{-19}	–1	9.11×10^{-31}
Proton	$+1.602\times10^{-19}$	+1	1.672×10^{-27}
Neutron	0	0	1.674×10^{-27}

are free to travel throughout the material. In insulators, the electrons are so tightly bound to the atoms that they are not free to flow though the material. This is why the electrons rubbed off your head onto the balloon stay at the spot on the balloon where they were put. The outermost electrons in semiconductors are not free to move, but are not tightly bound either. A small amount of electrical force on the electrons in a semiconductor is enough to pull the electrons out of their atoms so they are free to move about the material like the free electrons in a metal. The ability of semiconductors to conduct can be adjusted by injecting different atoms with slightly different electrical characteristics into the material. This process is called doping and is the key to the semiconductor industry.

5.3 COULOMB'S LAW

Coulomb's Law states that the magnitude F_{12} of the electric force exerted on one charged particle by another is proportional to the product of the absolute values of the charges q_1 and q_2 of the particles, and inversely proportional to the square of the separation r of the particles. It is given here in equation (5.1) as:

$$F_{12} = \frac{k|q_1|\cdot|q_2|}{r^2} \tag{5.1}$$

where k is the Coulomb constant, which is approximately $9.0 \times 10^9\ \frac{\text{Nm}^2}{\text{C}^2}$. The law can also be applied to objects as long as the charge distribution on each object is spherically symmetric and the center-to-center distance is used as the separation of the objects. Note that, given two particles 1 and 2, the magnitude of the electric force exerted by particle 2 on 1 is equal to the magnitude of the electric force exerted by 1 on 2. $F_{21} = F_{12}$.

The direction of the force is determined by the types of charges interacting. It is always directed along the line passing through both particles or, when dealing with objects, through their centers. In Figure 5.2, the objects have charge of the same sign so the forces are repulsive.

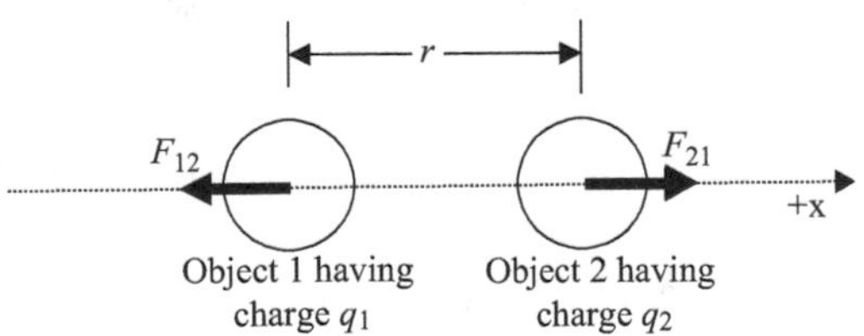

FIGURE 5.2 Forces on like charges $\frac{q_1}{|q_1|} = \frac{q_2}{|q_2|}$.

In Figure 5.3, the objects have charges of opposite signs so the forces are attractive.

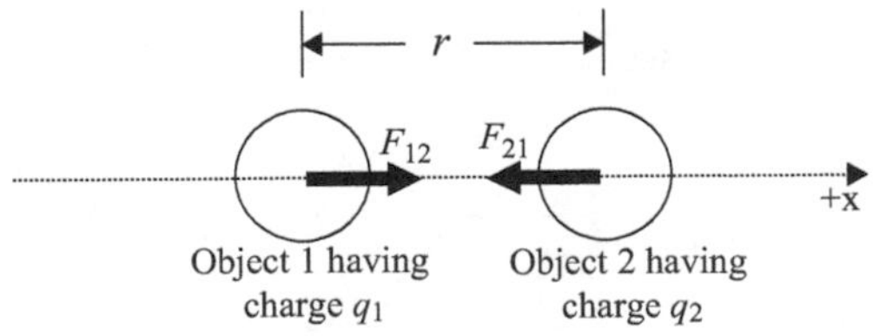

FIGURE 5.3 Forces on unlike charges $\frac{q_1}{|q_1|} = -\frac{q_2}{|q_2|}$.

To find the Coulomb force vector, calculate the magnitude of the force from $F_{12} = \frac{k|q_1| \cdot |q_2|}{r^2}$ and determine the direction of the force using the vector diagram and the fact that unlike charges attract and like charges repel.

5.3.1 Examples of Calculations Using Coulomb's Law

Example 5.1

Find the Coulomb force on a proton, q_1, located at the origin, due to another proton, q_2, that is 1×10^{-15} m to the right of q_1. These dimensions are similar to the size of the nucleus of an atom, so the solution to this example gives a sense of the strength of the forces pushing the protons apart in the nucleus of atoms larger than hydrogen.

Step 1: As shown in Figure 5.4, sketch the diagram and label the Coulomb force F_{12} on q_1 due to q_2.

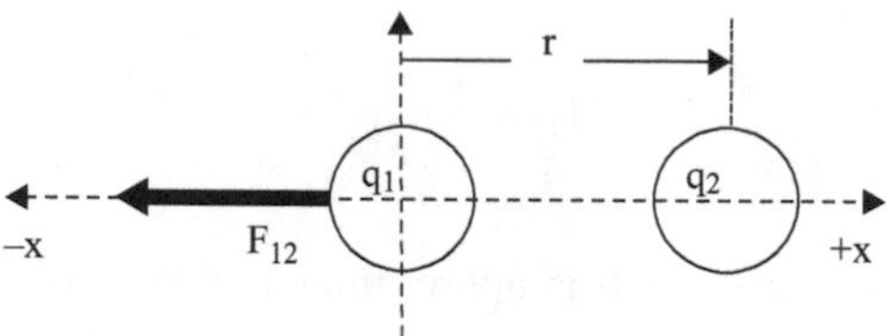

FIGURE 5.4 Diagram for Coulomb's Law Example 5.1.

Because both charges are positive, the Coulomb force on q_1 due to q_2 is repulsive and points to the left. The subscript used for this force in this diagram is F_{12} since this is the force on charge 1 due to charge 2. This is a technique used throughout this chapter to label the forces on a charge due to another charge.

Step 2: Compute the magnitude of the Coulomb force on charge 1 due to charge 2 using Coulomb's Law. Remember the charge on a proton is a positive value of $q = 1.602 \times 10^{-19}$C.

Start with Coulomb's Law: $F_{12} = \frac{k|q_1| \cdot |q_2|}{r^2}$

Plug in the values of k, charge, and r to find the magnitude of the force.

$$F_{12} = \frac{\left(9.0 \times 10^9 \ \frac{\text{Nm}^2}{\text{C}^2}\right) \left|1.602 \times 10^{-19} \ \text{C}\right| \cdot \left|1.602 \times 10^{-19} \ \text{C}\right|}{\left(1 \ \times 10^{-15} \ \text{m}\right)^2}$$

$$F_{12} = 231 \text{ N}$$

Example 5.2

One charge of +1.00C is located at the origin and another charge of −2.00C is located at a distance of 1 m away on the +x-axis. Compute the Coulomb force on Q_2 due to Q_1.

Step 1: As shown in Figure 5.5, sketch the diagram and label the Coulomb force F_{21} on Q_2 due to Q_1.

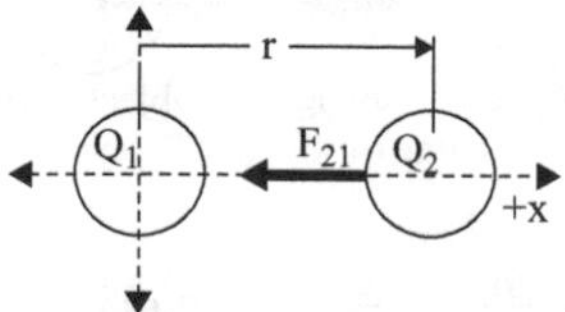

FIGURE 5.5 Diagram for Example 5.2.

Because Q_1 is positive and Q_2 is negative, the force on Q_2 due to Q_1 is attractive and is in the –x direction. Notice that the subscript used for the force is F_{21} since it is the force on Q_2 due to Q_1. Also, notice that Q was used instead of q since these are much bigger charges. This is not required but it is a common change.

Step 2: Compute the magnitude of the Coulomb force on charge 2 due to charge 1 using Coulomb's Law.

Start with Coulomb's Law: $F_{21} = \dfrac{k|Q_1|\cdot|Q_2|}{r^2}$

Plug in the values of k, charge, and r to find the magnitude of the force.

$$F_{21} = \frac{\left(9.0 \times 10^9 \ \dfrac{\text{Nm}^2}{\text{C}^2}\right)|1\text{ C}|\cdot|2\text{ C}|}{(1\text{ m})^2}$$

$$F_{21} = 1.8 \times 10^{10}\text{ N}$$

Step 3: Find the x- and y-components of the force.

$$F_{21x} = F_{21}\cos(180°) = -1.8 \times 10^{10}\text{ N and } F_{21y} = F_{12}\sin(180°) = 0$$

The magnitude of this force is huge, $F_C = 1.8 \times 10^{10}$N (1 lb/4.448 N) = 4,046,762,590 lb, because a Coulomb (C) is a large amount of charge. This is why, in most situations, the charge will be in micro-Coulombs, μC = 1×10^{-6}C, nano-Coulombs, nC = 1×10^{-9}C, or pico-Coulombs, pC = 1×10^{-12}C.

Example 5.3

Four charges are arranged, as shown in Figure 5.6.

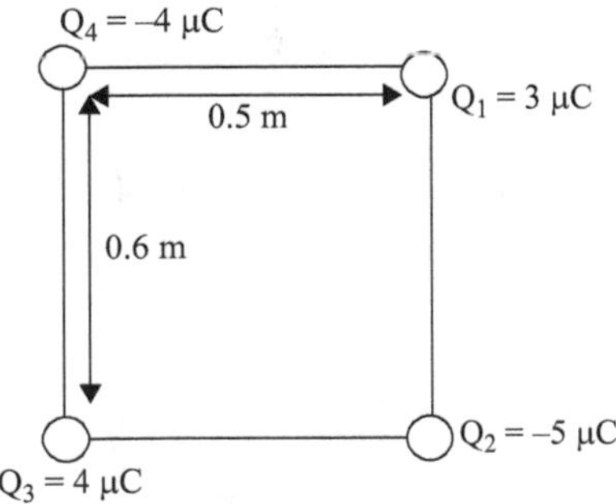

FIGURE 5.6 Example 5.3.

The horizontal side is 0.5 m in length and the vertical side is 0.6 m in length, as shown. The charges are labeled Q_1–Q_4 and their values are given on the diagram. Calculate the net force exerted on charge Q_1.

Step 1: Sketch a force diagram for the charges in this problem.

Shown in Figure 5.7, the steps are:

a. Center a coordinate system on Q_1, since the example asked us to find the force on this charge.
b. Sketch the forces for each charge on Q_1. Be careful to recognize if the force is repulsive or attractive and put it in the appropriate direction.
c. Sketch the distances, r, from Q_1 to the other charges and label using the same scheme. That is, for example, r_{12} is the distance from Q_1 to Q_2 and so on. As part of

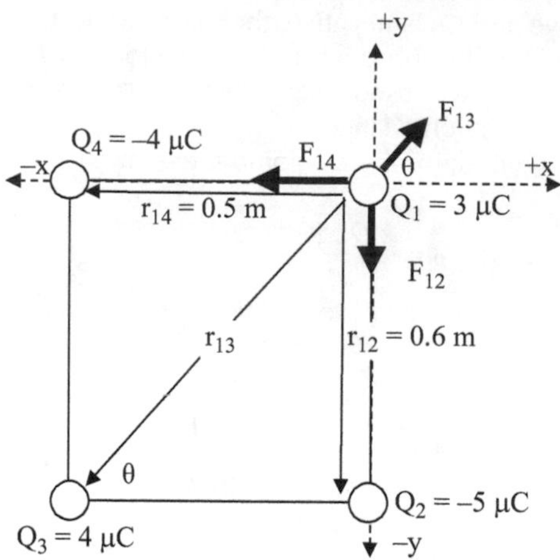

FIGURE 5.7 Force vector diagram for Example 5.3.

this step in the problem, the distances should be found. For r_{12} and r_{14}, the distances of 0.5 and 0.6 m, respectively, can be found by inspection of the diagram included with the original problem. The value of r_{13} requires the use of the Pythagorean theorem, $r_{13} = \sqrt{r_{13\_x}^2 + r_{13\_y}^2}$ so the distance between Q_1 and Q_3 is along the diagonal

$$r_{13} = \sqrt{[(0.5\ \text{m})2 + (0.6\ \text{m})2]} = 0.781\ \text{m}$$

Step 2: Calculate the absolute magnitude of each force separately:

$$F_{12} = \frac{k\ Q_1\ Q_2}{r_{12}^2} = \frac{9\times10^9\ \frac{\text{Nm}^2}{\text{C}^2}\ (3\times10^{-6}\text{C})\ (5\times10^{-6}\text{C})}{(0.6\ \text{m})^2} = 0.375\ \text{N}$$

$$F_{14} = \frac{k\ Q_1\ Q_4}{r_{14}^2} = \frac{9\times10^9\ \frac{\text{Nm}^2}{\text{C}^2}\ (3\times10^{-6}\ \text{C})\ (4\times10^{-6}\text{C})}{(0.5\ \text{m})^2} = 0.432\ \text{N}$$

$$F_{13} = \frac{k\ Q_1\ Q_3}{r_{13}^2} = \frac{9\times10^9\ \frac{\text{Nm}^2}{\text{C}^2}\ (3\times10^{-6}\ \text{C})\ (4\times10^{-6}\text{C})}{(0.781\ \text{m})^2} = 0.177\ \text{N}$$

Step 3: Find the x- and y-components of each force.
By observation of the Force diagram:

$$F_{12x} = 0\ \text{N and}\ F_{12y} = -0.375\ \text{N}$$

$$F_{14x} = -0.432\ \text{N and}\ F_{14y} = 0\ \text{N}$$

The angle θ of the force F_{13} can be found by the triangle with sides of 0.6 and 0.5 m.

$$\theta = \tan^{-1}(y/x) = \tan^{-1}(0.6/0.5) = 50.2°\ \text{to the x axis.}$$

The x- and y-components of F_{13} can be computed with the angle or with the ratios:

$$F_{13x} = (0.177\text{ N})\cos(50.2°) = 0.113\text{ N}$$

$$F_{13y} = (0.177\text{ N})\sin(50.2°) = 0.136\text{ N}$$

Step 4: Sum the components of the forces and find the magnitude and direction of the resulting force vector

$$F_x = F_{12x} + F_{14x} + F_{13x}$$

$$F_y = F_{12y} + F_{14y} + F_{13y}$$

$$F_x = -.319\text{ N}$$

$$F_y = -0.239\text{ N}$$

Find the magnitude ($\vec{F}$) and angle (θ_F) of the net force as shown in Figure 5.8

$$F = \sqrt{F_x^2 + F_y^2} = 0.4\text{ N},$$

$$\theta_F = \tan^{-1}\left(\frac{F_y}{F_x}\right) = 36.8°$$

From the diagram:

$$\theta_F = 180° + 36.8° = 216.8°.$$

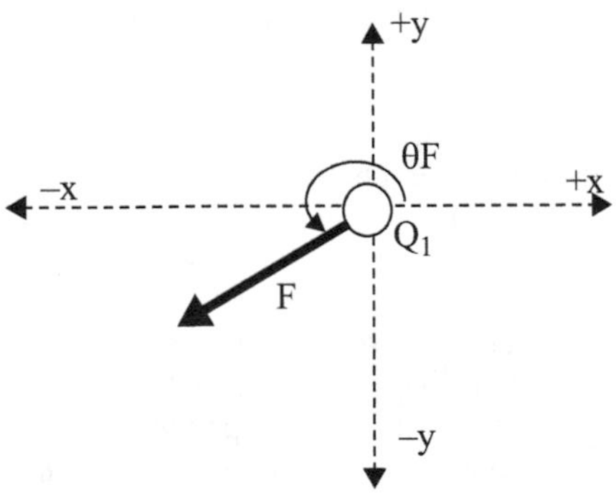

FIGURE 5.8 Resulting force vector for Example 5.3.

Notice the similarities between the analysis done in this solution and the solution to Example 5.1 of 2D Statics in Chapter 2. The electrostatic force vectors in Figure 5.7 are analyzed with the same techniques as the tension force vectors in Figure 2.9. The major difference is that in this chapter, it is assumed that the charges are fixed in place, so the calculation is only for the net electrostatic forces on the charges. The other difference is that the forces in Figure 5.7 are electrostatic and the forces in Figure 2.9 are tensions in the ropes during a tug-of-war.

5.4 ELECTRIC FIELDS

Like the gravitational field $(\vec{g})$ that is gravitational-force-per-mass in a region of space, the electric field $(\vec{E})$ is defined as the electric-force-per-charge in a region of space, given here in equation (5.2) as:

$$\vec{E} = \frac{\vec{F}}{q} \tag{5.2}$$

The electric field strength, the magnitude of the electric field at a specific point in space, is given as a value with units of newtons/Coulomb (N/C). If the electric field in a region of space is known, it is easy to compute the force on a charged particle placed in that region by multiplying the electric field vector $\vec{E}$ at the location of the charge q at that location, so equation (5.2) is rearranged to give: $\vec{F} = q\vec{E}$. The direction of the electric field at a point in space is defined as the direction of the force that the electric field would exert on a positively charged particle placed at that point in space.

5.4.1 Electric Field Due to a Particle

To compute the electric field $(\vec{E})$ at a location in space due to a charge (q_s), the following expression should be employed Since charged particles cause an electric field to exist in the region of space surrounding the said charged particle, the magnitude of the electric field due to a particle having charge q_s is given by equation (5.3) as:

$$E = \frac{F}{q} = \frac{k|q| \cdot |q_s|}{q \cdot r^2} = \frac{k \cdot |q_s|}{r^2} \tag{5.3}$$

where k is the Coulomb constant and r is the distance that the empty point in space is from the charged particle q_s, and $\hat{\mathbf{r}}$ is the unit vector in the direction of $\vec{\mathbf{r}}$. The direction of the electric field is directly away from the charged particle if the charge of the particle is positive, and directly toward the charged particle if the charge of the particle is negative.

5.4.2 Examples of Computing Electric Field Strength

Example 5.1

An object having a spherically symmetric charge distribution amounting to a total charge of −2 μC is located at the origin of a Cartesian coordinate system. Find the x-component of the electric field at an empty point in space at a distance of 2.00 cm away from the center of the object, in the +x direction (Figure 5.9).

Step 1: Draw the diagram: The electric field is in the −x direction because the electric field at an empty point in space, *p*, is in the direction of the force that would be experienced by a positively charged "test" particle if the positively charged particle were at that empty point in space, and a positively charged particle would be attracted by the given

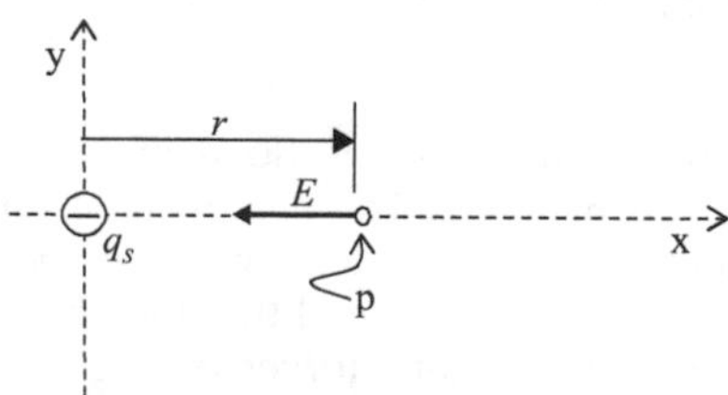

FIGURE 5.9 Diagram of electric field for Example 5.1.

negatively charged particle at the origin, meaning the positively charged particle would experience a force in the –x direction. So, that is the direction in which *E* points.

Step 2: Compute the magnitude of the electric field at this point of space

$$E = (kq_s) / r^2$$

$$E = (9.00 \times 10^9 \, \mathrm{Nm^2 / C^2}) \, (2 \times 10^{-6} \, \mathrm{C}) / (.02 \, \mathrm{m})^2$$

$$E = 4.5 \times 10^7 \, \mathrm{N/C}$$

Step 3: Compute the components of the electric field at the point of space.

$$E_x = E\cos(180°) = -4.5 \times 10^7 \, \mathrm{N/C}$$

Example 5.2

Suppose that a particle having charge +5 μC is in place at the location (2 cm, 0) at which the electric field was computed in Example 1. Find the magnitude and direction of the electric force that the particle having charge +5 μC would be experiencing.

Step 1: Draw the diagram for this example similar to Figure 5.10.

Remember that the electric field was found in the previous example and the particle is placed at the point given in the example.

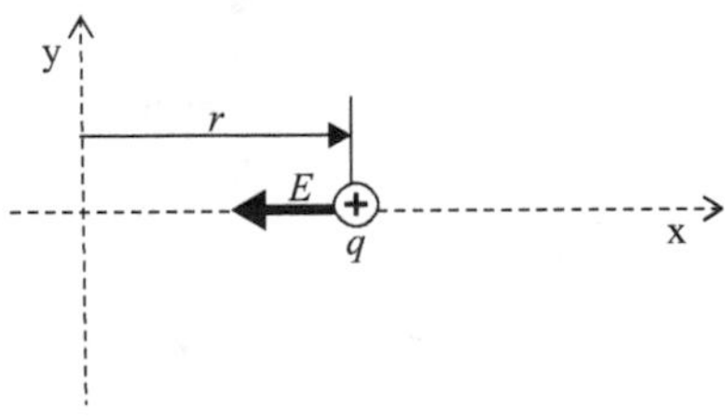

FIGURE 5.10 Diagram of electric field Example 5.2.

Step 2: Compute the magnitude of the electric force on this particle.

$$F = qE = (4.5 \times 10^7 \, \mathrm{N / C}) \, | \, 5 \times 10^{-6} \mathrm{C} = -225 \, \mathrm{N}$$

Step 3: Compute the components of the Electric field at the point of space.

$$F_x = F\cos(180°) = -225 \, \mathrm{N}$$

Example 5.3

Three charges are arranged as shown in Figure 5.11.

The horizontal side is 0.5 m in length and the vertical side is 0.6 m in length as shown. The charges are labeled Q_1–Q_4 and their values are given on the diagram. Calculate the net force exerted on charge Q_1.

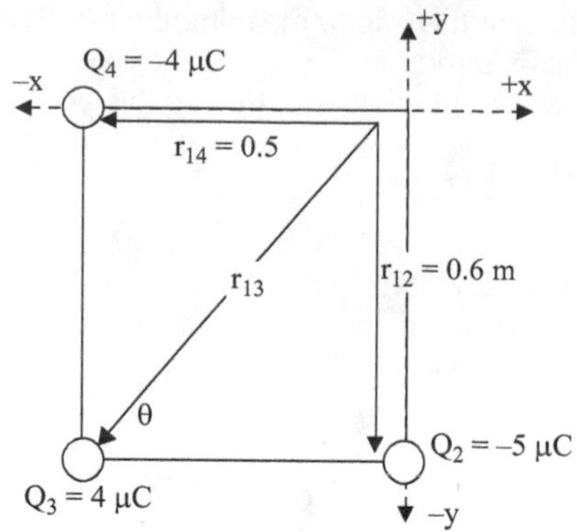

FIGURE 5.11 Electric fields Example 5.3.

Step 1: Sketch an electric field diagram for the charges in this problem. Shown in Figure 5.12, the steps are:

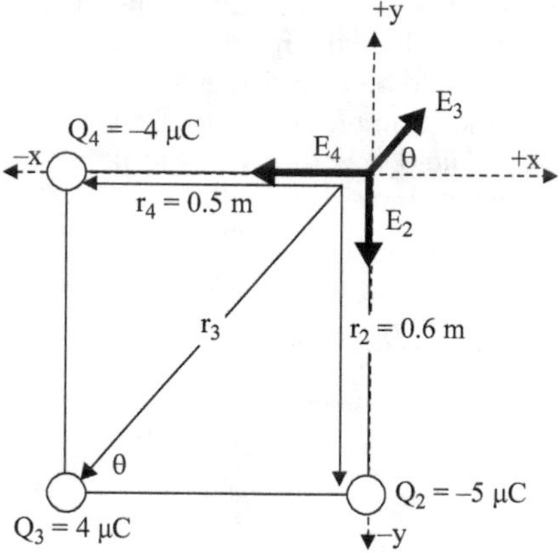

FIGURE 5.12 Electric field vector diagram for Example 5.3.

a. Sketch the electric field at the origin due to each charge. Be careful to recognize if the electric field points toward or away from the charge because the charge is either negative or positive, respectively.
b. Sketch the distances, r, from the origin to each of the charges and label using the same scheme. That is, for example, r_2 is the distance from the origin to Q_2 and so on. As part of this step in the problem, the distances should be found. For r_2 and r_4, the distances of 0.5 and 0.6 m, respectively, can be found by inspection of the diagram included with the original problem. The value of r_3 requires the use of the Pythagorean theorem, $r_3 = \sqrt{r_{13\_x}^2 + r_{13\_y}^2}$ so the distance between the origin and Q_3 is along the diagonal

$$r_3 = \sqrt{[(0.5\text{ m})2 + (0.6\text{ m})2]} = 0.781\text{ m}$$

Step 2: Calculate the absolute magnitude of each E-field separately:

$$E_2 = \frac{kQ_2}{r_2^2} = \frac{9\times10^9\ \frac{\text{Nm}^2}{\text{C}^2}\ (5\times10^{-6}\text{C})}{(0.6\text{ m})^2} = 125{,}000\ \frac{\text{N}}{\text{C}}$$

$$E_4 = \frac{kQ_4}{r_4^2} = \frac{9\times10^9 \frac{\text{Nm}^2}{\text{C}^2}\left(4\times10^{-6}\text{C}\right)}{(0.5\text{ m})^2} = 144{,}000\frac{\text{N}}{\text{C}}$$

$$E_3 = \frac{kQ_3}{r_3^2} = \frac{9\times10^9 \frac{\text{Nm}^2}{\text{C}^2}\left(4\times10^{-6}\text{C}\right)}{(0.781\text{ m})^2} = 59{,}000\frac{\text{N}}{\text{C}}$$

Step 3: Find the x- and y-components of each force.
By observation of the Force diagram:

$$E_{2x} = 0\text{ N and } E_{2\_y} = -125{,}000\ \frac{\text{N}}{\text{C}}$$

$$E_{4x} = -144{,}000\frac{\text{N}}{\text{C}}\text{ and } E_{4y} = 0\ \frac{\text{N}}{\text{C}}$$

The angle θ of the force F_{13} can be found by the triangle with sides of 0.6 and 0.5 m.

$$\theta = \tan^{-1}\left(y/x\right) = \tan^{-1}\left(0.6/0.5\right) = 50.2^\circ \text{ to the x axis.}$$

The x- and y-components of F_{13} can be computed with the angle or with the ratios:

$$E_{3x} = \left(59{,}000\frac{\text{N}}{\text{C}}\right)\cos(50.2^\circ) = 37{,}666.67\frac{\text{N}}{\text{C}}$$

$$E_{3y} = \left(59{,}000\frac{\text{N}}{\text{C}}\right)\sin(50.2^\circ) = 45{,}333.33\frac{\text{N}}{\text{C}}$$

Step 4: Sum the components of the E-fields and find the magnitude and direction of the resulting field vector.

$$E_x = E_{12x} + E_{14x} + E_{13x}$$

$$E_y = E_{12y} + E_{14y} + E_{13y}$$

$$E_x = -106{,}333.33\text{ N/C}$$

$$E_y = -79{,}666.67\text{ N/C}$$

Find the magnitude ($\vec{E}$) and angle (θ_F) of the electric as shown in Figure 5.13,

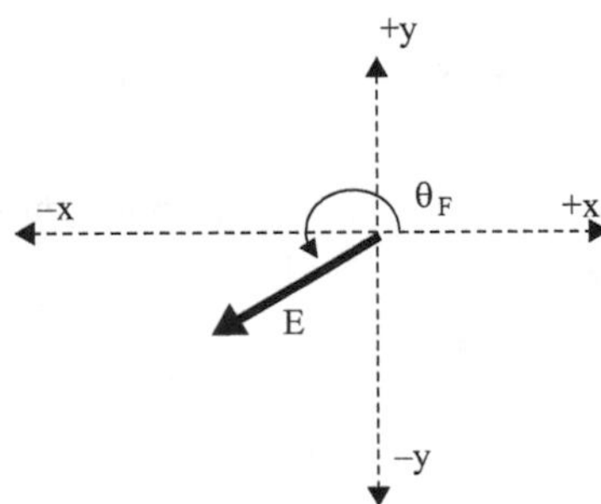

FIGURE 5.13 Resulting field vector for Example 5.3.

$$E = \sqrt{E_x^2 + E_y^2} = 133{,}333.33 \text{ N},$$

$$\theta_F = \tan^{-1}\left(\frac{F_y}{F_x}\right) = 36.8°$$

From the diagram:

$$\theta_F = 180° + 36.8° = 216.8°.$$

If you think this example looks like the one done earlier in this chapter, you are correct and that is the point. The electric field calculation is done the same way as the force is computed with the charge at the location of interest missing.

5.4.3 Electric Field Lines and Electric Field Diagrams

Electric field lines are used to represent the electric field in a region of space. They are drawn as a line or curved segments with arrowheads on them. The direction in which the arrowhead points represents the direction of the electric field at points on the line and the spacing between the lines represents the magnitude of the electric field; the closer together they are, the stronger the electric field.

The approximate direction of the electric field vectors at points in between electric field lines can be inferred from the directions of the electric field lines near those points. Notice the lines, all pointing outward from the particle having positive charge in Figure 5.14. Near the charged particle, the lines are close together, consistent with the fact that the electric field due to a charged particle is stronger at points closer to the particle than it is at points farther away. A diagram in which electric field lines are used to convey information about the electric field in a region of space is referred to as an electric field diagram. Here, two-dimensional diagrams giving the electric field in the plane of the diagram are demonstrated; in practice, the electric field extends outward in all directions.

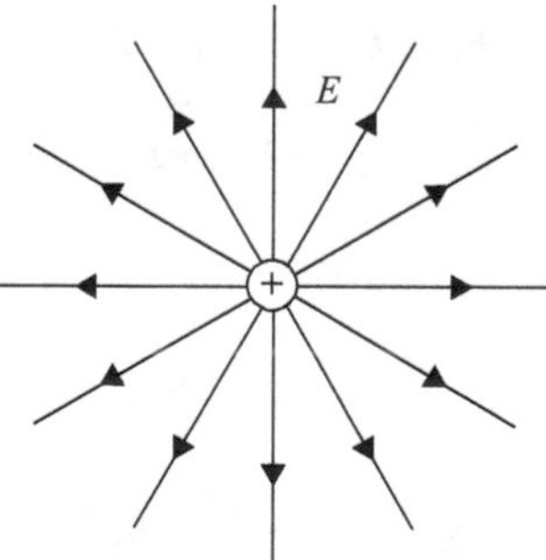

FIGURE 5.14 Electric field of a positive charge.

For the case of a negatively charged particle, the pattern is similar, but the arrows are all flipped so they point toward the particle, as shown in Figure 5.15.

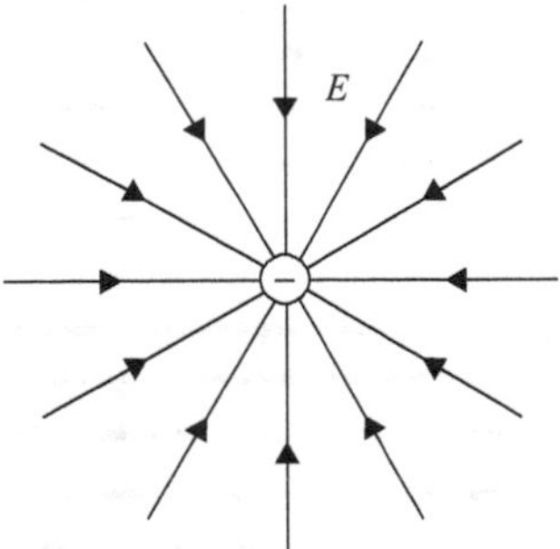

FIGURE 5.15 Electric field of a negative charge.

The electric field lines for an arrangement of a particle having a positive charge and another particle having an equal amount of negative charge are shown in Figure 5.16.

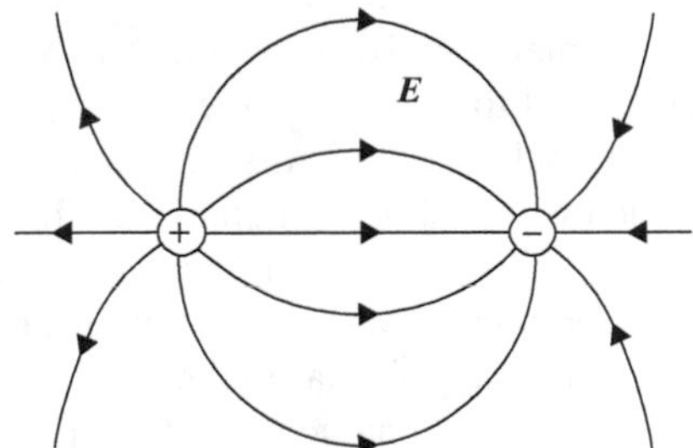

FIGURE 5.16 Electric field of a positive and negative charge in the same area.

Just like gravitational field lines, the electric field lines never cross and the relative number of field lines starting or ending at a charge is proportional to the charge. Figure 5.17 shows two positively charged particles, one with a charge of $+q$ and the other with a charge of $+2q$ (where $q>0$).

Note that there are twice as many field lines coming out of the particle having charge $+2q$ as there are coming out of the particle having charge $+q$.

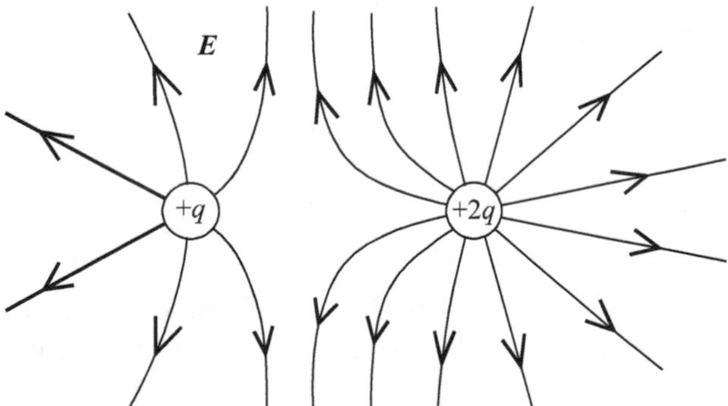

FIGURE 5.17 Electric field of two charges with different magnitudes of charge.

Given that each charged particle in a distribution of charged particles creates an electric field vector at all points in space around the charged particle, a distribution of charged particles also causes an electric field to exist in the region around that distribution. At any given point in space, the total electric field is the vector sum of all the individual contributions to the electric field, at that point in space, by all the charged particles making up the distribution. So, for example, the electric field between two lines of charges should look something like the E-field in Figure 5.18.

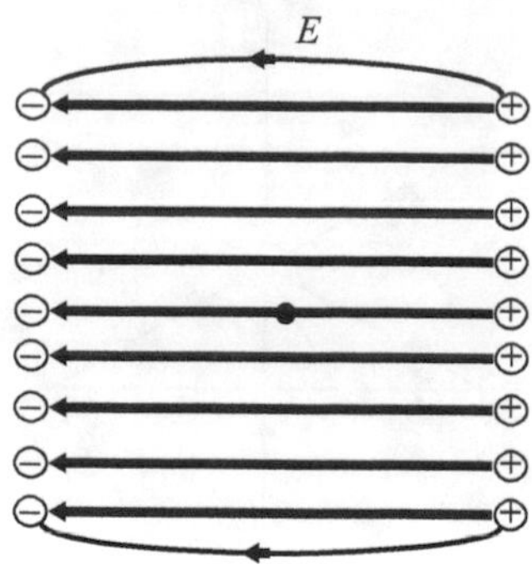

FIGURE 5.18 Electric field of two lines of opposite charge.

Looking at the solid dot in the middle of the field, it is obvious that if a positive charge that is small compared to the charges that make up the distribution is placed at that point, this positive "test" charge will be pushed down and to the left by the positive charge above it and pushed up and to the left by the positive charge below it in the distribution. The opposite will be the case for the negative charge to its left, which will pull the test charge to the left. Together, the positive and negative line of charges will cause positive charges in the region between the lines of charges to be pushed to the left. Near the top and bottom of the distributions, the field lines will curve up and down as demonstrated in Figure 5.18 in a way similar to the field lines in Figure 5.16, so it is important to use the middle of a line distribution of charge to get the most uniform field.

5.5 ELECTROSTATICS AND GRAVITY

An electric field characterizes how electric charge affects the region of space around it and a gravitational field characterizes how mass affects the region of space around it. A charged particle brought into a region with an electric field experiences a force, and a particle having mass brought into a region with a gravitational field experiences a force.

The magnitude of the *gravitational force* exerted by each of the two particles, having masses m_1 and m_2, respectively, on the other, when the particles are separated by a distance r, is given by Newton's Law of Universal Gravitation as:

$$F_g = \frac{Gm_1m_2}{r^2}, \text{ where } G = (6.67\times 10^{-11}\text{Nm}^2 / \text{kg}^2)$$

The *gravitational field strength* (magnitude of force-per-mass) at any point in the region of space around a particle having mass m_s is: $\mathbf{g} = \frac{Gm_s}{r^2}$. The SI units for the gravitational field are N/kg.

The *magnitude of the Coulomb force* exerted by each of two particles, having charges q_1 and q_2, respectively, on the other, when the particles are separated by a distance r is:

$$F_C = \frac{k|q_1q_2|}{r_{12}^2}$$

The electric field strength (magnitude of force-per-charge) at any point in the region of space around a particle having charge q_s is: $E = \frac{kq_s}{r^2}$. The SI units for electric field strength are $\left(\frac{N}{C}\right)$. It is obvious that the force and field equations are similar for gravity and electrostatics. The main differences are that gravitational forces are always attractive, while electrostatic Coulomb forces can be attractive or repulsive.

5.6 GAUSS' LAW

In the same way that Einstein's General Relativity gives us a geometric relationship for gravity, Gauss' Law gives us a geometric relationship for electric fields. Gauss' Law relates the geometry of the electric field to the arrangement of the quantities that generate the field, the charges. For example, a positive point charge, q, in Figure 5.19, will generate an electric field that points away from the charge so that if it is surrounded with a spherical Gaussian surface the electric field will be everywhere perpendicular to the surface.

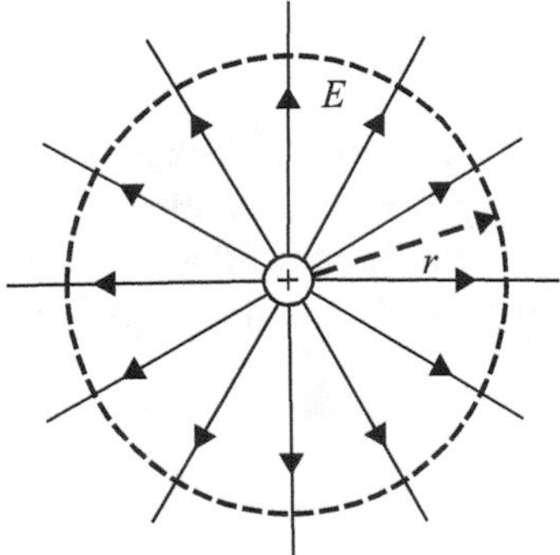

FIGURE 5.19 Gaussian surface around the electric field of a positive charge.

For this case, Gauss's Law gives the same expression of the electric field as found with Coulomb's Law. It turns out that Gauss' Law works for all geometries so it can be used to find the electric field around any arrangement of charge without knowing the forces. The only limitation is the mathematics needed to find the geometry to find the shape that encloses the field with the orientation. In most cases, to find the geometry, calculus needs to be used so this discussion will remain conceptual, but it is important to know that the electric field can be found for any distribution of charge.

5.7 ANSWER TO CHAPTER QUESTION

The role of the electric field in gel electrophoresis device is to apply a force to the molecules, such as proteins or DNA, to push them through the gel so they can eventually be grouped by their size. The experimenter starts by setting up an electric field across a container filled with a gel, as demonstrated in Figure 5.20.

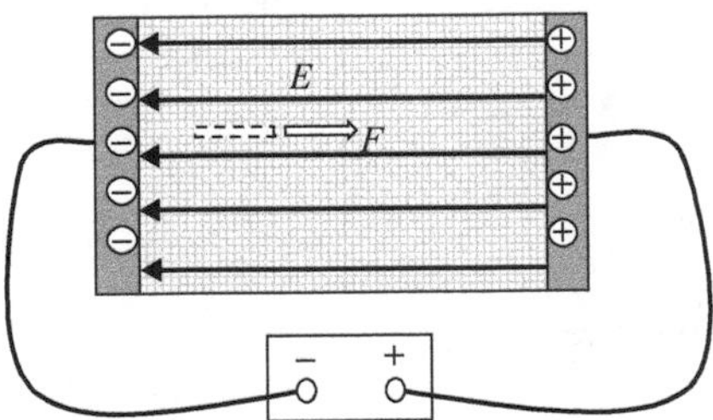

FIGURE 5.20 Gel electrophoresis device.

Negatively charged molecules, such as proteins or DNA, are inserted in the gel at the left.

In Figure 5.20, the molecule is negatively charged as it moves in a direction opposite the direction of the electric field ($\vec{E}$). The electric field exerts a force on charged molecules, which results in movement of the molecules through the gel. Force causes the molecules to move through the gel and separate out based on the size of the molecules of the same size. The forces involved and the resulting motion of the molecules will be the focus of a section in a later chapter, after all the necessary forces and descriptions of motion have been presented. Since gel electrophoresis is used in DNA analysis, it is worth understanding how this device works. In a few chapters, the analysis of this device will be presented after all the topics critical to understanding how it works are covered.

5.8 QUESTIONS AND PROBLEMS

5.8.1 Multiple-Choice Questions

1. As shown in Figure 5.21, two uncharged metal spheres, 1 & 2, are suspended from a non-conducting string and are in contact with each other. A negatively charged rod is brought close to the left side of sphere 1, without touching it, as shown in the diagram at right. The two spheres are then slightly separated (with the rod still close to sphere 1) and then the rod is withdrawn. As a result:
   A. both spheres are neutral.
   B. both spheres are positive.
   C. both spheres are negative.
   D. 1 is negative, and 2 is positive.
   E. 1 is positive, and 2 is negative.

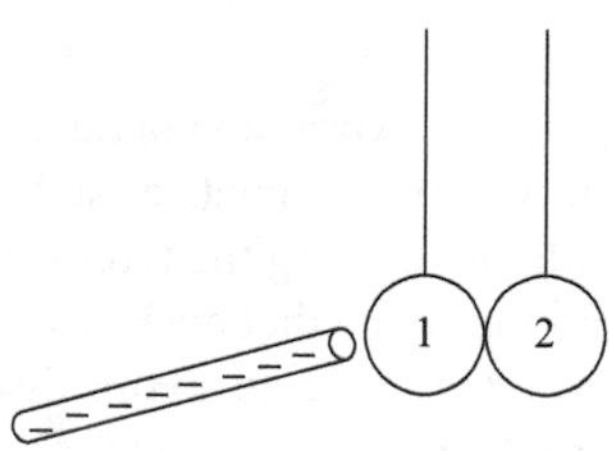

FIGURE 5.21 Multiple-choice questions 1 and 2.

2. As shown in Figure 5.21, two uncharged metal spheres, 1 & 2, are suspended from a non-conducting string and are in contact. A negatively charged rod is brought close to the left side of 1, but not touching it, as shown in in the diagram at right. Then someone, a person who is a good conductor and who is initially uncharged, touches the right side of sphere 2 while the rod is still close to sphere 1. The person withdraws her finger from sphere 2 and goes away. Then, the rod is withdrawn and finally (up until this point in time, the spheres have been in contact with each other) the spheres are separated from each other. As a result:
   A. both spheres are neutral.
   B. both spheres are positive.
   C. both spheres are negative.
   D. 1 is negative and 2 is positive.
   E. 1 is positive and 2 is negative.

3. As shown in Figure 5.22, an uncharged metal sphere is suspended from a nonconducting string, as shown in the diagram at right. Someone, who is a good conductor and is uncharged, places their finger on the right side of the sphere. With the finger in contact with the right side of the sphere, a positively charged rod is brought close to, but not into contact with, the left side of the sphere. The rod is removed from the region and then, after waiting for a few seconds, the finger is removed from the sphere, which results in a:
   A. neutral sphere. B. positive sphere. C. negative sphere.

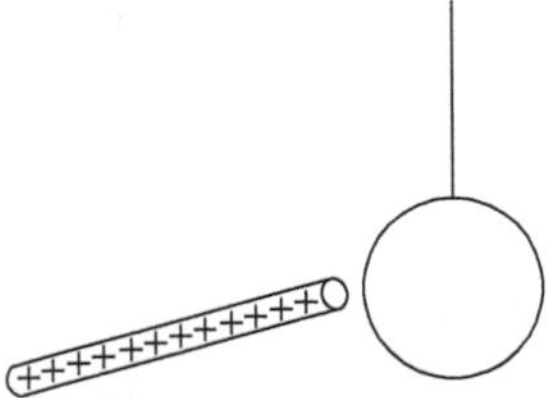

FIGURE 5.22 Multiple-choice questions 3 and 4.

4. An uncharged (neutral) metal sphere is suspended from a nonconducting string, as shown in Figure 5.22. A positively charged rod is brought close to the left side of the sphere and then someone who is a good conductor and is uncharged, places their finger on the right side of the sphere. With the rod still close to the left side of the sphere, the finger is removed from the right side of the sphere and after that, the rod is distanced from the sphere. This results in a:
   A. neutral sphere. B. positively charged sphere. C. negatively charged sphere.

5. Two particles, particle 1 and particle 2, having charges q_1 and q_2, respectively, are separated by the distance r. q_1 is twice q_2 and the magnitude of the electrostatic force being exerted by particle 1 on particle 2 is F. What is the magnitude of the electrostatic force being exerted by particle 2 on particle 1?
   A. $F/2$ B. F C. $2F$

6. Two particles, particle 1 and particle 2, having charges q_1 and q_2, respectively, are separated by the distance r. q_1 is twice q_2 and the magnitude of the electrostatic force being exerted by particle 1 on particle 2 is F. What would the magnitude of the electrostatic force being exerted by particle 1 on particle 2 be if the charge on particle 2 were twice what it is?
   A. $F/4$ B. $F/2$ C. F D. $2F$ E. $4F$

7. Is there a way to distribute charge on conducting spheres, so that appropriately drawn arrows representing forces, due only to the charge on the spheres, look like those given in the Figure 5.23?
   A. Yes B. No

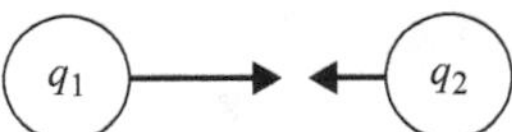

FIGURE 5.23 Multiple-choice question 7.

8. Is there a way to distribute charge on conducting spheres, so that appropriately drawn arrows representing forces, due only to the charge on the spheres, look like those given in Figure 5.24?
   A. Yes B. No

FIGURE 5.24 Multiple-choice question 8.

9. A particle having charge +1 nC charge is located at position (1 m, 0) and a particle having charge –1 nC is located at (–1 m, 0). In what direction is the electric field at the origin (0,0) due to the charge distribution?
   A. +x B. –x C. +y
   D. –y E. No direction ($E = 0$)

10. Particle 1 having charge $q_1 = .900$ μC is at the origin of a Cartesian coordinate system and particle 2 having charge $q_2 = -.100$ μC is at (5.00 cm, 0). Consider the +x direction to be rightward. In what region or regions is there a point that is not an infinite distance from the origin, where the electric field is 0?
    A. On the x-axis, to the left of particle 1.
    B. On the x-axis, between particles 1 and 2.
    C. On the x-axis, to the right of particle 2.
    D. Both a and b above.
    E. Both b and c above.
    F. Both a and c above.
    G. All of a, b, and c above.
    H. None of the above.
    I. There is no such region.

5.8.2 Problems

1. A 2 C charge is located 20 cm away from a 3 C charge. Compute the electrostatic force that the 2 C charge exerts on the 3 C charge.
2. A 5 C charge is located 10 cm away from a –4 C charge. Compute the electrostatic force that the 5 C charge exerts on the –4 C charge.
3. Compute the magnitude of the electric field 20 cm away from a 3 C charge.
4. Compute the magnitude of the electric field 10 cm away from a –5 C charge.
5. Two +4 nC charges are separated by 4 m. Find the magnitude of the electric field at a point midway between them.
6. In Figure 5.25, all three charges are fixed in their locations. The charge of q_1 is –300.0 μC and it is located at (+5 m, +4 m), the charge of q_2 is +40.0 μC and it is located at (+2 m, 0 m), and the charge of $q_3 = -20$ μC and it is located at (0, 1 m). Compute the magnitude of the net electric force on (q_3) due to q_1 and q_2.

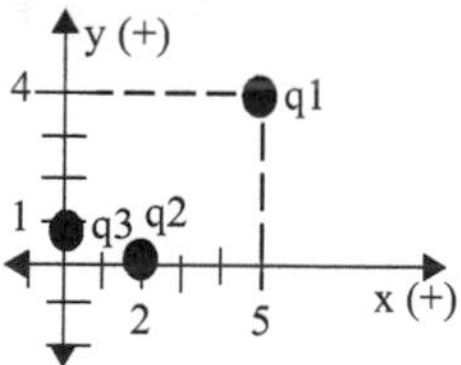

FIGURE 5.25 Problem 6.

7. In Figure 5.26, two charges are fixed in their locations. The charge q_1 is +2.0 μC and it is located at the origin (0 mm, 0 mm) and the charge of q_2 is –4.0 μC and it is located at (–3 mm, 4 mm). Compute the x & y coordinates of the force on q_1.

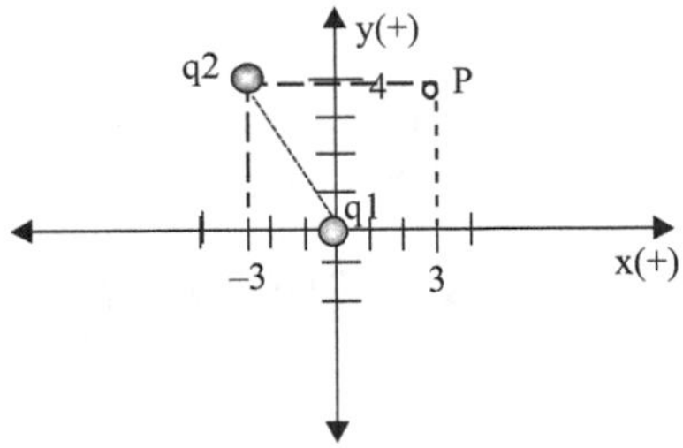

FIGURE 5.26 Problem 7.

8. Three charged particles are located at the following points, q_1 = .150 μC is at the origin, q_2 = .420 μC is at the (1.20 cm, 0), and q_3 = −.250 μC is at the (1.20 cm, .900 cm).
   Find the x-component of the net electrostatic force on q_1 due to q_2 and q_3.
9. There are two charges, q_1 = .900 μC at the origin and q_2 = −.100 μC is at (5.00 cm, 0). Find the electric field, due to the two charged particles, at the point (8.00 cm, 4.00 cm).
10. In Figure 5.27, two fixed charges, q_1 = 2.0 μC & q_2 = −4.0 μC, are located at (2 m, −1 m) and (−4 m, 3 m), respectively. Compute the electric field at point P, due to charges q_1 and q_2.

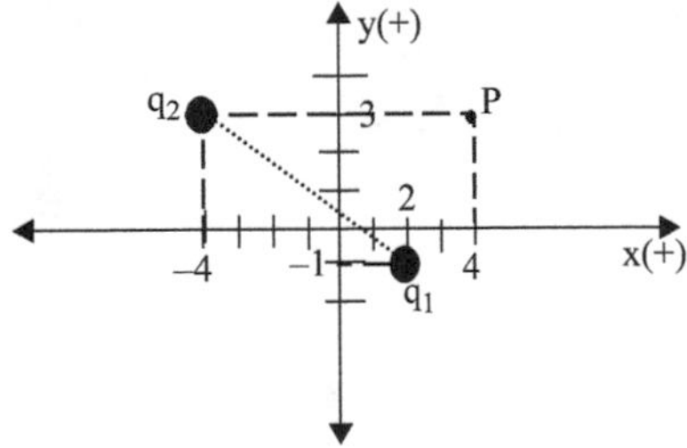

FIGURE 5.27 Problem 10.

APPENDIX: NUCLEAR FORCES

The Nuclear Strong Force

The nucleus of an atom, larger than hydrogen, is made up of protons and neutrons compressed in an area of approximately 1×10^{-15} m in diameter. Like charges repel and even with the tiny positive charge of each proton at this close proximity, the resulting electrostatic repulsive forces is hundreds of Newtons on these tiny particles. Given these repulsive forces, the nucleus of every atom should be ripped apart by its own internal electrostatic forces. Obviously, this doesn't happen, under normal conditions, so there must be another force of nature which holds a nucleus together against the enormous forces of electrostatic repulsion due to the protons. This force that holds the nucleus together is the Nuclear Strong Force.

A widely accepted expression for the Nuclear Strong Force was created by Hideki Yukawa. The force is very strong over a short distance, range, but its strength decreases rapidly outside this range. So, at short distances it is the strongest, but as the particles move further apart the strength of the force decreases to approximately zero. This lets the force dominate at the very small distances, within the nucleus, while having negligible effects at distances outside the nucleus.

The Nuclear Weak Force

Of the three common types of radioactivity, beta decay cannot be explained with the other forces of nature. Since the beta particle has a negative charge, this particle should be trapped inside the positive nucleus by electrostatic attraction. Thus, another force, with a format similar to the Nuclear Strong Force, was created to explain this decay. The base expression for the Nuclear Weak Force is the same as the expression for the strong force with different constants and range, so there is an intermediate region on the outside of larger nuclei that the weak force can overcome the strong force and eject a beta particle. At a distance between the nucleons of 6.5×10^{-15}m, the weak force becomes stronger than the strong force. This is the radius at the outside of the nucleus where an unpaired neutron will undergo B-decay, leaving a proton in its place.

Although humans do not have firsthand knowledge of these nuclear forces they are important to the structure and operation of all matter. The strong force is crucial to the formation of all nuclei and the weak force provides an understanding of a crucial step in the processes of the sun along with an explanation of beta decay.

6 Magnetic Forces

6.1 INTRODUCTION

If you have ever used a magnet to pick up a paper clip, you have experienced the magnetic force. This force of nature is normally combined with the electric force from the previous chapter into one force of nature called the electromagnetic force. As you read through this chapter, it will become apparent that both these forces, electric and magnetic, are associated with the electrical charge of an object. Therefore, in this chapter, the magnetic force will be computed with the quantities of electrical current and electric charge. Examples of each type of force calculation will be presented and the definition of the magnetic field will be established. The connection between the analyses employed in the study of torques will also be presented to complement and spiral with the sequence of analysis established in the first three chapters of the text.

Chapter question: An MRI (Magnetic Resonance Imaging) is a sophisticated piece of medical machinery that allows doctors to see soft tissue in a way that X-ray cannot. The word magnetic is part of the acronym by which these devices are known, so what is the role of the magnetic in MRI? This question will be answered at the end of this chapter after the concepts of magnetic forces and magnetic fields are studied.

6.2 MAGNETS

The simplest place to start a discussion about magnetic phenomena is a bar magnet, which is most likely some composite of iron, cobalt, and nickel that has been magnetized by placing it near another strong magnet. These metals that hold their magnetic field are called ferromagnets. If a bar magnet, like the one in Figure 6.1, is split in half, each half will have a north and a south pole.

If each half is split over and over again, each piece will still have both poles. In fact, atoms, electrons, protons, and neutrons have both a north and a south pole. If we split a magnet in half and then flip one of the magnets, the like poles will face each other, and the magnets will repel. See Figure 6.1. So, like electrical phenomena that have positive and negative properties, there are two types of magnetic poles known as north and south (N and S) that attract each other and repel like poles, but unlike the positive and negative charges of electricity that can be separated, the north and south poles cannot be separated. There is no such thing as free magnetic charge.

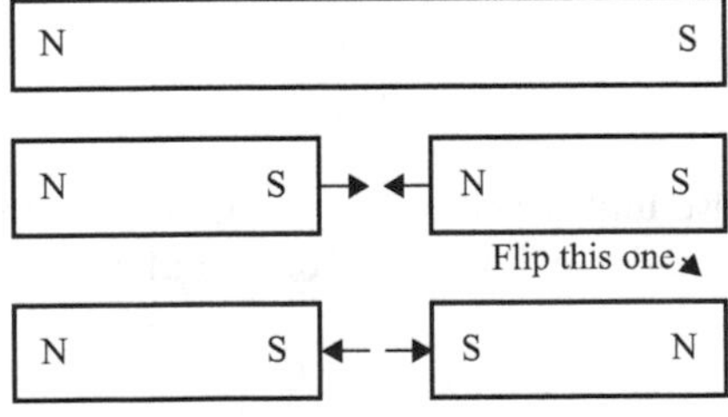

FIGURE 6.1 Magnet poles (N's and S's) cannot be separated.

As with gravity and electrostatics, since there is a force there must be a magnetic field, which is a vector denoted with the symbol, $\vec{\mathbf{B}}$.

This field points from North to South between two magnets, as shown in Figure 6.2.

DOI: 10.1201/9781003308065-6

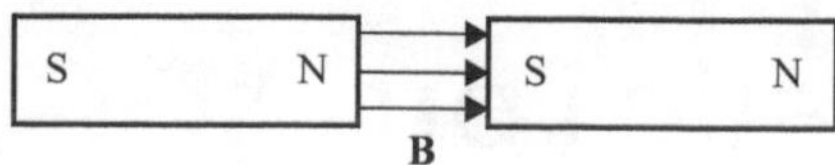

FIGURE 6.2 Magnetic field between two permanent magnets.

If you have played with magnets, you would know that the magnetic force is similar to the forces of gravity and the Coulomb electrostatic force, in that it can act over a distance. A strong enough magnet can pick up a paper clip off a desk without initially touching the paperclip as long as it is brought close enough. Like the Coulomb electrostatic force, it can be both attractive and repulsive. Unlike the forces of gravity and the Coulomb (electrostatic) force, which depend on the mass and the charge of a particle, respectively, the characteristic of a particle that mediates the magnetic force is not obvious since the individual poles (N's & S's) cannot be separated. Thus, since any object must always contain the same amount of north and south poles, the magnetic force must be based on another quantity. It turns out that this quantity is the electrical current (I or i).

6.3 MAGNETIC FORCE ON AN ELECTRICAL CURRENT

Analogous to the flow of water through a pipe, the electrical current, **I**, is the follow rate of electric charge past a point in space but more commonly past a point in a wire. For charge to flow through a wire, an electric field must be present in the wire to push the charge in the direction of the field, Figure 6.3. The field is established by creating an excess of negative charge on one side of the wire and leaving excess positive charge on the other side of the wire. This can be done with a battery or a DC power supply.

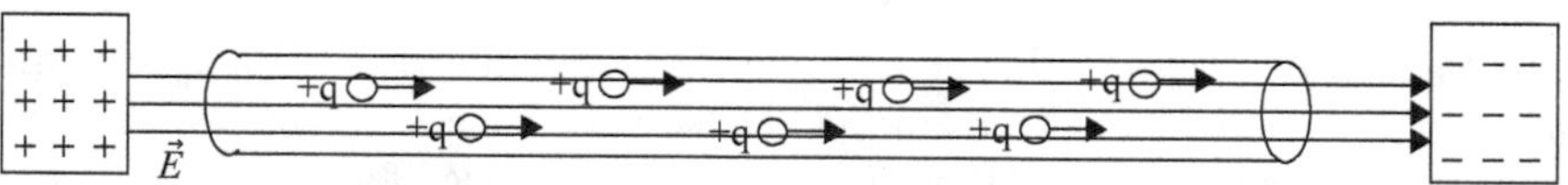

FIGURE 6.3 The flow of electric charge is electrical current.

Once the electric field is established, the current can flow. For variables that are changing, the capital delta (Δ) is used in front of the variable so the current is expressed in terms of the change in the amount of charge, Δq, flowing through a specific region of space in a specific amount of time, Δt, given by equation (6.1) as:

$$I = \frac{\Delta q}{\Delta t} \tag{6.1}$$

As stated in Chapter 1, the base unit for electrical phenomena is the unit of electric current, the Ampere (Amp = A), which is equal to a flow rate of one Coulomb per second.

$$1 \text{ Amp} = 1 \text{ C/s}$$

The current flows in the direction of the electric field from the positive to the negative of a power supply, like a battery. Since a Coulomb is a large amount of charge, an Amp (A) is a lot of current. Thus, in many situations, the current is given in milliamps mA (1×10^{-3} A) or microamps μA (1×10^{-6} A).

The connection between electrical current and magnetic forces was made in 1820 by Hans Christian Oersted, when he observed an electrical current twisting the needle of a compass. In his

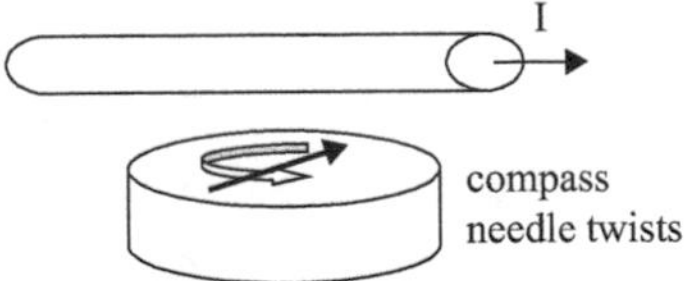

FIGURE 6.4 Oersted's experiment showing the effect of electrical currents on magnets.

experiment, as depicted in Figure 6.4, he aligned a compass so it pointed north, stretched a wire so it was parallel to the compass needle, then he connected a circuit so a current flowed through the wire.

The compass needle twisted and the connection between electrical current and the magnetic field was established.

Soon after this discovery, Andre-Marie Ampere, the French scientist after whom the unit of electrical current is named, observed the force of attraction between two parallel wires carrying electrical currents in the same direction. Ampere reasoned that since the current in the wire exerts a force on a magnetic compass needle, the force between the wires is also magnetic. He set up an experiment with two parallel wires, as shown in Figure 6.5, with the current running through them in the same direction.

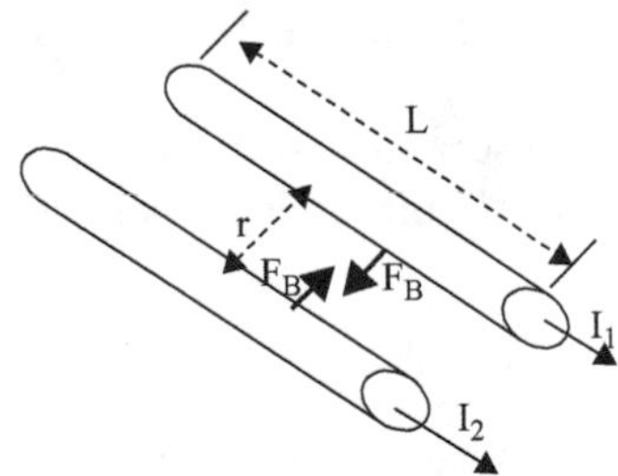

FIGURE 6.5 Ampere's experiment with two electrical currents exerting forces on each other.

He measured the force, **F**, between the wires at a series of currents, I_1 and I_2, and separation distances, **r**. He reasoned that the force was of a form similar to the force of gravity and electrostatic force, but the $1/r^2$ dependence did not match the data. Instead, the equation has $1/r$ dependence.

Ampere also switched one of the currents so the currents were still parallel, but flowing in the opposite direction. In this orientation, the force had the same magnitude, but it was repulsive instead of attractive. Because in practice it is difficult to measure the flow rate of charge through a wire, the Amp is defined as the amount of current in each of the two long straight wires separated by 1 m, which results in a force of 2×10^{-7} N, on each meter of the wires. The force between wires with even a substantial amount of current is very small, thus it is negligible for most situations.

The magnetic field is defined by the force on an electric current in a magnet field. As demonstrated in Figure 6.6, the magnetic field ($\vec{B}$) is defined as flowing from the north (N) to the south (S) pole of the horseshoe magnet and the current flows from left to right across the page.

The magnitude of the magnetic force (F), which the magnetic field exerts on the segment of the current (I) with a length (L), is perpendicular to the plane defined by the magnetic field and the current is given by equation (6.2) as:

$$F = B\ I\ L|\sin(\theta)| \tag{6.2}$$

where:

F is the force exerted on the wire-segment,
I is the current in the wire,

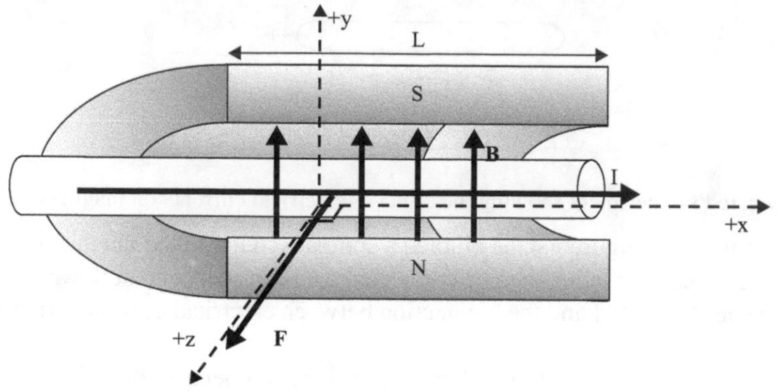

FIGURE 6.6 Magnetic force on a current-carrying wire.

L is the length of that segment of the wire, which is *in* the magnetic field,
B is the uniform magnetic field along the entire length of the wire,
θ is the angle between the magnetic field and the direction of the current.

These relationships can be more easily seen in Figure 6.7, which includes the forces, currents, and the magnetic field. The dot in the center of the circle is a simple way to represent the force coming out of the page.

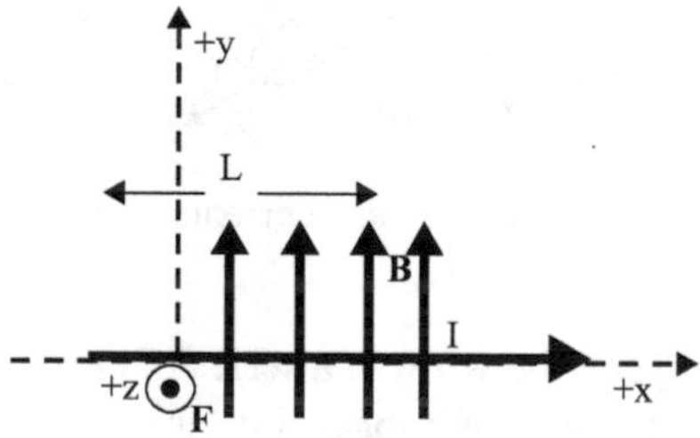

FIGURE 6.7 The magnetic force on a current-carrying wire.

Thus, if the field and the current are in the same direction ($\theta = 0$) the $\sin(0°) = 0$, there is no force. This is consistent with experimental evidence that the force is zero on a current in a magnetic field, which is parallel or antiparallel. The force is maximum when $\theta = 90°$ or $\theta = 270°$.

Another common orientation for a magnetic force problem is given in Figure 6.8; the same force presented in a different orientation and the magnetic field is generated between two separate magnets. The dot in the center of the circle is a simple way to represent a current flowing out of the page. The dot represents the tip of an arrow pointing out of the page.

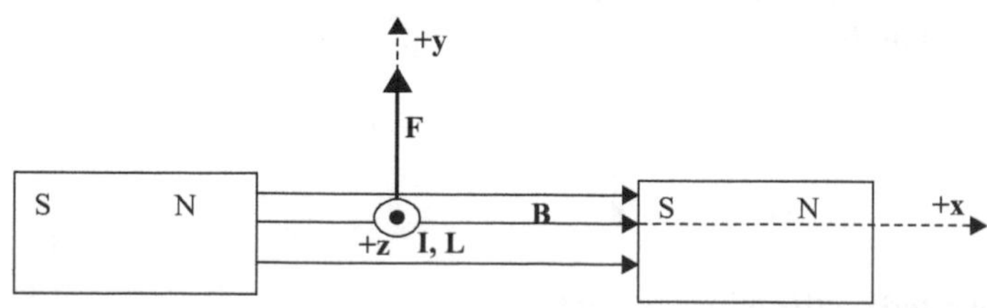

FIGURE 6.8 A current flowing through a region of space that is filled with a magnetic field.

If on the other hand the current was flowing into the page, there would be an (**x**) in the center of the circle to represent the tail feathers of an arrow. As shown in Figure 6.9, the force resulting from the interaction of the current and magnetic field is perpendicular to the plane created by the current and the magnetic field.

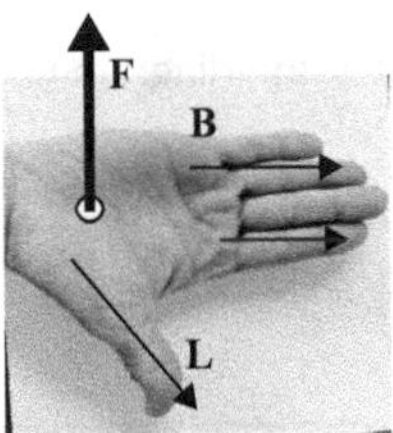

FIGURE 6.9 Magnetic right-hand rule.

The force resulting from the interaction of an electrical current and a magnetic field is perpendicular to the plane created by the current and the magnetic field. Thus, the direction of the force on a current-carrying wire can be found with a right-hand rule (RHR).

Magnetic force right-hand rule, as shown in Figure 6.9, is:

1. First, put the fingers of your right hand in the direction of the magnetic field.
2. Second, put the thumb of your right hand in the direction of the length of the wire in the direction in which the current is flowing.
3. Third, the palm of your hand will be pointing in the direction of the force on the current-carrying wire in the magnetic field.

6.3.1 Units of Magnetic Field

Since the magnetic field is defined by the force on a length of current-carrying wire, equation (6.1) can be solved for B to give:

$$B = \frac{F}{I\ L\ \sin(\theta)}$$

So, the units of magnetic field are $\left(\frac{\text{N}}{\text{mA}}\right)$, which is called a Tesla (T).

$$1\ \text{T} = 1\frac{\text{N}}{\text{m A}}$$

Since many common magnetic fields are a very small number of Tesla, for example, the average magnetic field of the Earth is $B_{\text{Earth}} = 0.00005$ T, a smaller unit of magnetic field was created. This is the Gauss (G):

$$1\ \text{G} = 1 \times 10^{-4}\text{T}$$

The magnetic field is defined as the force per current lengths in the region, since it is an electrical current upon which a force is applied if it is in a magnetic field:

Unlike the electric field strength, in which $E = \frac{F}{q}$, and the gravitational field strength, where $g = \frac{F}{m}$, there is no free "magnetic charge" since the poles cannot be separated. Instead, the magnetic field is defined as *Force per length of current*, $B = \frac{F}{(I\ L)}$. Thus, it is electrical currents that "feel" magnetic forces.

6.3.2 Example of Computing a Magnetic Force on a Current-Carrying Wire

Compute the magnitude and direction of the force on a 12 cm long wire carrying 3.0 A of current in the +x-direction in a region with a constant magnetic field of 4.0×10^{-7} T in the +z-direction.

Solution:

Step 1: As shown in Figure 6.10, sketch the diagram paying close attention to the directions of the current and the magnetic field.

The dots represent the magnetic field coming out of the page in the +z-direction and the current is moving to the right in the +x-direction.

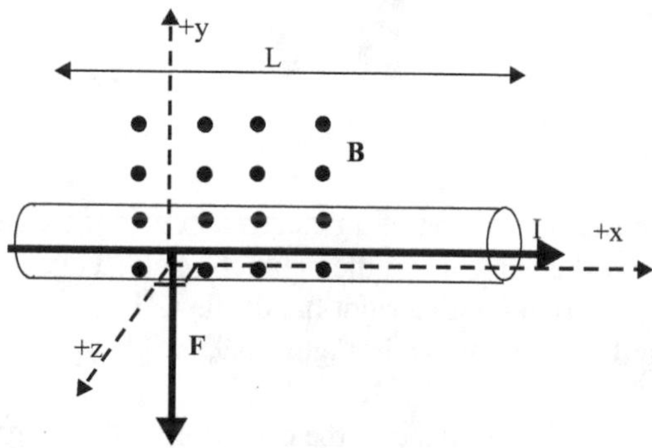

FIGURE 6.10 Example of the magnetic force on a current-carrying wire.

Step 2: Compute the magnitude of the magnetic force on the current-carrying wire with equation (6.2),

$$\mathbf{F = B\,I\,L\,sin(\theta)}$$

$$F = B\ I\ L\sin(\theta) = \left(4.0 \times 10^{-7}\text{T}\right)(3.0\text{ A})(12\text{ m}) = 1.4 \times 10^{-7}\text{ N}$$

Step 3: Use the RHR to find the direction of the magnetic force on the current-carrying wire and draw the force vector on the diagram. Notice that the x- and z-components of the force are zero and the magnetic force on the current-carrying wire is in the –y-direction.

$$F_x = 0, \quad F_y = -1.4 \times 10^{-7}\text{ N}, \quad F_z = 0$$

6.4 MAGNETIC FORCE ON A MOVING CHARGED PARTICLE

As demonstrated in Figure 6.11, an electrical current can be thought of as the flow of electrical charge moving through a wire. If this wire is in a magnetic field, the force due to the magnetic field can be thought of as acting on the individual charges moving at a given speed.

Given that the force due the magnetic field is $F = B\ I\ L\ \sin(\theta)$ and the electrical current is the flow rate of electric charge, $\text{I} = \frac{\Delta q}{\Delta t}$, the product of the current and a length is equivalent to the charge moving a specific speed. For example, a current of 1 A is flowing through a 2 m length of wire that is the equivalent flow of charge to every (C) Coulomb of charge moving at 2 m/s. Using this relationship, the force on a charged particle moving at with a velocity through a magnetic field can be written as:

$$F = \frac{\Delta q}{\Delta t} L\ B\sin(\theta) = q\frac{\Delta L}{\Delta t} B\sin(\theta) = q\ v\ B\sin(\theta)$$

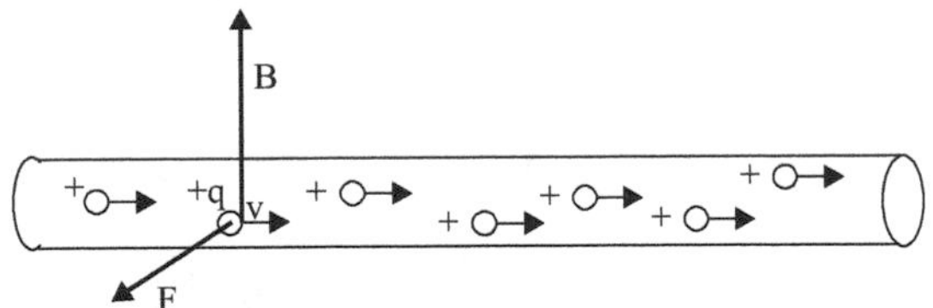

FIGURE 6.11 Positively charged particle moving through a wire in a magnetic field.

which can be simplified in equation (6.3) as:

$$F = q\ v\ B\left|\sin(\theta)\right| \tag{6.3}$$

with q and v as the electrical charge and speed of the moving particle in a magnetic field B. The angle θ is between the direction of the speed (v) and the magnetic field (B).

The direction of the magnetic force on a moving positively charged particle can be determined by the RHR.

1. Point your fingers in the direction of the magnetic field.
2. Point your thumb in the direction of the positive charge's motion.
3. The palm of your hand points in the direction of the force on the charge.

If the charge that is moving through a magnetic field is negative the magnitude is still computed with the same equation and the direction is opposite. So, you can either flip the direction given by the RHR or use your left hand. The magnetic force on a moving charged particle depends on the magnetic field, the amount of charge on the particle, the speed of the particle, and the angle between the direction the particle is moving and the magnetic field.

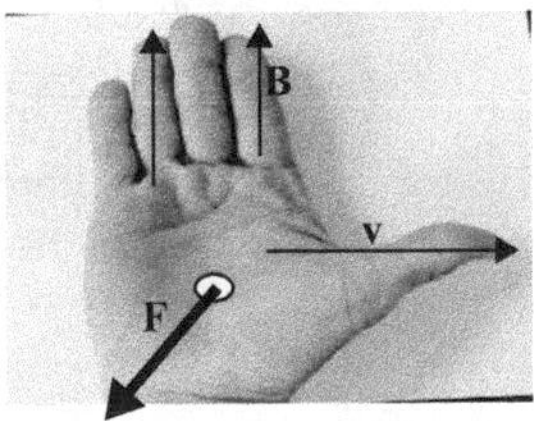

FIGURE 6.12 RHR for a moving positively charged particle.

If the charge is at rest ($v = 0$), there is no magnetic force. Also, since the $\sin(0°) = \sin(180°) = 0$, a charged particle moving in the direction or in the opposite direction of the magnetic field does not experience a magnetic force. This is the case for any charged particles moving in magnetic fields.

6.4.1 Examples of Computing the Force on a Moving Charged Particle

Example 6.1

A proton travels along the +z-axis with a speed of 1,000,000 m/s in a region of space with a magnetic field of 0.5 T in the +y-direction. Find the magnitude and direction of the force on the proton.

Step 1: As shown in Figure 6.13, sketch the diagram by first putting a circle at the origin to represent the proton, and then add the vectors for the velocity and the magnetic field.

Step 2: Compute the magnitude of the magnetic force on the proton with equation (6.3)

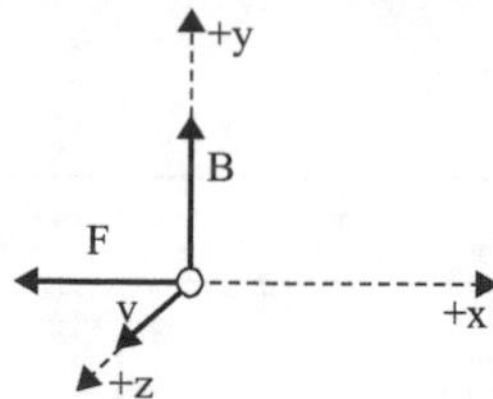

FIGURE 6.13 Diagram for the example of a proton moving through a magnetic field.

$$F = B\ q\ v\ \sin(\theta)$$

$$F = B\ q\ v\sin(\theta) = (0.5\ \text{T})\left(1.602\ \times10^{-19}\text{C}\right)\left(1{,}000{,}000\ \text{m/s}\right) = 8.01 \times 10^{-14}\ \text{N}$$

Step 3: Find the direction of the force with the RHR, by pointing the fingers of your right hand in the direction of the magnetic field and pointing your right thumb in the direction of the velocity. The palm of your right hand will point in the negative x-direction. This results in a final answer for this example of:

$$F_x = -8.01 \times 10^{-14}\ \text{N}, \quad F_y = 0, \quad F_z = 0$$

For the force on the proton as it moves through this magnetic field.

Example 6.2

Radioactivity is a natural process in which particles are ejected out of some isotopes of some elements. How can a magnetic field be used to discover the sign of the particles ejected in a radioactive process?

Answer: In many cases, radioactive elements have a large number of protons and neutrons in their nucleus, like Uranium-238 ($_{92}U^{238}$) (α) or Thorium-234 ($_{90}TH^{234}$) (β). In some cases, the isotopes of relatively light elements like Carbon-14 ($_6C^{14}$) are radioactive. When a radioactive sample is dug out of the ground, it will likely contain several different elements, such as Uranium and Thorium. In some cases, these elements will contain some percentage of radioactive isotopes that eject different types of particles. Information about the types of particles ejected from the sample can be gathered by using electric and magnetic fields.

If the sample is placed in a lead box, with thick enough sides, none of the radioactive particles will escape. If a second lead box is placed around the box containing the sample and a hole is drilled in the top of the two boxes, as sketched in Figure 6.14, a stream of particles will come straight out of the top of the boxes.

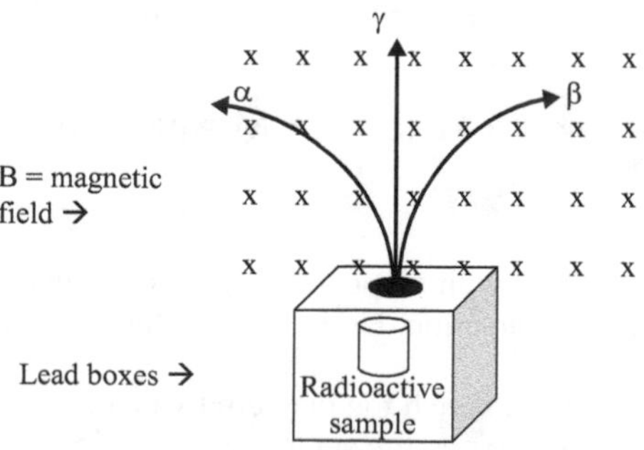

FIGURE 6.14 The three types of radioactive particles.

In this way, the sign of the charge of the particles can be discovered by observing the way the particles bend and thus finding the locations at which they are detected. The three common types of particles are named for the first three letters of the Greek alphabet, alpha (α), beta (β), and gamma (γ).

Using the RHR, with the fingers of your right hand pointing to the page (direction of the magnetic field) and your thumb pointing to the top of the page (direction of the velocity), the palm of your right hand will point left, the path of the α particles. So α particles are positive. You can then point the fingers of your left hand to the page (direction of the magnetic field) and your thumb pointing to the top of the page (direction of the velocity) such that the palm of your left hand will point right, the path of the β-particles. So, β particles are negative. Since the γ particles pass through a magnetic field without bending, they are neutral.

6.5 AMPERE'S LAW

Like Einstein's General Relativity for gravitational fields and Gauss' Law for electric fields, there is a geometric relationship for magnetic fields known as Ampere's Law. This law relates the magnitude of magnetic field (B) in a space around an electric current (I). These currents can be huge like circulating charge that produces the magnetic field of the sun or tiny like the "orbiting" electrons around the nucleus of atoms. When the magnetic force was first introduced, the magnetic field ($\vec{\mathbf{B}}$) was defined in terms of the force on electrical currents or moving charged particles. Since electrical currents, which are moving charged particles, generate magnetic fields, it makes sense that magnetic fields should generate a force on moving charges or electrical currents. The moving charges generate their own magnetic fields, which interact with the magnetic field in the region resulting in a force.

For example, the magnetic field generated by a straight electrical current is in the form of symmetrical, concentric, circles around the wire, as depicted in Figure 6.15.

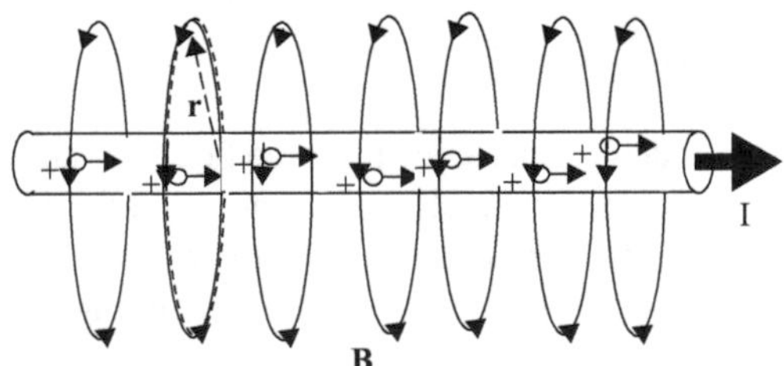

FIGURE 6.15 Magnetic field around a straight electrical current.

So, the magnetic field around a wire with a current, I, flowing through it can be found with Ampere's Law and information about the geometry of the arrangement. In the same way that Gauss' Law provides deeper understanding to the Coulomb's Force expression, Ampere's Law can be used to explain the force Ampere measured between two parallel current-carrying wires. Referring to the two current-carrying wires in Figure 6.16, a magnetic field, B_1, generated by the current one, I_1, at the location of current two, I_2, results in a force.

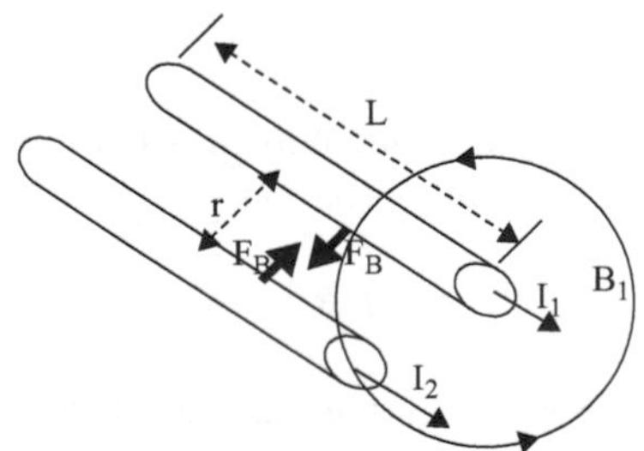

FIGURE 6.16 Magnetic field and resulting force between two parallel electrical currents.

Therefore, it is not simply that two electrical currents have a force on each other, it is that each wire produces a magnetic field that can be found with Ampere's Law and the resulting force is simply the force on the current-carrying wire in a magnetic field.

6.6 MAGNETIC MOMENT

If a rectangular coil of wire with a length, **L**, width, **W**, and a current, **I**, flowing through each of the N loops of wire is placed in a magnetic field, **B**, the loop will experience a torque, $\boldsymbol{\tau}$. For the loop described in Figure 6.17, there is a force, F_1, into the page on the left side of the loop and a force, F_2, out of the page on the right side.

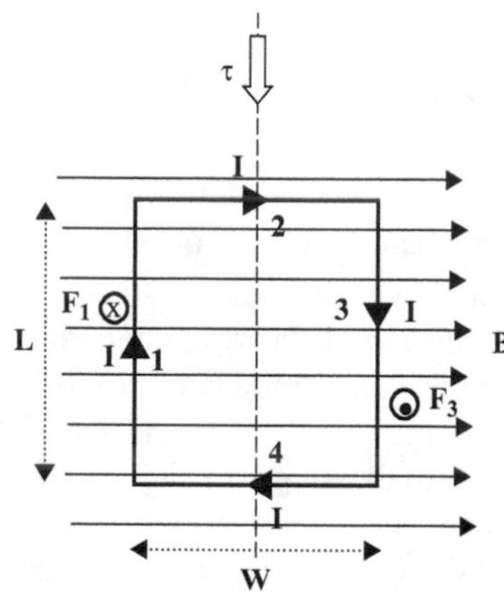

FIGURE 6.17 Current loop in a magnetic field.

The top, side 2, and bottom, side 4, experience no force because the angle between the magnetic field and the current is 0° and 180°, respectively. Since the sine of each of these angles is zero the force is zero for these sides. The pair of forces on the left and right side of the coil will generate a torque about an axis that bisects sides 2 and 4, the dashed line in Figure 6.20.

$$\tau = F_1\left(\frac{W}{2}\right) + F_3\left(\frac{W}{2}\right) = F\ W = (B\ I\ L)W = B\ I\ (L\ W) = B\ I\ A$$

where A represents the area of the coil. If there are N number of loops of wire in the coil, the current experiencing the force will be ($N\,I$), so the torque will be:

$$\tau = B\ N\ I\ A$$

The number of loops, the current, and the area of the loop are put together in equation (6.4) as

$$\mu = N\ I\ A \tag{6.4}$$

which is the magnetic moment, μ, of the loop. With this definition of the magnetic moment, the torque on a current loop is defined in equation (6.5) as

$$\tau = \mu\ B\ \sin(\theta) \tag{6.5}$$

where θ is the angle between the direction of the magnetic field and the direction of the magnetic dipole, which is defined as perpendicular to the current loop and given by a RHR in which the fingers of your right hand and your thumb point in the direction of the magnetic dipole. For example, in Figure 6.17, the magnetic dipole points to the page.

Thus, in this figure, the angle between the magnetic field and the magnetic dipole is 90°, since the magnetic field is from left to right across the page and the magnetic dipole is into the page. As explained in the description of the forces on the individual lines of current in the loop and the resulting torque on the loop, the net result will be that the external magnetic field will tend to align the magnetic dipole of the loop with itself. This magnetic moment of a particle is the quantity that is assigned to the concept of particle spin.

6.7 ANSWER TO THE CHAPTER QUESTION

In an MRI, the large external magnet aligns the magnetic moments of the hydrogen nuclei in the patient's body in the same or opposite direction as the magnetic field. The hydrogen nuclei begin to precess around the magnetic field lines, as described by the dashed line in Figure 6.18, in a direction that is in line or in the opposite direction of the magnetic field. These two orientations of the hydrogen nuclei have different signatures in the radio spectrum that can help doctors learn about the structure of the tissue in the device. So, the giant magnetic field in an MRI changes the orientation of the magnetic moments in a patient's body to help generate the image.

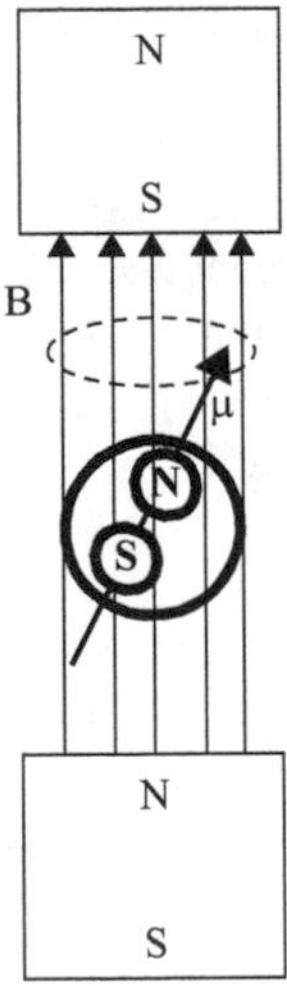

FIGURE 6.18 Particle in a magnetic field.

6.8 CHAPTER QUESTIONS AND PROBLEMS

6.8.1 Multiple Choice Questions

1. A wire extends along the x-axis of a Cartesian coordinate system. In that wire, there is a current in the +x-direction. The wire is in a uniform magnetic field that is parallel to the x-y plane and makes an angle of 75° with the x-axis. What is the direction of the force being exerted on the wire by the magnetic field?
   A. $\hat{\mathbf{i}}$ B. $-\hat{\mathbf{i}}$ C. $\hat{\mathbf{j}}$ D. $-\hat{\mathbf{j}}$ E. $\hat{\mathbf{k}}$ F. $-\hat{\mathbf{k}}$ G. There is no force.

2. In Figure 6.19, what direction is the force on the wire with current I flowing through it?
   A. +x B. −x
   C. +y D. −y
   E. +z F. −z
   G. There is no force.

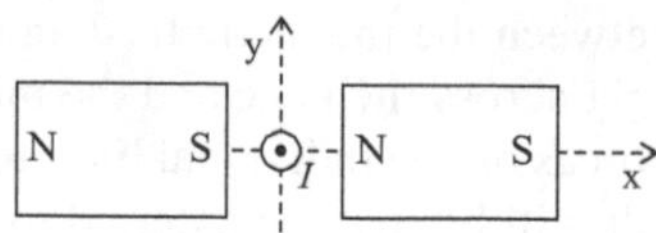

FIGURE 6.19 Multiple choice question 2.

3. A straight horizontal wire carries a current flowing northeastward through a region in space in which, there is a uniform upward magnetic field. What is the direction of the force, if any, being exerted on the wire by the magnetic field?
A. Northward B. Northeastward C. Eastward D. Southeastward
E. Southward F. Southwestward G. Westward H. Northwestward
I. Upward J. Downward K. There is no force

4. A straight horizontal wire carries a current flowing northward through a region in space in which, there is a uniform eastward magnetic field. What is the direction of the force, if any, being exerted on the wire by the magnetic field?
A. Northward B. Northeastward C. Eastward D. Southeastward
E. Southward F. Southwestward G. Westward H. Northwestward
I. Upward J. Downward K. There is no force

5. A straight horizontal wire carries a current flowing northward through a region in space in which, there is a uniform southward magnetic field. What is the direction of the force, if any, being exerted on the wire by the magnetic field?
A. Northward B. Northeastward C. Eastward D. Southeastward
E. Southward F. Southwestward G. Westward H. Northwestward
I. Upward J. Downward K. There is no force

6. A positively charged particle is moving due east, in a region of space which has a magnetic field that points directly south. What is the direction of the force, if any, exerted on the charged particle by the magnetic field?
A. Upward B. Downward C. Eastward D. Westward
E. Northward F. Southward G. No direction since $F = 0$

7. A negatively charged particle is moving due north in a region of space, which has a magnetic field that points directly west. What is the direction of the force, if any, exerted on the charged particle by the magnetic field?
A. Upward B. Downward C. Eastward D. Westward
E. Northward F. Southward G. No direction since $F = 0$

8. A positively charged particle is moving downward, in a region of space with a magnetic field which points directly east. What is the direction of the force, if any, exerted on the charged particle by the magnetic field?
A. Upward B. Downward C. Eastward D. Westward
E. Northward F. Southward G. No direction since $F = 0$

9. A proton, moving with a velocity in the +x-direction, enters a region with a magnetic field that points in the –z-direction. The magnetic force on the proton as it first enters the magnetic field points:
A. in the +x-direction. E. in the +z-direction.
B. in the –x-direction. F. in the –z-direction.
C. in the +y-direction. G. In no direction whatsoever, because it is 0.
D. in the –y-direction. H. None of the other answers is correct.

10. An electron, moving with a velocity in the +y-direction, enters a region with a magnetic field that points in the +x-direction. The magnetic force on the electron as it first enters the magnetic field points:

A. in the +x-direction.
B. in the –x-direction.
C. in the +y-direction.
D. in the –y-direction.
E. in the +z-direction
F. in the –z-direction.
G. In no direction whatsoever, because it is 0.
H. None of the other answers is correct.

6.8.2 Problems

1. A current of 1.2 A flows through a wire in the +x-direction. The wire is in a uniform magnetic field of magnitude 350 mT that is parallel to the x-y plane and makes an angle of 75° with the x-axis. What is the magnitude and direction of the force being exerted on the wire by the magnetic field?
2. A straight horizontal wire carries a northeastward 850 mA current through a region in space in which, there is a uniform upward 650 mT magnetic field. What is the magnitude and direction of the force being exerted on a 1 m segment of the wire by the magnetic field?
3. A particle with a charge of 3 C is moving in the +y-direction with a velocity of 2 m/s through a region of space in which there is a magnetic field pointing in the –z-direction with a magnitude of 5 T. Find the magnitude and direction of the force on the charged particle due to the magnetic field.
4. A particle with a charge of 2 C is moving in the +x-direction with a velocity of 5 m/s through a region of space in which there is a magnetic field pointing in the +y-direction with a magnitude of 3 T. Find the magnitude and direction of the force on the charged particle due to the magnetic field.
5. A particle with a charge of –26.4 μC is moving in the –x-direction with a velocity of 45.7 m/s through a region of space in which there is a magnetic field pointing in the –z-direction with a magnitude of 1.36×10^{-5} T. Find the magnitude and direction of the force on the charged particle due to the magnetic field.
6. A particle with a charge of –3.0 C is moving with a velocity $\vec{\mathbf{v}} = 3.0\text{ m/s }\hat{\mathbf{i}} + 4.0\text{ m/s }\hat{\mathbf{j}}$ in a uniform magnetic field $\vec{\mathbf{B}} = 1.0\text{ T }\hat{\mathbf{i}} + 2.0\text{ T }\hat{\mathbf{j}}$. Calculate the force on this particle due to the magnetic field.
7. A particle having charge 77.7 μC is moving in a region of space in which there is a uniform magnetic field of magnitude 1.50 T in the +x-direction. At the instant in question, the velocity of the particle is 222 m/s at 122°. What is the magnitude and direction of the force being exerted on the charged particle by the magnetic field in which the particle is moving?
8. At an instant of time, a 2.0 C charge is passing through the origin (0,0) moving at a velocity of 5 m/s at an angle of $\phi = 40°$ relative to the +x-axis. In the region of space, which includes the origin (0,0), there is a magnetic field with a magnitude of 2 T directed perpendicular to the path of the particle and in the –z-direction. Find the magnitude and direction of the magnetic force on this charged particle at the instant described.
9. A particle with a charge of +5 μC travels with a velocity given by $\mathbf{v} = (-3\text{ m/s})\,\boldsymbol{i} + (4\text{ m/s T})\,\boldsymbol{j} + (0\text{ m/s})\,\boldsymbol{k}$, in a region in which there is a magnetic field, which can be expressed as: $\mathbf{B} = (0.2\text{ T})\,\boldsymbol{i} + (0.3\text{ T})\,\boldsymbol{j} + (0\text{ T})\,\boldsymbol{k}$. Compute the x-, y-, and z-components of the magnetic force on the charged particle at the instant described in the problem.
10. A particle with a charge of +2 μC travels with a velocity 45 m/s up and to the left so that its trajectory makes an angle of $\phi = 20°$ with the left pointing toward the horizontal axis. In the region in which the particle is traveling there is a magnetic field, which can be expressed as: $\mathbf{B} = 0\text{ T }\boldsymbol{i} + (-0.3\text{ T})\,\boldsymbol{j} + (+0.4\text{ T})\,\boldsymbol{k}$. Compute the x-, y-, and z-components of the magnetic force on the charged particle at the instant described.

APPENDIX: CROSS-PRODUCT FORM OF THE MAGNETIC FORCE EQUATION

Since the force on a moving charged particle in a magnetic field is perpendicular to the plane established by the velocity and magnetic field, the equation that represents the relationship of the magnetic force is the cross-product of the velocity and the magnetic field, which can be written as given in equation (6.6) as

$$\vec{F} = q\vec{v} \times \vec{B} \tag{6.6}$$

This expression can be used to find the magnitude and direction of the magnetic force on a moving charged particle.

For example, consider a positively charged particle in Figure 6.20 moving with velocity v at angle θ in the x-y plane of a Cartesian coordinate system in which there is a uniform magnetic field in the +x-direction.

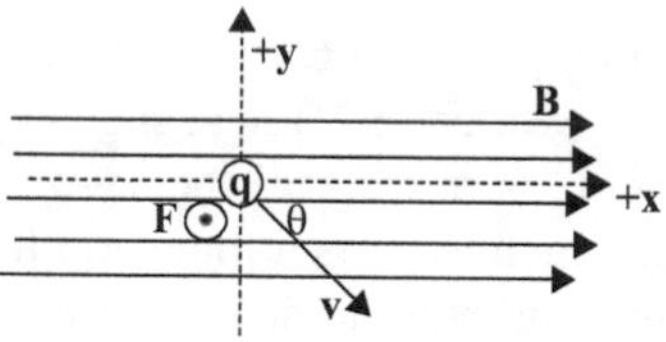

FIGURE 6.20 Positive charge moving in the +x and –y directions through a magnetic field in the +x-direction.

For this example:

$$\vec{F} = q\ \vec{v} \times \vec{B} = q*\left[\left(v_y B_z - v_z B_y\right)i + \left(v_z B_x - \ v_x B_z\right)j + \left(v_x B_y - v_y B_x\right)k\right]$$

$$\vec{F} = q\ \vec{v} \times \vec{B} = q*\left[\left(-v_y(0) - (0)(0)\right)i + \left((0)B - \ v_x(0)\right)j + \left(v_x(0) - -v_y B\right)k\right] = +\ q\ \left(v_y B\right)k$$

$\vec{F} = q\ (v_y B)\ k$ *(the circle with the dot in it represents the point of an arrow coming out of the page and thus the +z-direction of the force vector)*

The y-component of the velocity is just $v\ \sin(\theta)$, so the magnitude of the magnetic force is:

$$\left|\vec{\mathbf{F}}\right| = \left|qvB\sin\theta\right|$$

and the direction can be determined by the RHR and labeled as the circle with a dot in it to indicate the tip of an arrow coming out of the page. Therefore, this cross-product equation gives both the magnitude and direction of the force on a charged particle moving though a constant magnetic field.

7 Kinematics

7.1 INTRODUCTION

Thus far in this volume, the motion of the objects analyzed was simply at rest or in the case of the magnetic force, moving at a constant speed at one instant of time. The techniques applied to analyze a situation were centered around the analysis of vectors. A diagram was produced, vector components were found, an expression was applied, and solutions were found.

At this point in the volume, the topic of kinematics, which is the terminology and the analysis employed to study the motion of objects, is introduced. The solutions to the problems in this chapter are associated with the definitions of different types of motion and not focused on the components of force vectors. Therefore, the pattern of analysis established in the previous chapters is not applied in this chapter, but it will return in the next chapter after the terminology and techniques of kinematics are studied.

Chapter question: A family makes an 800 km trip, strictly eastward, as described in Figure 7.1. The family travels the first half of the trip distance at 80 km/h, quickly speeds up, and then maintains a speed of 100 km/h for the second half of the trip distance. For this entire trip, is the average speed 90 km/h? This question will be answered at the end of the chapter using the formal mathematical study of motion, called Kinematics, which is the topic of this chapter.

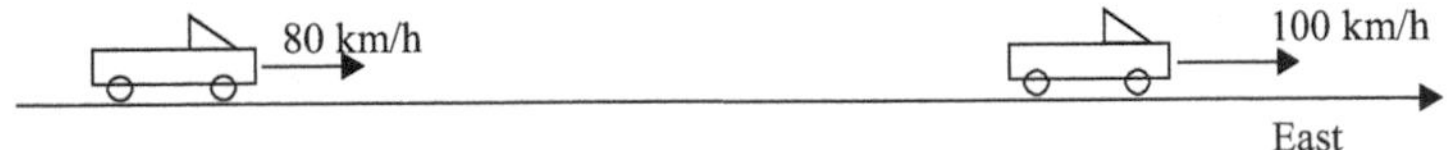

FIGURE 7.1 Family trip.

7.2 KINEMATIC DEFINITIONS

Kinematics, first developed by Galileo, is the formal study of the motion of objects using diagrams, graphs, and equations. In the previous chapters, the analysis employed assumed the objects were either at rest or moving at a constant speed. Situations under these conditions are known as a static, because the forces are balanced. When the forces become unbalanced, accelerated motion results. In this chapter, the terminology used to describe all types of motion is formalized in preparation for subsequent chapters in which the resulting motion of unbalanced force will be studied. This chapter will start with a definition of displacement and then explore velocity and acceleration in both graphical and analytical formats. It is important to realize that the problem-solving techniques employed in this chapter are different than those used in the previous and subsequent chapters, and this is only a brief aside to introduce the terminology and practices of kinematics.

Displacement: The *displacement* ($\vec{\mathbf{d}}$) of an object is the change in position of an object relative to a fixed origin, as described in Figure 7.2.

Displacement is a vector. Suppose that at time t_0, a particle is at position $\vec{\mathbf{r}}_0$ (relative to the origin of an x-y coordinate system), and at time t_1, the particle is at position $\vec{\mathbf{r}}_1$. The displacement that the particle undergoes from time t_0 to time t_1 is then given in equation (7.1) as:

DOI: 10.1201/9781003308065-7

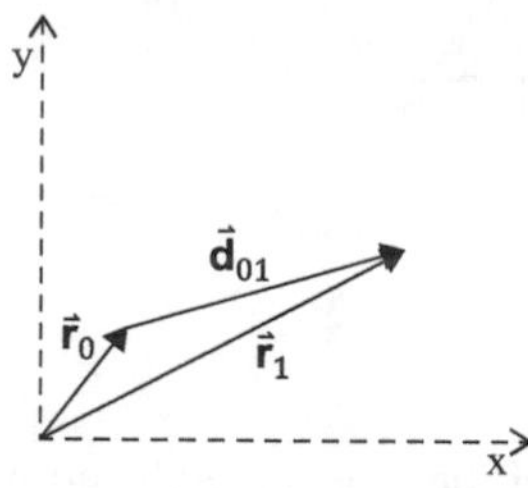

FIGURE 7.2 Definition of displacement.

$$\vec{\mathbf{d}}_{01} = \vec{\mathbf{r}}_1 - \vec{\mathbf{r}}_0 \tag{7.1}$$

The SI unit for the magnitude of the displacement vector is the meter (m).

As an example of a displacement, as depicted in Figure 7.3, at time t_0 a man is 5 m east of a lamppost and is walking directly away from the lamppost so that he is 15 m east of the lamppost at time t_1. His displacement $\vec{\mathbf{d}}_{01}$ from time t_0 to time t_1 would be 10 m eastward.

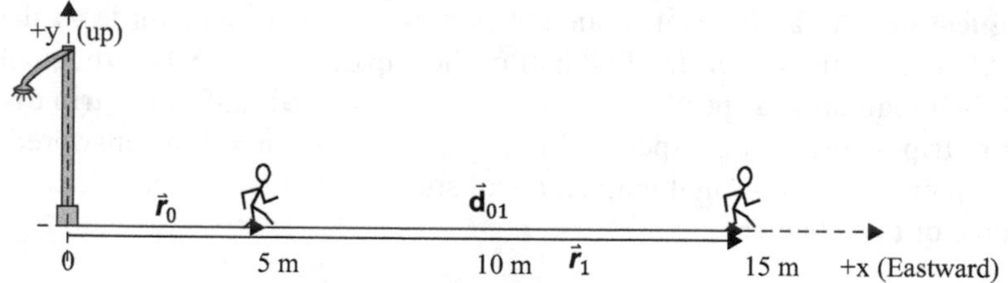

FIGURE 7.3 Displacement example.

Velocity: In previous chapters, the *velocity* $(\vec{\mathbf{v}})$ of an object was defined as the speed of the object and the direction in which the object is moving. This definition is more precisely known of the instantaneous velocity of an object, to distinguish it from the average velocity over a known time interval. The average value-with-direction of the velocity $\vec{\mathbf{v}}$ over some time interval, from time t_0 to time t_1, is shown in equation (7.2) and obtained by dividing the total displacement $\vec{\mathbf{d}}_{01}$ that occurs during that time interval by the duration of the time interval $t_{01} = t_1 - t_0$:

$$\vec{\mathbf{v}}_{\text{Avg01}} = \vec{\mathbf{d}}_{01}s\ /t_{01} \tag{7.2}$$

The SI unit for velocity is the meter per second (m/s).

For example, if the man in the previous example, shown in Figure 7.3, took 5 seconds to move the 10 minutes of displacement the average velocity of the man is:

$$\vec{\mathbf{v}}_{\text{Avg01}} = \frac{\vec{\mathbf{d}}_{01}}{t_{01}} = \frac{10\text{ m}}{5\text{ s}} = 2\frac{\text{m}}{\text{s}}, \text{ Eastward or just in the + x-direction.}$$

Another common unit for velocity is miles per hour (mph). As sketched in Figure 7.4, if a car traveled 100 miles due north in 2 hours, the average velocity of the car for the trip is, $\vec{\mathbf{v}}_{\text{Avg01}}$ = (100 miles northward)/(2h) = 50 mph northward.

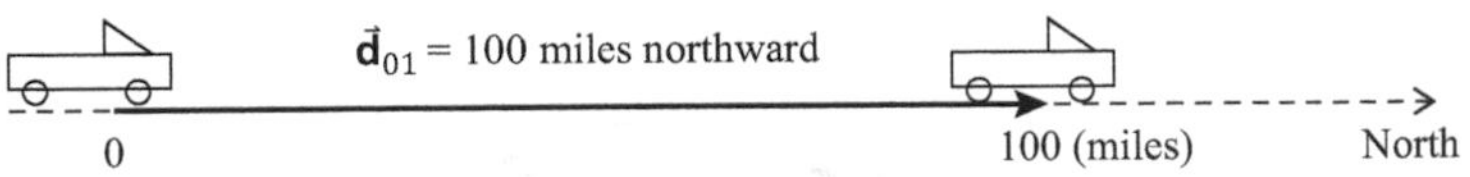

FIGURE 7.4 Average velocity example.

Acceleration: The *acceleration* of an object is how fast and in which direction the velocity of the object is changing. It is represented by a vector ($\vec{\mathbf{a}}$) whose magnitude is how fast the velocity is changing and the direction associated with the way the velocity is changing. Like velocity, acceleration can be both instantaneous and average. The *instantaneous acceleration* is at one particular point in time and the *average acceleration* of the object over some time interval. The average value-with-direction of the acceleration $\vec{a}$ over some time interval, from time t_0 to time $t_{1,}$ is given in equation (7.3) and obtained by dividing the total change in velocity $\vec{v}_1 - \vec{v}_0$ that occurs during that time interval by the duration of the time interval $t_{01} = t_1 - t_0$:

$$\vec{\mathbf{a}}_{\text{Avg01}} = (\vec{\mathbf{v}}_1 - \vec{\mathbf{v}}_0)/t_{01} \tag{7.3}$$

The SI unit for acceleration is (m/s)/s, which is usually written m/s^2.

For example, as shown in Figure 7.5, if a car is traveling northward at 10 m/s and speeds up to 30 m/s northward in 5 seconds, the average acceleration during those 5 seconds is:

$$\vec{\mathbf{a}}_{\text{Avg01}} = (30 \text{ m/s northward} - 10 \text{ m/s northward})/(5 \text{ seconds}) = 4 \text{ m/s}^2 \text{ northward.}$$

If, during the 5 seconds in which the car is speeding up, the car increases its speed (without changing the direction of its velocity) *at a constant rate*, the acceleration is said to be constant and the acceleration at any instant in time during those 5 seconds will be the same as the average acceleration.

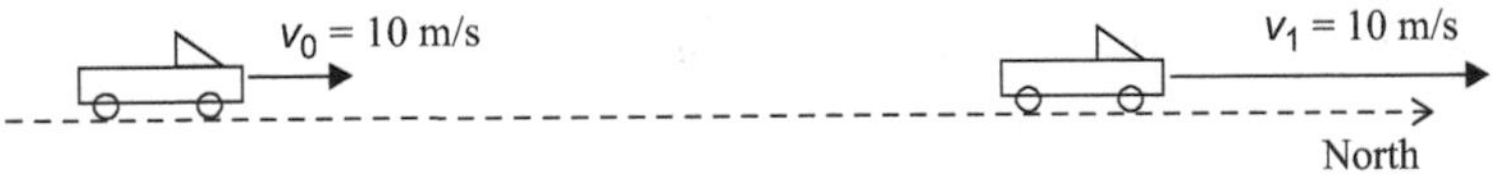

FIGURE 7.5 Example of constant acceleration.

When a rigid body undergoes purely translational motion, every part of the body undergoes the same motion, but different parts move along different curves or lines. When studying the motion of a rigid body, to determine the motion of the object, it is common to track the motion of the rigid body's center of mass.

7.2.1 One-Dimensional Motion

For this and the next few chapters, the motion studied will be along a line in one-dimension. It is common to choose our x-y-z coordinate system so that the x-axis or y-axis is along the line in which the motion occurs. Then, the motion will only have one component of displacement $\vec{\mathbf{d}}$, velocity $\vec{\mathbf{v}}$, and acceleration $\vec{\mathbf{a}}$, such as the x-components, d_x, v_x, a_x, or the y-components, d_y, v_y, a_y, respectively.

As shown in Figure 7.6, the x-component d_x of the displacement $\vec{\mathbf{d}}$ of an object will be positive if the displacement of that object is in the +x-direction.

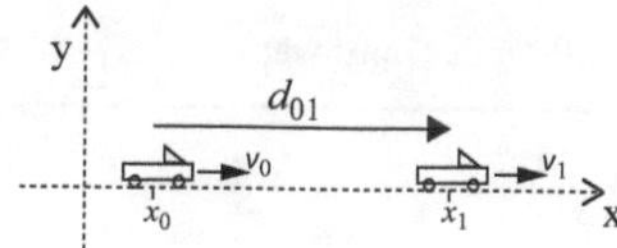

FIGURE 7.6 The vector $\vec{\mathbf{d}}_{01}$ is in the +x-direction, so d_{01x} is positive.

This is the case when the object moves from a position with a smaller value of x to a position with a larger value of x.

As shown in Figure 7.7, the x-component d_x of the displacement $\vec{\mathbf{d}}$ of an object will be *negative* if the displacement of that object is in the −x-direction. Note that in the diagram at right, $x_0 > x_1$.

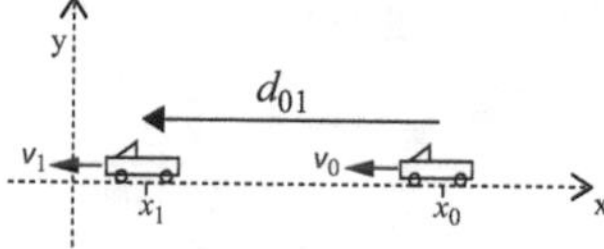

FIGURE 7.7 The vector $\vec{\mathbf{d}}_{01}$ is in the −x-direction, so d_{01x} is negative.

As shown in Figure 7.8, the x-component v_x of the velocity $\vec{\mathbf{v}}$ of an object will be positive if the velocity of that object is in the +x-direction.

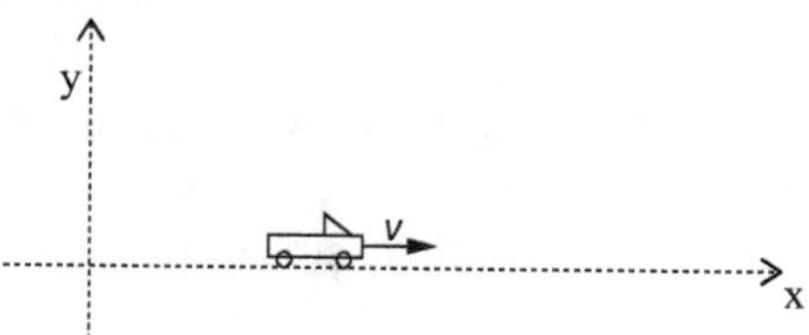

FIGURE 7.8 The vector $\vec{\mathbf{v}}$ is in the +x-direction, so v_x is positive.

As shown in Figure 7.9, the x-component v_x of the velocity $\vec{\mathbf{v}}$ of an object will be *negative* if the velocity of that object is in the −x-direction.

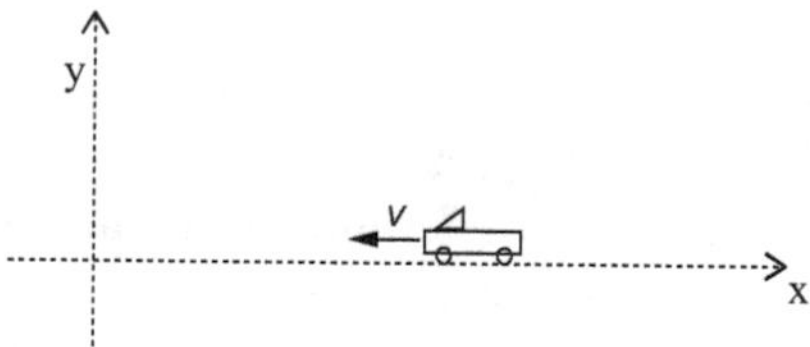

FIGURE 7.9 The vector $\vec{\mathbf{v}}$ is in the −x-direction, so v_x is *negative.*

The acceleration vector for an object that is traveling along a line and is speeding up is in the same direction as the velocity vector, but the acceleration vector for an object that is traveling along a line and is slowing down is in the direction *opposite* that of the velocity vector.

As shown in Figure 7.10, when an object is moving in the +x-direction and speeding up, its acceleration vector $\vec{\mathbf{a}}$ is in the +x-direction so the x-component a_x of the object's acceleration is positive.

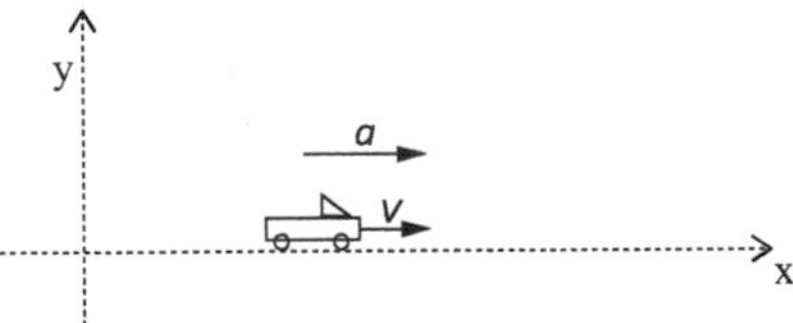

FIGURE 7.10 Moving in the +x-direction and speeding up, so $\vec{\mathbf{a}}$ is in the +x-direction and a_x is positive.

As shown in Figure 7.11, when an object is moving in the +x-direction and slowing down, its acceleration vector $\vec{\mathbf{a}}$ is in the –x-direction so the x-component a_x of the object's acceleration is *negative.*

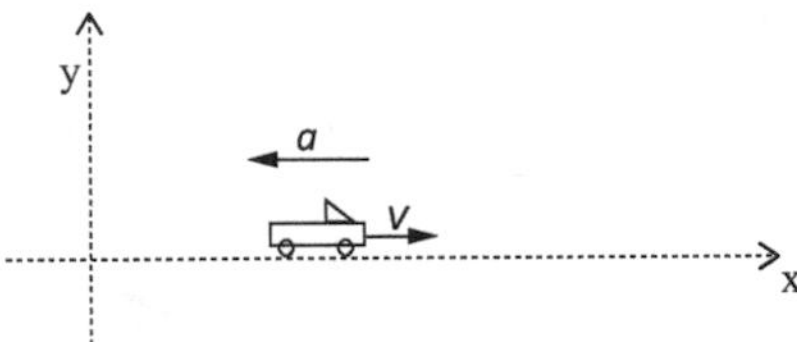

FIGURE 7.11 Moving in the +x-direction and slowing down, so $\vec{\mathbf{a}}$ is in the –x-direction and a_x is negative.

As shown in Figure 7.12, when an object is moving in the –x-direction and speeding up, its acceleration vector $\vec{\mathbf{a}}$ is in the –x-direction so the x-component a_x of the object's acceleration is negative.

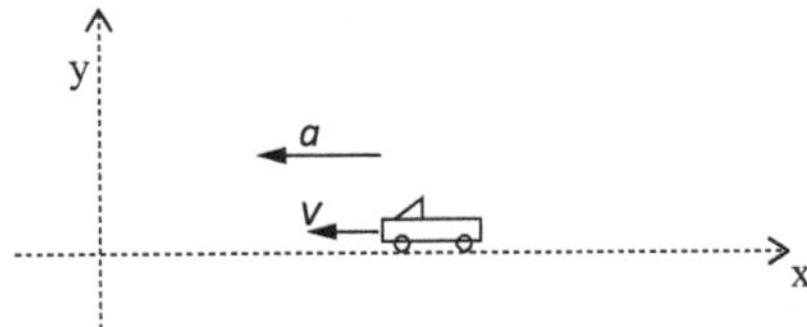

FIGURE 7.12 Moving in the –x-direction and speeding up, so $\vec{\mathbf{a}}$ is in the –x-direction and a_x is negative.

As shown in Figure 7.13, when an object is moving in the –x-direction and slowing down, its acceleration vector $\vec{\mathbf{a}}$ is in the +x-direction so the x-component a_x of the object's acceleration is positive.

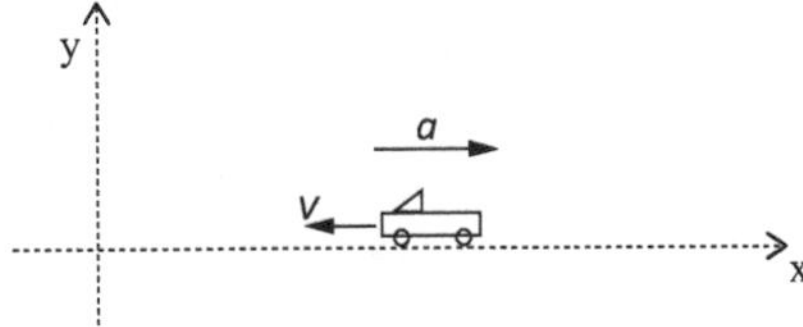

FIGURE 7.13 Moving in the –x-direction and slowing down, so $\vec{\mathbf{a}}$ is in the +x-direction and a_x is positive.

7.3 KINEMATIC GRAPHS

7.3.1 Constant Velocity Graphs

Consider a cart moving along a straight track with a constant velocity $\vec{\mathbf{v}} = 2$ m/s $\hat{\mathbf{i}}$. (Recall that $\hat{\mathbf{i}}$ indicates the vector is in the +x-direction.) In Figure 7.14, a cart is shown in a motion diagram, which is a series of images at a constant time interval, which in this case is each second. Since the cart is moving with a constant speed of 2 m/s, the cart moves 2 m in the +x-direction every second.

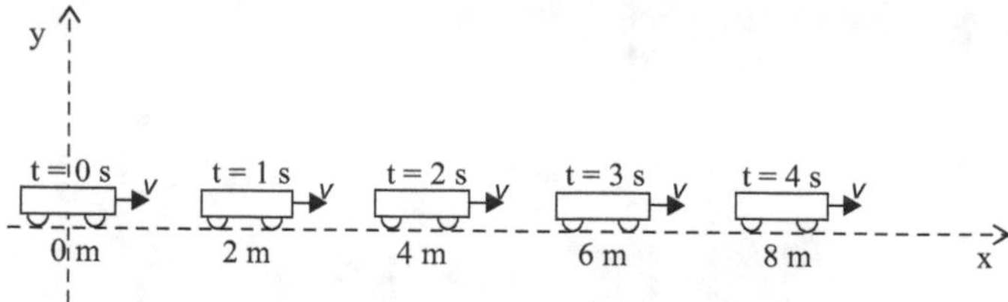

FIGURE 7.14 Motion dot diagram of a cart moving down a track at a constant speed.

A graph of the *x-component of the position of the cart relative to the origin vs. time,* shown in Figure 7.15, is a straight line with a slope of 2 m/s.

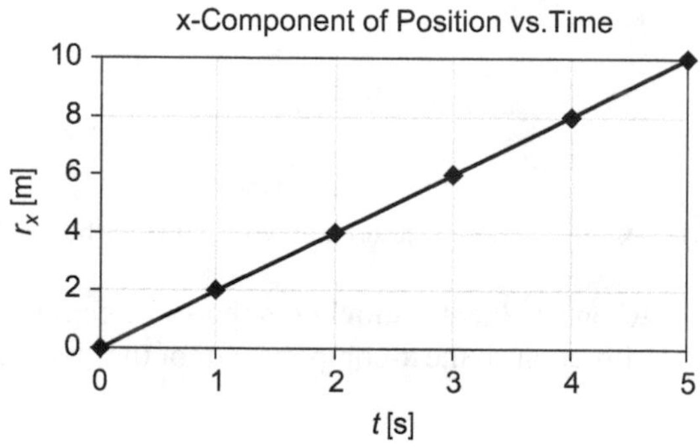

FIGURE 7.15 Graph of the x-component of displacement of the cart vs. time.

The slope of the line is the x-component of the velocity of the cart. Since, a straight line has the same slope everywhere along the line, the x-component of the velocity, v_x, is constant. Note that, since $\vec{\mathbf{r}}$ is the relative position vector from the origin to the object, the x-component of $\vec{\mathbf{r}}$, namely r_x, is the same as the x-coordinate of position, x, and hence, a graph of r_x vs. t is the same as a graph of x vs. t, where x is the x-coordinate of the position (x, y, z) of the object.

As shown in Figure 7.16, since v_x has a constant value of 2 m/s, the *x-component of velocity vs. time* graph for this motion is a horizontal line at the value of $v_x = 2$ m/s.

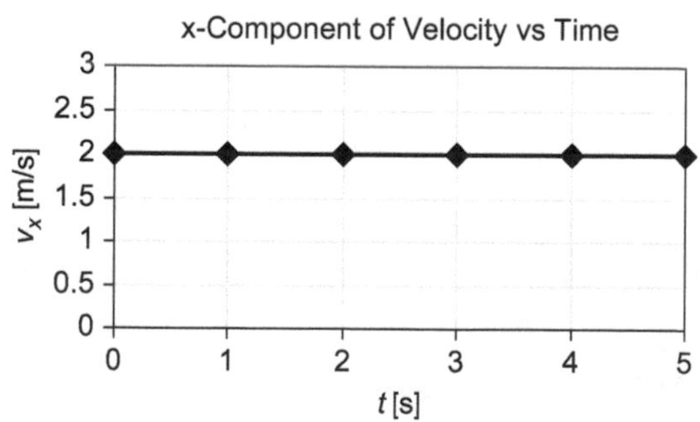

FIGURE 7.16 Graph of the x-component of velocity of the cart vs. time.

As shown in Figure 7.17, since the velocity is not changing, the acceleration is zero, so the *x-component of acceleration vs. time* graph for this situation is a horizontal line at $a_x = 0$.

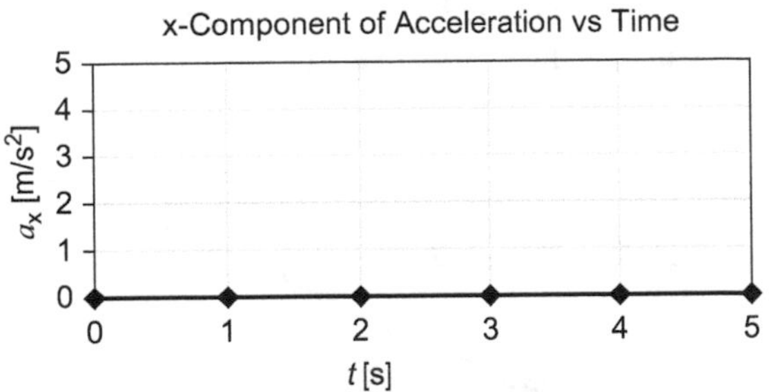

FIGURE 7.17 Graph of the x-component of velocity of the cart vs. time.

7.3.2 Constant Acceleration Graphs

Consider a cart starting from rest and moving along a straight track with a constant acceleration of 5 m/s² $\hat{\imath}$. A motion diagram for this situation is shown in Figure 7.18.

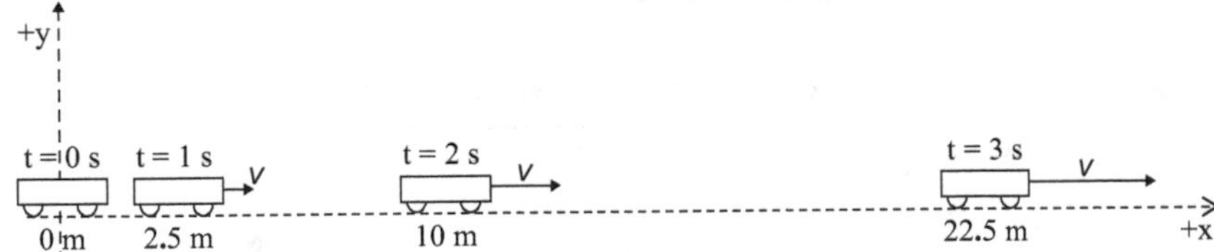

FIGURE 7.18 Motion diagram of a cart moving down a track with a constant acceleration.

As shown in Figure 7.18, the cart is speeding up, so the magnitude of the velocity at each second is greater than that at the preceding second. The distance covered during each subsequent second is also greater since the cart is moving faster during each subsequent second.

A graph of the *x-component of the position vs. time*, shown in Figure 7.19, of the cart, is a curve with an increasing slope, since the slope is the x-component of the velocity, which is increasing every second.

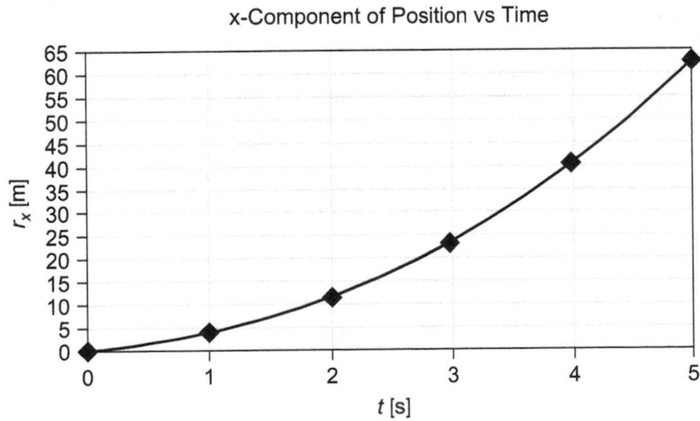

FIGURE 7.19 Graph of the x-component of displacement of the cart vs. time.

As shown in Figure 7.20, the *x-component of velocity vs. time* graph for this situation is a straight line with a slope of 5 m/s², which is the x-component of the acceleration.

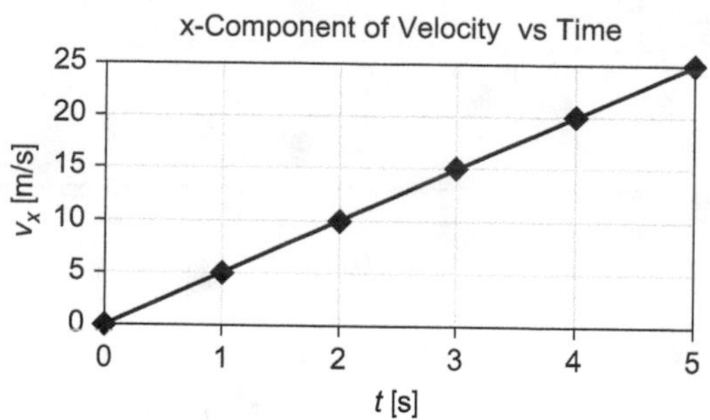

FIGURE 7.20 Graph of the x-component of velocity of the cart vs. time.

As shown in Figure 7.21, the *x-component of the acceleration vs. time* graph for this situation is a horizontal line at 5 m/s², since that is the (given) x-component of the acceleration.

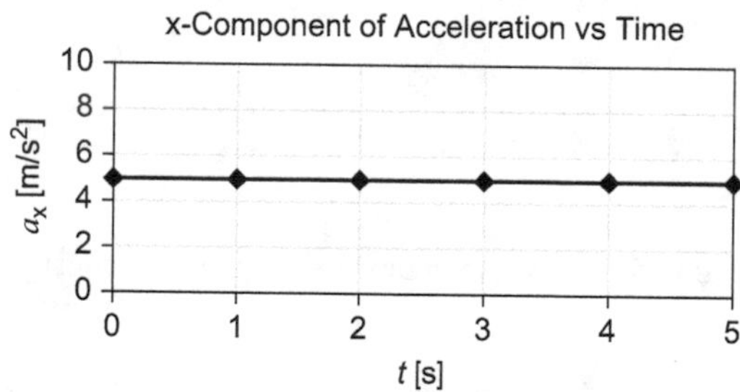

FIGURE 7.21 Graph of the x-component of acceleration of the cart vs. time.

7.3.3 Combined Motion Graphs

Example 7.1

Consider a cart starting from rest, accelerating in the +x-direction at 5 m/s² for 3 seconds, then moving along for the next 2 seconds at a constant velocity equal to that at the end of the first-time interval. The motion diagram of the cart described in the previous sentence is shown in Figure 7.22. The displacement of the cart that occurs during a 1-second time interval increases for the first 3 seconds then stays constant for the last 2 seconds.

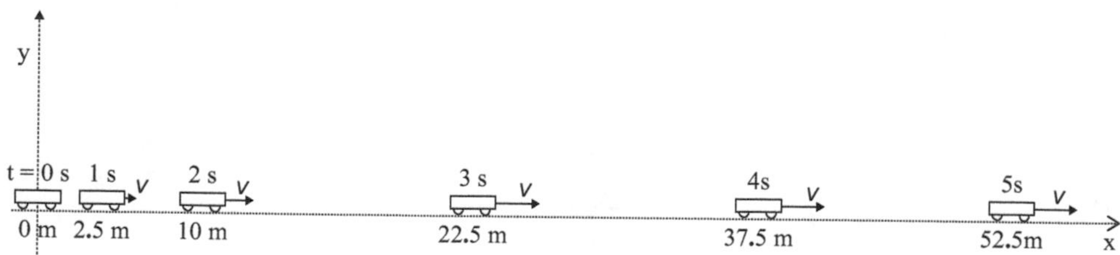

FIGURE 7.22 Motion diagram of a cart accelerating from rest and then maintaining a constant speed.

A graph of the *x-component of the position of the cart relative to the origin vs. time,* given in Figure 7.23, is a curve with an increasing slope for the first 3 seconds, and then a straight line for the next 2 seconds.

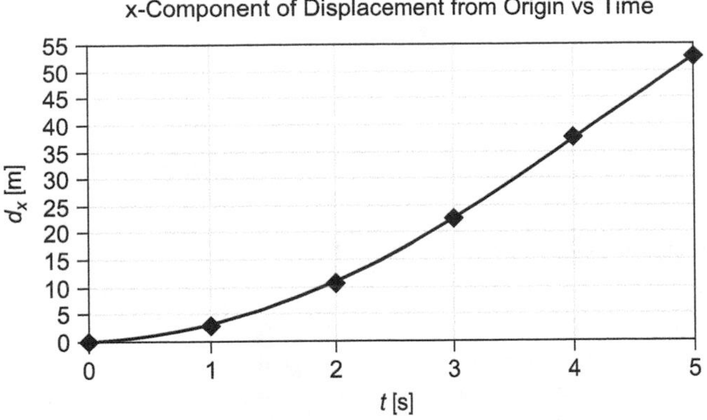

FIGURE 7.23 Graph of the x-component of displacement of the cart vs. time.

The *x-component of the velocity vs. time* graph, shown in Figure 7.24, for this motion is a straight line with a slope of 5 m/s^2 for the first 3 seconds, then a horizontal line for the last 2 seconds.

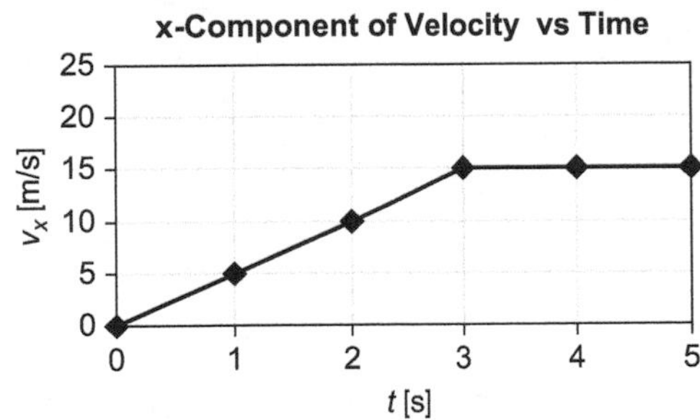

FIGURE 7.24 Graph of the x-component of velocity of the cart vs. time.

Given in Figure 7.25, the *x-component of the acceleration vs. time* graph for this motion is a horizontal line at 5 m/s^2, for the first 3 seconds, then a horizontal line at 0 m/s^2 for the last 2 seconds.

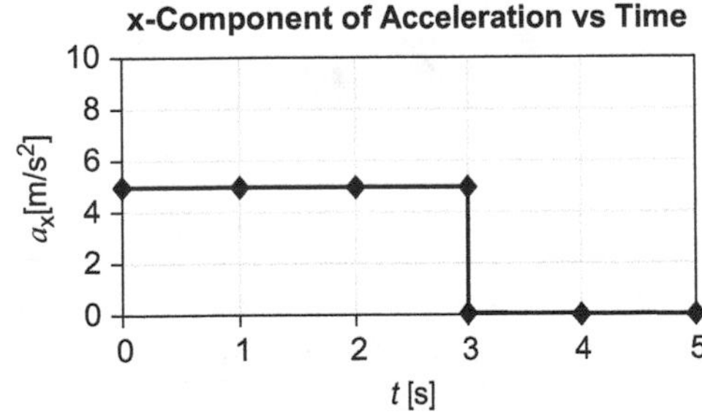

FIGURE 7.25 Graph of the x-component of the acceleration of the cart vs. time.

Example 7.2

Consider a cart starting from rest, accelerating in the +x-direction at 5 m/s^2 for 2 seconds, then slowing at constant acceleration until it comes to rest 2 seconds later. A motion diagram for this situation is shown in Figure 7.26.

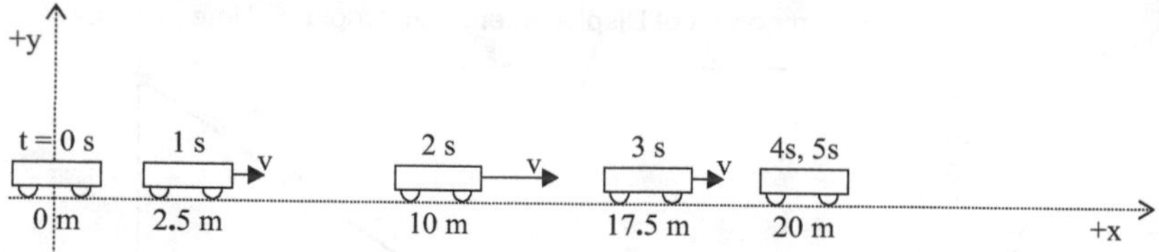

FIGURE 7.26 Motion diagram of a cart accelerating from rest and then slowing down to a stop.

Figure 7.27 is a graph of the *x-component of the position of the cart relative to the origin vs. time* of the cart, which would be a curve with an increasing slope for the first 2 seconds, then a curve with a decreasing slope for the next 2 seconds, and finally a horizontal line for the last second. Notice that the x-component of the position vector is still increasing from 2 to 4 seconds, but at a decreasing rate.

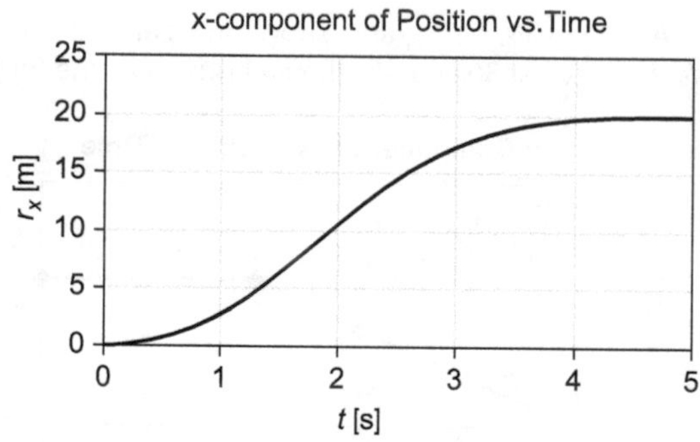

FIGURE 7.27 Graph of the x-component of displacement of the cart vs. time.

Figure 7.28 is a graph of the *x-component of velocity vs. time* graph for this motion, which is a straight line with a slope of 5 m/s² for the first 2 seconds, then a straight line with a slope of –5 m/s² for the next 2 seconds, and finally a horizontal line at zero for the last second.

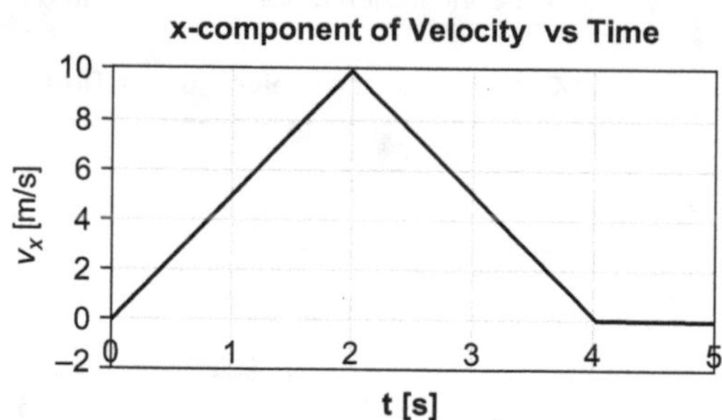

FIGURE 7.28 Graph of the x-component of displacement of the cart vs. time.

Figure 7.29 is a graph of the *acceleration vs. time* for this motion, which is a horizontal line at 5 m/s² for the first 2 seconds, then a horizontal line at –5 m/s² for 2 seconds, and finally a horizontal line at zero for the last second.

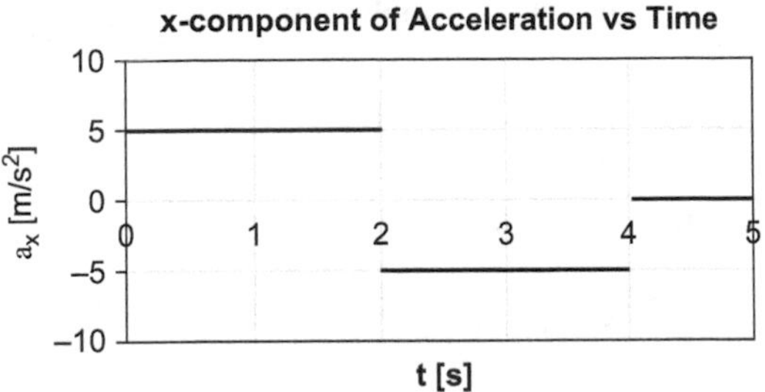

FIGURE 7.29 Graph of the x-component of displacement of the cart vs. time.

7.4 KINEMATIC EQUATIONS

From the definitions for velocity and acceleration, the following relations can be established assuming a *constant acceleration* and using the letter t to indicate an elapsed time: $t = t_1 - t_0$. Start by rearranging the x-component of equation (7.3), which is the definition of acceleration, $\bar{\mathbf{a}}_{Avg01} = (\bar{\mathbf{v}}_1 - \bar{\mathbf{v}}_0) / t_{01}$, so it gives an expression of the x-component of the velocity, $v_{1x}(t)$, at a specific time t. Equation (7.4) is one of the two fundamental equations of kinematics.

$$v_{1x}(t) = v_{0x} + a_x\ t \tag{7.4}$$

The next fundamental equation for kinematics starts with an x-component version of equation 7.2 with the definition of the average velocity as the change in displacement over a given time: $v_{ave\_x} = (d_{01x} - d_{0x})/t$, this can be rearranged to give an expression for calculating the displacement from t_0 to t_1 in the x-direction (d_{01x}) at a given time:

$$d_{01x}(t) = d_{0x} + v_{ave\_x}\ t$$

Given the common definition of an average in which you add the start and finish values and then divide by 2: $v_{ave} = (v + v_o)/2$, the equation for displacement becomes:

$$d_{01x}(t) = d_{0x} + \tfrac{1}{2}(v_{1x} + v_{0x})t$$

Which can be rearranged to give equation (7.5) as:

$$d_{01x}(t) = d_{0x} + \tfrac{1}{2}\, v_{1x}\ t + \tfrac{1}{2}\, v_{0x}\ t. \tag{7.5}$$

Inserting equation (7.4) into equation (7.5), the expression for displacement from t at time zero to time at t is given in equation (7.6) as:

$$d_{01x}(t) = d_{ox} + v_{0x}\ t + \tfrac{1}{2}\, a_x t^2 \tag{7.6}$$

Equations (7.4) and (7.6) are the two kinematic equations that together are the key to solving any kinematic problem. Another equation that comes in handy is the combination of equations (7.4) and (7.6) that eliminates the time (t) from the equations. Solving equation (7.4) for time gives: $t = \left[v_x(t) - v_{ox}\right]/a_x$. If this expression is substituted for the t's in equation (7.6)

$$d_{01x}(t) = d_{x0} + v_{x0}\ \left\{\left[v_x(t) - v_{ox}\right]/\ a_x\right\} + \tfrac{1}{2}\, a_x\ \left\{\left[v_x(t) - v_{ox}\right]/a_x\right\}^2\ ,$$

Next, subtract d_{xo} from each side and then multiply both sides by $\mathbf{a_x}$

$$a_x\left[d_{01x}(t) - d_{x0}\right] = v_{ox}\left[v_x(t) - v_{ox}\right] + \tfrac{1}{2}\left[v_x(t) - v_{ox}\right]^2,$$

Multiplying through by v_{ox} and expand the square:

$$a_x\left[d_{01x}(t) - d_{x0}\right] = v_{ox}\, v_x(t) - v_{ox}^2 + \tfrac{1}{2}\, v_x(t)^2 - v_{ox}\, v_x(t) + \tfrac{1}{2}\, v_{ox}^2,$$

Cancel the + and – $v_{ox}\, v_x(t)$ terms on the right-hand side and then multiply through by 2

$$2\, a_x\left[d_{01x}(t) - d_{x0}\right] = v_x(t)^2 - v_{ox}^2,$$

This is often rearranged and written in equation (7.7) as:

$$v_x(t)^2 - v_{ox}^2 = 2\, a_x\left[d_{01x}(t) - d_{x0}\right] \qquad (7.7)$$

Equations (7.4), (7.6), and (7.7) are the commonly used kinematic equations used to analyze situations with a constant acceleration, of any value including zero. If the acceleration of the object is not constant, these equations cannot be used and a function must be used for the acceleration along with the techniques of calculus.

Examples: (The examples in this chapter are to illustrate the use of kinematic equations and not the process of dynamics. So, they don't follow the same format of examples in previous chapters. We will return to the original format in the next chapter.)

Example 7.1

A car is traveling on a long straight road. A coordinate system has been established for which the origin is at the location of the car at the start of observations (t = 0 second) and the +x-direction is the direction in which the car is going at the start of observations. The velocity of the car is monitored for 10 seconds and a graph of the x-component of the car's velocity for time 0 second to time 10 seconds is given in Figure 7.30.

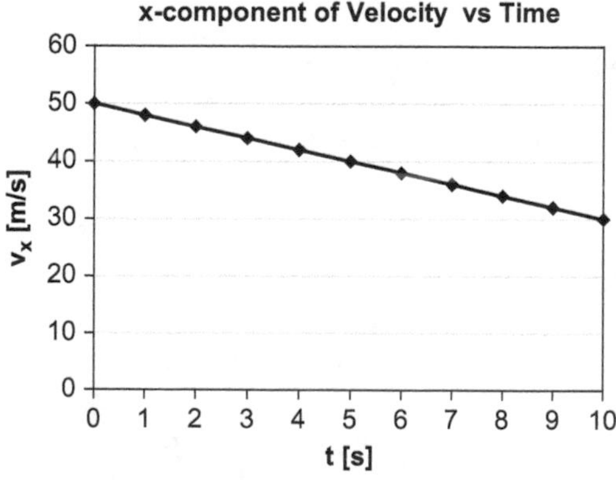

FIGURE 7.30 Graph of the x-component of velocity of a car vs. time for Example 7.1.

A. Find the acceleration of the car valid for every instant in time from 0 second to 10 seconds.
B. Calculate the displacement of the car from time 0 second to time 10 seconds.

C. Make a graph of the position *x* of the object vs. time.

Solution:

A. Find the acceleration of the car valid for every instant in time from 0 second to 10 seconds.

Define $t_0 = 0$ second and $t_1 = 10$ seconds.

Then, $t = t_1 - t_0 = 10$ seconds − 0 second = 10 seconds.

From the question: $dx_o = 0$.

From the graph, the initial velocity, $v_{ox} = 50$ m/s. The fact that the graph of v_x vs. t is a straight line means that the x-component of the acceleration is a constant. That one constant is the x-component of the acceleration of the car at every instant from 0 second to 10 seconds. That constant is equal to the slope of the line, and it is equal to the average acceleration during any time interval, including the time interval from 0 second to 10 seconds:

$$a_x = \frac{v_{1x} - v_{0x}}{t_{01}} = \frac{30 \text{ seconds} - 50 \text{ seconds}}{10 \text{ seconds}} = -2\frac{\text{m}}{\text{s}^2}$$

Because the motion is strictly along the x-axis, the x-component is the only component of the vector. A coordinate system was given in the problem statement so we can use that in stating our answer. Here are three acceptable ways of stating the final answer.

$$\vec{\mathbf{a}} = -2\frac{\text{m}}{\text{s}^2}\hat{\mathbf{i}} = 2\frac{\text{m}}{\text{s}^2} \text{ in the -x direction } = 2\frac{\text{m}}{\text{s}^2} \text{ at } 180°$$

Note that each answer includes both the magnitude and the direction of the acceleration vector.

B. Calculate the displacement of the car from time 0 second to time 10 seconds.

At this point, we can solve for the x-component of the displacement of the car from 0 second to 10 seconds using equation (7.6) and the x-component of the acceleration that we found in solving part *a*:

$$d_{01x} = v_{0x}t_{01} + \frac{1}{2}a_x t_{01}^2 = \left(50\frac{\text{m}}{\text{s}}\right)10 \text{ seconds} + \frac{1}{2}\left(-2\frac{\text{m}}{\text{s}^2}\right)(10 \text{ seconds})^2 = 400 \text{ m}$$

Displacement is a vector. We have found the x-component of that vector and the x-component is the only component the vector has, so it is easy to write down the magnitude and direction of the displacement vector. Again, there are several ways to communicate the direction, but in this part only one format is given:

$$\vec{\mathbf{d}}_{01} = 400 \text{ m } \hat{\mathbf{i}}$$

C. Make a graph of the position *x* of the object vs. time.

Let t_n represent any arbitrary time from 0 second to 10 seconds. Then, t_{0n} represents the time interval from t_0 to t_n. The x-component of the total displacement from time t_0 to t_n is given by the constant acceleration kinematic equation that, in this context, reads:

$$d_{0nx} = v_{0x}t_{0n} + \frac{1}{2}a_x t_{0n}^2$$

Given that the clock used was a stopwatch that was started at time t_0, the time interval t_{0n} from time t_0 (which is 0 second) to time t_n is equal in value and units to the current clock reading which we shall call t. This allows us to write the x-component of our displacement as:

$$d_{0nx} = v_{0x}t + \frac{1}{2}a_x t^2$$

The position of the car at t_n is just the position it had at time t_0 plus the displacement it has undergone from time t_0 to time t_n.

$$x_n = x_0 + d_{0nx} = 0 + v_{0x}t + \frac{1}{2}a_x t^2$$

At this point, to simplify the notation, we let x represent the position of the car at time t_n, in other words we define $x = x_n$. Substituting that and our values for v_{0x} and a_x into the expression just above, we obtain: $x = 0 + 50\frac{\text{m}}{\text{s}}t + \frac{1}{2}\left(-2\frac{\text{m}}{\text{s}^2}\right)t^2$ which simplifies to:

$$x = 50\frac{\text{m}}{\text{s}}t - 1\frac{\text{m}}{\text{s}^2}t^2.$$

A graph of x vs. t is given in Figure 7.31. Note that the slope of the curve decreases with increasing time. The slope of a graph of the x-coordinate of the position vs. time is equal to the x-component of the velocity. Thus, given in the problem that the x-component of the velocity is decreasing, the slope of x vs. t must decrease as well. So, the displacement is a curve that flattens as time progresses.

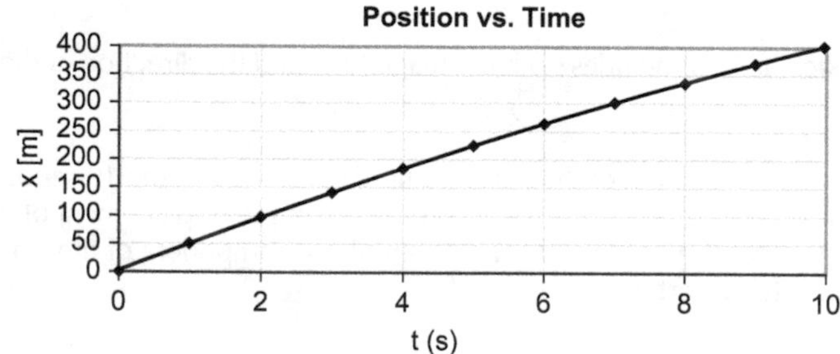

FIGURE 7.31 Graph of the x-component of displacement of a car vs. time for Example 7.1.

Example 7.2

A car starting from rest accelerates forward at a constant rate of 2.00 m/s² along a straight road. How far along the road has the car traveled when it reaches a speed of 100 km/h?

Solution:

Step 1: Start with a diagram in which we define our coordinate system and some variables, as shown in Figure 7.32:

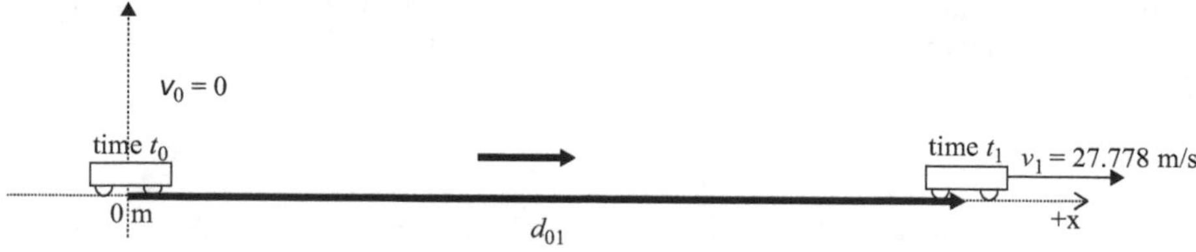

FIGURE 7.32 Diagram for Example 7.2 of kinematics.

Convert the velocity from km/h to m/s. $v_1 = 100\frac{\text{km}}{\text{h}} = 100\frac{\text{km}}{h}\frac{1{,}000\text{ m}}{1\text{ km}}\frac{1\text{ hour}}{3{,}600\text{ seconds}}$ $= 27.778\frac{\text{m}}{\text{s}}$ has already been included in our diagram.

All our vectors are in the +x-direction or 0 so as far as the x-components go:

$$v_{0x} = v_0 = 0$$

$$v_{1x} = v_1 = 27.778\text{ m/s}$$

$$a_x = a = 2\text{ m/s}^2$$

Since the acceleration is a constant and displacement is given, we can try to employ equation (7.7) to solve the problem:

$$v_{1x}^2 = v_{0x}^2 + 2a_{01x}d_{01x}$$

First, rearrange equation (7.7) to find the x-component of the displacement:

$$v_{1x}^2 - v_{0x}^2 = 2a_{01x}d_{01x}$$

$$d_{01x} = \frac{v_{1x}^2 - v_{0x}^2}{2a_{01x}}$$

Substituting in values with units:

$$d_{01x} = \frac{(27.778\text{ m/s})^2 - (0\text{ m/s})^2}{2\left(2\text{ m/s}^2\right)} = 192.90\text{ m}$$

The x-component of the displacement is the only component of the displacement, so the magnitude of the displacement is just the absolute value of the x-component of the displacement.

$$d_{01} = |d_{01x}| = |192.90\text{ m}|$$

$$d_{01} = 192.90\text{ m}$$

7.5 ANSWER TO THE CHAPTER QUESTION:

If a family makes an 800 km trip, strictly eastward. The family travels the first half of the trip distance at 80 km/h, quickly speeds up, and then maintains a speed of 100 km/h for the second half of the trip distance. For this entire trip, is the average speed 90 km/h? The answer is no, because they will spend more time at a lower speed.

The diagram for the situation with the specific displacements is given in Figure 7.33:

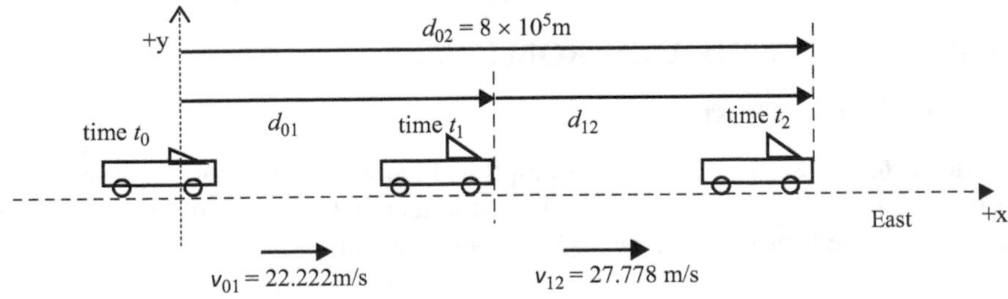

FIGURE 7.33 Diagram for the chapter question.

Assign the variables to the values in the problem:

$$d_{02x} = d_{02} = 8 \times 10^5 \text{m}$$

$$v_{01x} = v_{01} = 22.222 \text{ m/s} \qquad v_{12x} = v_{12} = 27.778 \text{ m/s}$$

$$d_{01x} = d_{01} \qquad d_{12x} = d_{12}$$

Split the two parts of the distance into the first and second halves.

$$d_{01x} = d_{02x}/2 = (8 \times 10^5 \text{ m})/2 = 4 \times 10^5 \text{ m}$$

and

$$d_{12x} = d_{02x}/2 = (8 \times 10^5 \text{ m})/2 = 4 \times 10^5 \text{ m}$$

Given that the velocity during each half of the trip is constant, we can write:

$$d_{01x} = v_{01x} t_{01} \Rightarrow t_{01} = d_{01x}/v_{01x} = (4 \times 10^5 \text{m})/22.222 \text{ m/s} = 18{,}000 \text{ seconds}$$

and

$$d_{12x} = v_{12x} t_{12} \Rightarrow t_{12} = d_{12x}/v_{12x} = (4 \times 10^5 \text{m})/27.778 \text{ m/s} = 14{,}400 \text{ seconds}$$

The total duration of the trip is: $t_{02} = t_{01} + t_{12} = 18{,}000 \text{ seconds} + 14{,}400 \text{ seconds} = 32{,}400 \text{ seconds}$.

Substitute this into the definition of velocity to find the x-component of the average velocity for each half of the trip.

$$v_{\text{Avg}02x} = \frac{d_{02x}}{t_{02}} = \frac{8 \times 10^5 \text{m}}{32{,}400 \text{ s}} = 24.691 \frac{\text{m}}{\text{s}}$$

In terms of the original units, this is:

$$v_{\text{Avg}02x} = 24.691 \frac{\text{m}}{\text{s}} \frac{\text{km}}{1{,}000 \text{ m}} \frac{3{,}600 \text{ seconds}}{\text{h}} = 88.887 \frac{\text{km}}{\text{h}}$$

The average speed is the absolute value of the x-component of the average velocity because the x-component is the only component:

$$v_{\text{Avg}02} = \left| v_{\text{Avg}02x} \right| = \left| 88.887 \frac{\text{km}}{\text{h}} \right| = 88.887 \frac{\text{km}}{\text{h}}$$

The average speed is not 90 km/h, and it is less because the car spends more time at the lower speed to travel the same distance.

7.6 CHAPTER QUESTIONS AND PROBLEMS

7.6.1 Multiple-Choice Questions

For questions 1–6, choose the letter of the graph in Figure 7.34 that best represents the motion described in the question. You will not use all eight letters and you may use the same letter more than once. Consider the forward direction to be the +x-direction.

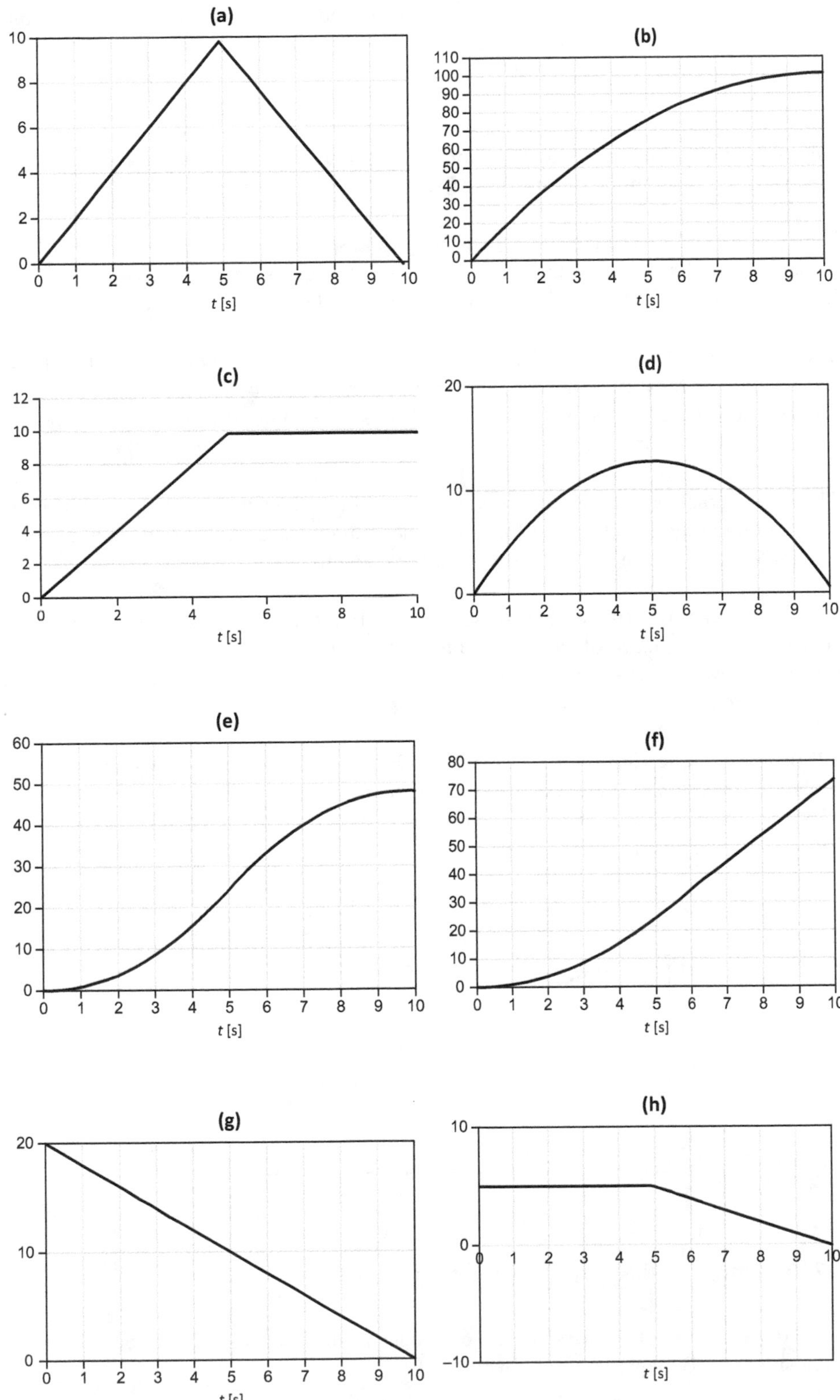

FIGURE 7.34 (a–h) Graphs for multiple-choice questions 1–6.

1. The d_x vs. t graph of a runner that crosses the start line with a nonzero velocity and continues running forward along the straight track, slowing down at a constant rate for the entire 10 seconds until she comes to rest.

2. The d_x vs. t graph of a runner starting from rest and speeding up in the forward direction at a constant rate for approximately 5 seconds, then continuing along the straight track slowing down at a constant rate for the next 5 seconds until she comes to rest.

3. The v_x vs. t graph of a runner starting from rest and speeding up in the forward direction at a constant rate for approximately 5 seconds, then continuing along the straight track slowing down at a constant rate for the next 5 seconds until she comes to rest.

4. The d_x vs. t graph of a runner that crosses the start line with a nonzero velocity and continues running forward along the straight track at the same speed for approximately 5 seconds, then turns around very quickly and runs back to the start at the same speed she maintained for the first 5 seconds.

5. The v_x vs. t graph of a runner that crosses the start line with a nonzero velocity and continues running away from the start line along the straight track, slowing down at a constant rate for the entire 10 seconds it takes for her to come to rest.

6. The v_x vs. t graph of a runner that crosses the start line with a nonzero forward velocity and continues running forward along the straight track at the same speed for 5 seconds, then she slows down at a constant rate for the next 5 seconds, coming to rest at the end of that 5-second time interval.

For multiple-choice questions 7–9, please refer to Figure 7.35 and the following description. The displacement of a person moving along a straight track with a constant acceleration is plotted in Figure 7.35. The starting line is at $x = 0$. The positive x-direction is defined to be the forward direction. The speed of the person at time 0 second is + 3.00 m/s.

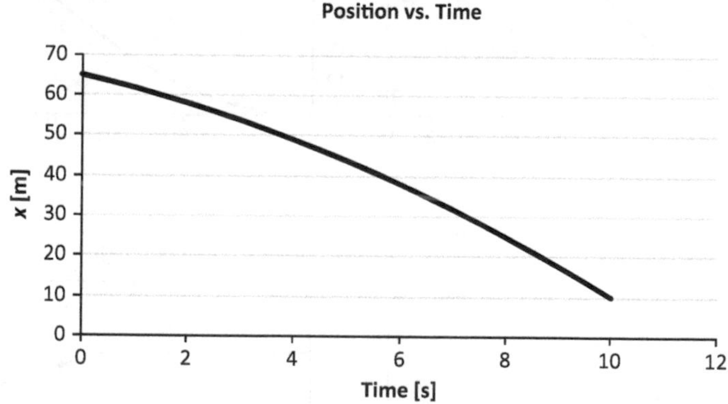

FIGURE 7.35 Position vs. time graph for multiple-choice questions 7–9.

7. Is the displacement of the person at $t = 6$ seconds positive, negative, or zero?
   A. positive B. negative C. zero

8. Is the velocity of the person at $t = 4$ seconds positive, negative, or zero?
   A. positive B. negative C. zero

9. Is the acceleration of the person from $t = 0$ second to $t = 8$ seconds positive, negative, or zero?
   A. positive B. negative C. zero

10. What condition must be met in order for the constant acceleration equations to be valid?
   A. The acceleration must be constant.
   B. The initial velocity of the object whose motion is under study must be 0.
   C. The acceleration must be 0.
   D. The x-component of the velocity must be a quadratic function of time.
   E. The duration of the time interval during which the equations are supposed to apply must be negative.

7.6.2 Problems

1. At the start of the recording time, $t = 0$, a runner crosses a start line with a velocity of 3 m/s forward. The runner continues at that velocity for 5 seconds, she then accelerates at a constant rate obtaining a velocity of 6 m/s forward after 5 seconds. Then, the runner slows at a constant rate to rest in 5 seconds. All motion of the runner is along one straight line. Sketch a graph of d_x vs. t, v_x vs. t, and a_x vs. t and use the start line as $x = 0$, and use forward direction as the +x-direction.
2. A car is at rest at a position 25 m ahead of the start line, at $t = 0$. The car accelerates uniformly in the forward direction, obtaining a speed of 25 m/s at the end of 10 seconds. It continues at that speed for 20 seconds at which point the driver applies the brakes and the car slows down smoothly to rest at the end of 5 seconds. Then, the car accelerates uniformly in the backward direction obtaining a speed of 10 m/s in reverse after 5 seconds. The driver immediately applies the brakes and the car steadily slows down and comes to a stop after another 5 seconds. All motion of the car is along one straight line. Use the start line as $x = 0$ and use forward direction as the +x-direction. Sketch a graph of d_x vs. t, v_x vs. t, and a_x vs. t.
3. A jogger is running at a velocity of 1.5 m/s forward. A dog on a long leash comes out from behind a house and begins to chase her. At the instant she realizes the dog is after her, she begins to accelerate at a constant rate of 0.5 m/s^2 forward. She maintains this acceleration for the 8.0 m she must travel to get out of reach of the dog.
   a. What is her velocity after this period of acceleration?
   b. How long (how much time) did it take her to travel the 8.0 m, while she was accelerating?

4. A car moves along a straight road with constant acceleration. At time 0, the car is 122 m ahead of the start line and moving forward at 80.0 km/h. 5.00 seconds later, the car is 398 m ahead of the start line.
   a. Find the acceleration of the car. After you calculate your answer, state your answer by giving a magnitude (a positive value with units) and a direction (write the word "forward" or the word "backward").
   b. Find the velocity of the car at time $t = 5.00$ seconds. State your answer by giving a magnitude and a direction.

5. A train travels along a straight track at a constant acceleration of 0.550 m/s^2. At the start of observations, the train is already moving forward at 15 m/s and the nose of the engine is already 826 m past a railroad crossing sign that you, as an observer, are using as a reference position. How fast is the train going when the nose of the engine is 1.000 km past the same sign?
6. A cart on a straight, horizontal air track is accelerating forward with a magnitude of 2.25 m/s^2. At the start of observations, the cart is already moving forward at 1.02 m/s. How much time does it take, beginning at the start of observations, for the cart to achieve a speed of 2.26 m/s?

7. A rocket-sled starts from rest on a horizontal, straight track, with the direction of motion labeled as the positive direction. Upon ignition, the rocket-sled accelerates at 50 m/s^2 for 3 seconds, while the rocket fuel burns. At this same instant, a brake is applied so that the rocket-sled slows to a stop at a constant rate in 5 seconds from the time the brake is applied. Compute the displacement of the rocket-sled during its deceleration, from the instant the brake is applied at the 3-second point to the instant at which the rocket-sled comes to a stop.
8. When a golf ball is hit off a tee, it experiences an acceleration of (1,000 g) = (1,000)(9.8 m/s^2) 9,800 m/s^2 for a time of 3.6 ms = 0.0036 second. Assuming the ball is initially at rest, compute the speed of the ball at the instant it leaves the club?
9. Sam is practicing his high-speed acceleration on a bicycle on a long straight track. Sam builds up some speed and at the point he passes his coach, his initial speed is 15.6 m/s, and at the end of a 30 m long straight-away another coach measures his speed to be 21.1 m/s in the same direction.
   a. How long did it take Sam to travel the 30 m between coaches?
   b. What was his acceleration?

10. A car decelerates uniformly from 42 to 34 m/s in 2.4 seconds. Determine the deceleration of the car and the distance traveled during the deceleration.

8 Dynamics 1

8.1 INTRODUCTION

In the first few chapters of this volume, the techniques associated with vector components and solving static problems were established. These techniques were applied with forces ranging from tension, to gravity, to those associated with the electromagnetic forces. In the previous chapter, the vocabulary and expressions associated with the motion of an object was formalized through kinematics. At this point in the text, these previous concepts and techniques will be integrated through Newton's Laws in a branch of physics known as dynamics. After an introduction to Newton's Laws, several examples will be presented that combine the concepts from the previous chapters to emphasize that these laws can be applied to analyze the motion of objects under the influence of all forces including gravity, tension, and even the electromagnetic forces.

Chapter question: When an arrow, which has been shot from a bow, flies through the air, does it need a force applied in the forward direction to keep it moving forward? Your answer to this question illustrates whether you understand the motion of objects like Aristotle or Galileo (Figure 8.1).

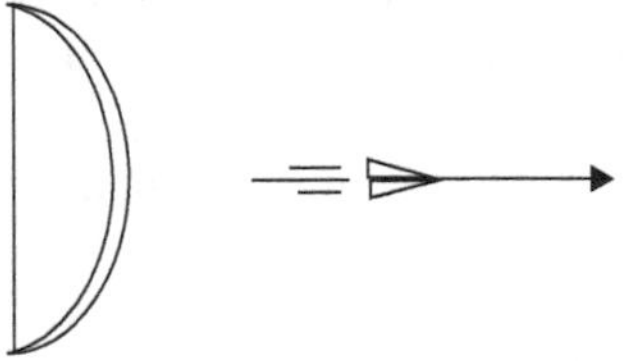

FIGURE 8.1 Arrow flying through the air moments after it leaves the bow.

8.2 KINEMATICS REVIEW

The kinematic equations from the previous chapter provide a method to calculate the displacement and velocity of an object undergoing a constant acceleration. These equations are given here again for convenience as equations (8.1)–(8.3) as:

$$v(t) = v_0 + a\, t \tag{8.1}$$

$$d(t) = d_0 + v_0\, t + 1/2\, a\, t^2 \tag{8.2}$$

$$v^2 - {v_0}^2 = 2\, a\, [d] \tag{8.3}$$

are all that is needed to make predictions of motion. Unfortunately, they provide no understanding of how and why a motion is occurring.

DOI: 10.1201/9781003308065-8

8.3 NEWTON'S LAWS

In the late 1600s, Sir Isaac Newton provided the logical framework for the study of forces and their effect on the motion of an object in one of his great works called *Principia*. In this document, he revealed his three laws and a structure to understand and predict the motion of an object. Although these laws are known as Newton's Laws, Galileo Galilei did quite a bit of work to set the stage for Newton. Learning from a set of experiments in which he rolled a set of spheres down long ramps of different steepness, Galileo observed that a sphere released from rest from the top of a ramp steadily sped up on the way down the ramp. As long as the ramps were not too steep, he could observe that the ball doesn't just roll down the ramp at some fixed speed, it accelerates the whole way down. This is all very impressive if you consider he didn't have a clock capable of making these measurements. Galileo further noted that the steeper the ramp was made the faster the ball would speed up on the way down. He did trial after trial, starting with a slightly inclined plane and gradually making it steeper and steeper. Each time he made it steeper, the ball would, on the way down the ramp, speed up faster than it did before, until the ramp got so steep that he could no longer see that it was speeding up on the way down the ramp—it was simply happening too fast to be observed. Galileo concluded that, as he continued to make the ramp steeper, the ball's speed was still increasing on its way down the ramp and the greater the angle, the faster the ball would speed up. Galileo hypothesized that if the ramp is tilted at 90°, the ball will actually fall as opposed to rolling down the ramp, and the object, for which air resistance is negligible, will speed up the whole way down, until it hits the Earth. So, it is clear Galileo did quite a bit to set the stage for Sir Isaac Newton, who was coincidentally born the same year that Galileo died.

Building upon the work of Galileo, it was Newton who recognized the relationship between force and motion. He formalized the link between force and acceleration, more specifically, that whenever an object is experiencing a net force that object experiences an acceleration in the same direction as the force. Newton's work on forces and motion are summarized in three laws of motion.

Newton's First Law states that an object will remain at rest or move with constant velocity unless it is acted upon by unbalanced forces. This law defines the property of mass known as *inertia*, which is a measure of an object's resistance to changes in its motion. An object doesn't need anything to keep it going if it is already moving. It can continue to move at a constant velocity as long as there is no net force acting on it. In fact, it takes a force to *change* the velocity of an object.

Newton's Second Law states that an object will accelerate at a rate directly proportional to the net force on the object and inversely proportional to its mass. This law is best known by equation (8.1) as:

$$\vec{F}_{\text{net}} = m\vec{\mathbf{a}} \tag{8.1}$$

where $\vec{\mathbf{a}}$ is the acceleration of the object, m is the mass of the object, and $\vec{F}_{\text{net}}$ is the net force on the object. The net force was first introduced in Chapter 2 in equation (2.1) as the vector sum of all the forces acting on the object. In Chapter 2, for systems at rest, the starting point of our analysis of these static systems was to set the net force to zero. This is consistent with Newton's Second Law since for systems at rest the acceleration is zero, so the right side of equation (8.1) is zero for systems at rest. Thus, statics is a subset of dynamics. Like the analysis of static systems, Newton's Second Law is the fundamental equation of dynamics.

In addition, this law relates the SI units for force to those of the mass and acceleration of an object, so: $1\ \text{N} = 1\ \text{kg}\frac{\text{m}}{\text{s}^2}$. It is important to note that this is not the definition of a Newton but an equivalence of the units involved in Newton's Second Law. There doesn't need to be a definition of a Newton beyond a quantity of force, any more than there needs to be a definition of a pound, beyond an amount of force. Remember, a Newton (N) is an amount of force that does not require an acceleration to be present for the force to be applied. This connection between the units of Newton, kg, and m/s^2 is based upon Newton's Second Law.

Newton's Third Law states that when one object applies a force on a second object, the second object applies a force back on the first object that has an equal magnitude and in the opposite direction. This law is commonly referred to as the law of **action–reaction**, since for every action there is an equal and opposite reaction.

Newton's Third Law is a statement of the fact that any force is just one half of an interaction where an *interaction* is the mutual pushing or pulling of objects on each other. These two forces are called a **force pair**. In some cases, where the effect is obvious, the validity of Newton's Third Law is evident. For instance, two people who have the same mass and are wearing skates on ice and are facing each other, as shown in Figure 8.2. If skater-1 pushes off against the stationary hands of skater-2, both skaters go backward away from each other, even if skater-2 only holds their arm still and does not push back. It might at first be hard to accept the fact that skater-2 is pushing back on the hands of the first skater, but we can tell that both skaters experience a backward acceleration. In fact, while the pushing is taking place, the force exerted on skater-2 must be equal in magnitude and in the opposite direction as the force that skater-1 exerts on the other skater because we see that they both have the same speed in opposite directions.

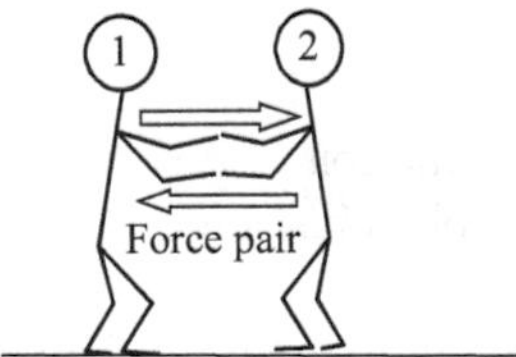

FIGURE 8.2 Action–reaction of two skaters.

There are also cases where the effect of at least one of the forces in the interaction pair is not at all evident. For example, as shown in Figure 8.3, when you kick a soccer ball, the ball applies an equal and opposite force back on your foot.

FIGURE 8.3 Action–reaction of kicking a ball.

As you apply force on the soccer ball, the ball begins to accelerate, which limits the magnitude of the force that you can apply to the ball and thus it limits the force that the ball applies back to your foot. The force pair is still the force you apply to the ball and the force that the ball applies to your foot. The acceleration of the ball limits the force you can apply to the ball since it accelerates away from your foot, but remember the force that you apply to the ball is still equal and opposite to the force that the ball applies back to your foot. Since there is an acceleration, apply Newton's Second Law for the ball: $F_{app} = m_b a_b$, where F_{app} is the applied force from your foot, m_b is the mass of the ball, and a_b is the acceleration of the ball in the direction of the applied force. The force that is associated with the acceleration of the ball is just the force that you apply to the ball; the reaction force is on your foot not on the ball. If on the other hand, you try to kick a wall as hard as you can, (please don't try this), the force you kick the wall with is accompanied by an equal but opposite force that the wall applies back to your foot. This reaction force is the one you feel. Since the wall will not accelerate no matter how hard you kick it, the wall pushes back on your foot as hard as you push on the wall. The force you apply to the wall is a force pair with the force the wall applies to your foot, since these forces are always equal.

In summary, *Newton's Three Laws of Motion* state:

I. Inertia is the property of an object to resist its change in motion, so if there is no net force acting on an object, then the velocity of that object is not changing.
II. $\vec{F}_{net} = m\vec{a}$ states that if there is a net force on an object of mass (m), then that object is experiencing acceleration $\vec{a}$.
III. Action–reaction, so any time one object is exerting a force on a second object, the second object is exerting a force of equal magnitude but opposite direction back on the first object.

8.4 DYNAMICS

Applying Newton's Laws to model the motion of object is the study of dynamics. The process of generating a solution to a dynamics problem has a few key steps:

- **Step 1**: Draw a complete free-body diagram of the object whose motion is under study. Only include the forces on the object that is the focus of the problem and the acceleration of this same object. This is an important clarification that sometimes is more difficult than it seems, so pay attention to only include the forces on the object you are analyzing.
- **Step 2**: Find the components of the vectors involved in the problem.
- **Step 3**: Expand term-by-term Newton's Second Law, $\vec{F}_{net} = m\vec{a}$, in each of the vector directions present in the problem.
- **Step 4**: Solve the dynamic equations generated in step 2 and apply kinematic equations if necessary.

8.4.1 Dynamics Examples

Example 8.1

A sled of mass of 30 kg is being pulled forward over a horizontal frictionless surface by pulling on a horizontal rope attached to the front of the sled. If the tension in the rope is 90 N, what is the acceleration of the sled? (Figure 8.4)

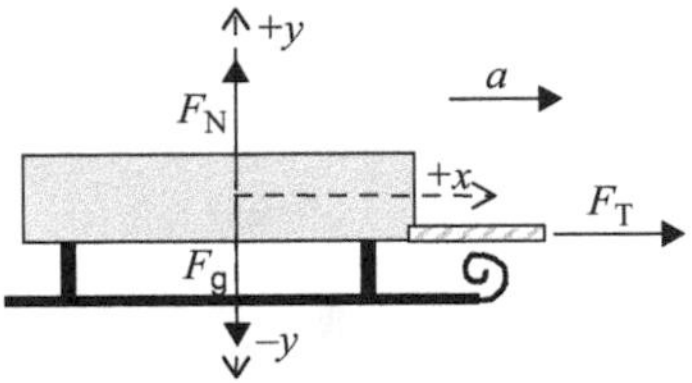

FIGURE 8.4 Free-body diagram for dynamics Example 8.1.

Step 1: Draw the free-body diagram of the sled. The upward force on the sled due to the snowy ground is called the *normal force* and is commonly denoted as F_N. This is because the term normal in mathematics means perpendicular and this is the force perpendicular to the surface on which an object is located. This force balances out the weight so there is no acceleration downward for objects on a surface. In this chapter, the normal force will be included to balance out the weight of the object on flat surfaces. In the next chapter, the normal force will be investigated further in the context of friction, in which it plays an important role in computing frictional forces.

Step 2: In the free-body diagram, an x-y coordinate system is defined and in this case the acceleration and all of the forces lie along either the x- or y-axis. Breaking the vectors up into their x- and y-components can be done by inspection:

$$F_{Tx} = F_T = 90 \text{ N} \quad \text{and} \quad F_{Ty} = 0$$

$$F_{Nx} = 0 \quad \text{and} \quad F_{Ny} = +F_N$$

$$F_{gx} = 0 \quad \text{and} \quad F_{gy} = -F_g = -(\text{mg}) = -(30 \text{ kg})\left(9.8\frac{\text{N}}{\text{kg}}\right) = -294 \text{ N}$$

$$a_x = a \quad \text{and} \quad a_y = 0$$

The acceleration along one line is independent of any forces perpendicular to that line so each axis is considered one at a time.

Step 3: Write Newton's Second Law for the x-components as:

$$\Sigma F_x = ma_x$$

$$F_{Tx} = ma_x$$

$$F_T = ma$$

$$90 \text{ N} = (30 \text{ kg})a$$

Write Newton's Second Law for the y-components as:

$$\Sigma F_y = ma_y$$

$$F_{Ny} + F_{gy} = ma_y$$

$$F_N - F_g = 0$$

$$F_N = F_g$$

Step 4: Solve for the x- and y-component expressions from Newton's Second Law to find the acceleration of and the normal force on the box:

x-components:

$$90\ \text{N} = (30\ \text{kg})a$$

$$3\ \frac{\text{m}}{\text{s}^2} = a$$

y-components:

$$F_N = 294\ \text{N}$$

Example 8.2

A 2 kg rock is thrown up into the air. Find the acceleration of the rock on the way up (Figure 8.5).

Step 1: Generate a free-body diagram: Notice that the acceleration is down since the rock is moving upward but is slowing down, so the acceleration is in the opposite direction of the motion. Also, notice that the velocity of the rock is not part of the free-body diagram.
Step 2: Find the components of the forces and acceleration:

$$F_{gx} = 0 \quad \text{and} \quad F_{gy} = -F_g = -(2\ \text{kg})\left(9.8\ \frac{\text{N}}{\text{kg}}\right) = -19.6\text{N}$$

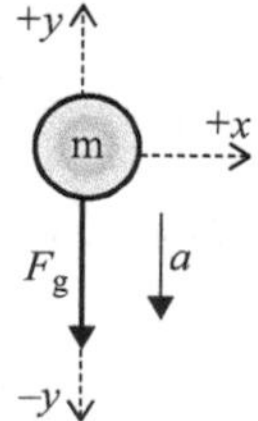

FIGURE 8.5 Free-body diagram for dynamics Example 8.2.

$$a_x = 0 \quad \text{and} \quad a_y = -a$$

Step 3: Newton's Second Law for the y-components:

$$\Sigma F_y = ma_y$$

$$F_{gy} = ma_y$$

$$-F_g = m(-a)$$

$$-19.6\text{ N} = (2\text{ kg})(-a)$$

The x-components for force and acceleration are zero.

Step 4: Solve the y-components for acceleration.

$$-a = \frac{-19.6\text{ N}}{2\text{ kg}} = -9.8\ \frac{\text{m}}{\text{s}^2}$$

$$a = 9.8\ \frac{m}{s^2}$$

This example illustrates why the gravitational field strength ($g=9.80\frac{\text{N}}{\text{kg}}$) is sometimes referred to as the acceleration due to gravity and given the value, $g = 9.8\ \frac{\text{m}}{\text{s}^2}$, downward. This is true for objects that are traveling through the air in which gravity is the only force on the object. It becomes a bit misleading when there are other forces such as air friction on the object.

Example 8.3

A 2-kg box is moving along a flat horizontal table with an applied force of 20 N in the +x-direction and another force of 2 N is applied in the –x-direction. Find the magnitude of the acceleration of the object in the x-direction.

Step 1: Generate a free-body diagram, see Figure 8.6:

Step 2: In the free-body diagram, an x-y coordinate system is defined and in this case the acceleration and all of the forces lie along either the x- or y-axis. Breaking the vectors up into their x- and y-components can be done by inspection in such a case.

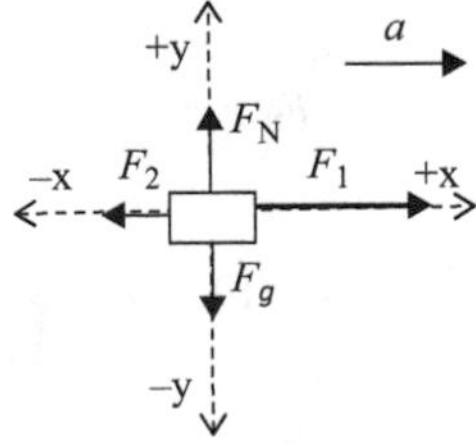

FIGURE 8.6 Free-body diagram for dynamics Example 8.3.

$$F_{1x} = +F_1 = 20\text{ N} \quad \text{and} \quad F_{1y} = 0$$

$$F_{2x} = -F_2 = -2\text{ N} \quad \text{and} \quad F_{2y} = 0$$

$$F_{gx} = 0 \quad \text{and} \quad F_{gy} = -F_g = -19.6\text{N}$$

$$F_{Nx} = 0 \quad \text{and} \quad F_{Ny} = +F_N$$

$$a_x = a \quad \text{and} \quad a_y = 0$$

The acceleration along one line is independent of any forces perpendicular to that line, so we can consider one axis at a time.

Step 3: Write Newton's Second Law for the x-components as:

$$\Sigma F_x = ma_x$$

$$F_{1x} + F_{2x} + F_{Nx} + F_{gx} = ma_x$$

$$F_1 - F_2 + 0 + 0 = ma$$

$$F_1 - F_2 = ma$$

$$20\text{ N} - 2\text{ N} = (2\text{ kg})a$$

Write Newton's Second Law for the y-components as:

$$\Sigma F_y = ma_y$$

$$F_{1y} + F_{2y} + F_{Ny} + F_{gy} = ma_x$$

$$0 + 0 + F_N - F_g = m\ 0$$

$$F_N = F_g$$

$$F_N = 19.6\text{ N}$$

Step 4: Solve for the x-component expressions from Newton's Second Law for the acceleration of the box:

$$20\text{ N} - 2\text{ N} = (2\text{ kg})a$$

$$18\text{ N} = (2\text{ kg})a$$

$$9\,\frac{\text{m}}{\text{s}^2} = a$$

Example 8.4

Two boxes with masses of $m_1 = 1$ kg and $m_2 = 2$ kg are connected by a string, of negligible mass, and placed on a smooth surface, with negligible friction. As shown in Figure 8.7., a horizontal tension force of 6 N is applied to mass 2 (m_2) by pulling on the string in the +x-direction. Find the tension in the string between the masses and the speed of each mass after the tension is applied for 4 seconds.

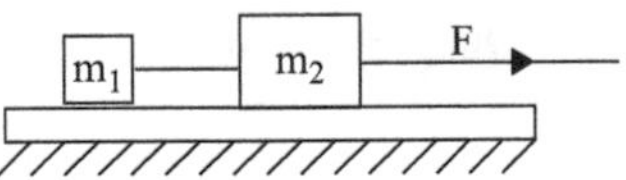

FIGURE 8.7 Diagram for dynamics Example 8.4.

Step 1: Draw a free diagram for each box, separately (Figure 8.8).

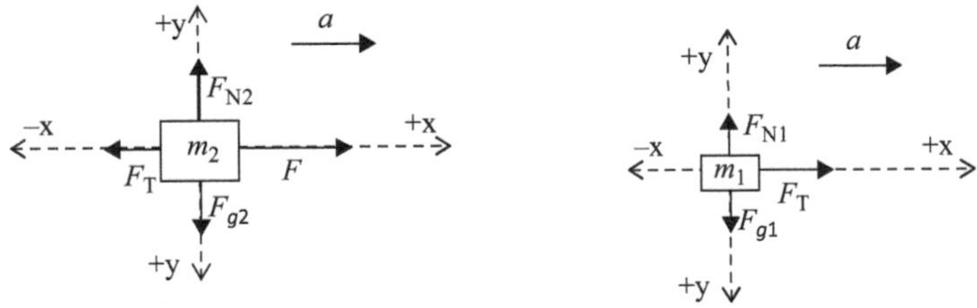

FIGURE 8.8 Free-body diagrams for dynamics Example 8.4.

Step 2: In the free-body diagram, an x-y coordinate system is defined and in this case the acceleration and all of the forces lie along either the x- or y-axis. Breaking the vectors up into their x- and y-components can be done by inspection in such a case. For mass-1, the forces are:

$$F_{Tx} = F_T \quad \text{and} \quad F_{Ty} = 0$$

$$F_{N1x} = 0 \quad \text{and} \quad F_{N1y} = +F_N$$

$$F_{g1x} = 0 \quad \text{and} \quad F_{g1y} = -F_{g1} = -(1\text{ kg})\left(9.8\,\frac{\text{N}}{\text{kg}}\right) = -9.8\text{N}$$

$$a_{1x} = a \quad \text{and} \quad a_y = 0$$

For mass-2, the forces are:

$$F_x = +F = 6\text{ N} \quad \text{and} \quad F_y = 0$$

$$F_{Tx} = -F_T \quad \text{and} \quad F_{Ty} = 0$$

$$F_{N2x} = 0 \quad \text{and} \quad F_{N2y} = +F_N$$

$$F_{gx} = 0 \quad \text{and} \quad F_{g2y} = -F_{g2} = -(2\text{ kg})\left(9.8\ \frac{\text{N}}{\text{kg}}\right) = -19.6\text{N}$$

$$a_{2x} = a \quad \text{and} \quad a_y = 0$$

The acceleration along one line is independent of any forces perpendicular to that line, so we can consider one axis at a time.

Step 3: Write Newton's Second Law for the x-components of the forces as:

$$\begin{aligned} \Sigma F_{1x} &= m_1 a_x & \Sigma F_{2x} &= m_2 a_x \\ F_{Tx} &= m_1 a_x & F_x + F_{Tx} &= m_2 a_x \\ F_T &= m_1 a & F - F_T &= m_2 a \\ F_T &= (1\text{ kg})a & 6N - F_T &= (2\text{ kg})a \end{aligned}$$

Write Newton's Second Law for the y-components as:

$$\begin{aligned} \Sigma F_{1y} &= m_1 a_y & \Sigma F_{2y} &= m_2 a_y \\ F_{N1} - F_{g1} &= m_1\ 0 & F_{N2} - F_2 &= m_2\ 0 \\ F_{N1} - F_{g1} &= m_1\ 0 & F_{N2} - F_{g2} &= m_2\ 0 \\ F_{N1} - 9.8\text{ N} &= 0 & F_{N2} - 19.6\text{ N} &= 0 \end{aligned}$$

Solve these two expressions to find the normal forces:

$$F_{N1} = 9.8\text{ N} \quad F_N = 19.6\text{ N}$$

Step 4: The final two expressions from Newton's Second Law provide a way to find the acceleration and the tension in the string between the masses.

Start with the expressions at the end of the x-component analysis of the Second Law:

$$F_T = (1\text{ kg})a \quad 6N - F_T = (2\text{ kg})a$$

Solve both expressions for F_T and set them equal:

$$F_T = (1\text{ kg})a \quad F_T = 6\text{N} - (2\text{ kg})a$$

$$(1\text{ kg})a = 6\text{N} - (2\text{ kg})a$$

$$(3 \text{ kg})a = 6\text{N}$$

$$a = 2 \frac{\text{m}}{\text{s}^2}$$

Since the acceleration and the time for which the force is applied is now known, the speed of the masses after 4 seconds of applied tension can be calculated by using kinematic equation (8.1) :

$$v(t) = 0 + 2 \frac{\text{m}}{\text{s}^2} 4 \text{ seconds} = 8 \frac{\text{m}}{\text{s}}$$

Example 8.5

Newton's Laws work for all forces of nature, not just the forces of gravity or tension. For example, as depicted in Figure 8.9, if an electron is located in a region of space in which an electric field of 3 N/C exists, the electron will accelerate.

To find the speed of the electron if it accelerates across a region which is 2 cm long, we follow a similar procedure once we have determined the force on the charged particle.

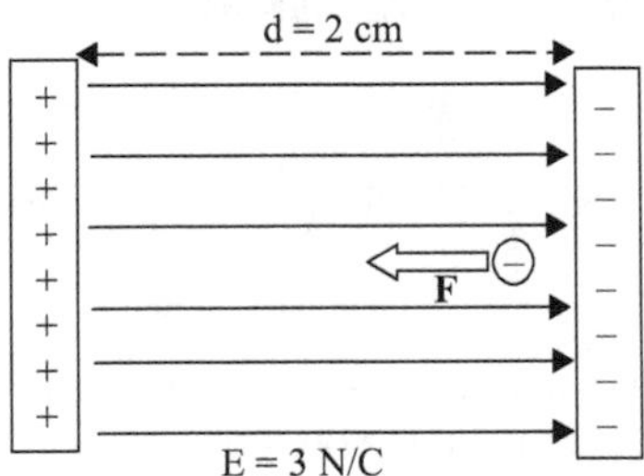

FIGURE 8.9 Dynamics Example 8.5 of an electron in an electric field.

Step 1: Generate a free-body diagram of the electron in the electric field, as shown in Figure 8.10.

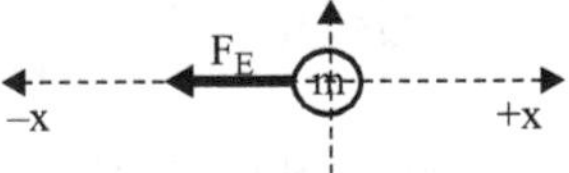

FIGURE 8.10 Free-body diagram for dynamics Example 8.5.

Remember that the direction of the electric field is the direction that a positive charge would be pushed so because an electron is negative, the direction of the electric force $\mathbf{F_E}$ will be in the –x-direction.

Step 2: In the free-body diagram, the only force and acceleration is in the –x-direction, so only those components will be found. Since an electron has a charge of $q_e = 1.602 \times 10^{-19}$C and the electric field in the region between the two plates in Figure 8.9, the force on the electron comes from the definition of the electric field defined in Chapter 5.

Given that the electric field is $E = \dfrac{F}{q}$ therefore,

$$F = E\,q = \left(3\frac{\text{N}}{\text{C}}\right)\left(1.602\times10^{-19}\ \text{C}\right) = 4.806\times10^{-19}\ \text{N}$$

$$F_{\text{Ex}} = -F_{\text{E}} = -4.806\times10^{-19}\ \text{N}$$

$$a_{\text{x}} = -a$$

The acceleration along one line is independent of any forces perpendicular to that line, so we can consider one axis at a time.

Step 3: Write Newton's Second Law for the x-components as:

$$\Sigma F_{\text{x}} = ma_{\text{x}}$$

$$F_{\text{Ex}} = ma_{\text{x}}$$

$$-F_{\text{E}} = m(-a)$$

$$-4.806\times10^{-19}\ \text{N} = \left(9.11\times10^{-31}\ \text{kg}\right)a$$

Step 4: Solve for the x- and y-component expressions from Newton's Second Law to find the acceleration of and the normal force on the box:

x-components:

$$90\ \text{N} = (30\ \text{kg})a$$

$$5.2755\times10^{11}\ \frac{\text{m}}{\text{s}^2} = a$$

The speed of the electron can then be found using kinematics. The best choice of an equation is (8.3):

$$v^2 - v_o^2 = 2(a)(d)$$

Since the initial velocity v_o is zero, the acceleration was just computed, and the distance is known, the velocity can be found by solving for v and plugging in the values.

$$v = \sqrt{(2\ a\ d)} = \sqrt{\left(2\left(5.2755\times10^{11}\ \frac{\text{m}}{\text{s}^2}\right)(0.02\ \text{m})\right)} = 1.45\times10^5\,\frac{\text{m}}{\text{s}}$$

8.5 INERTIAL FRAME (THE FINE PRINT)

Newton's Laws of Motion only apply in an inertial reference frame, which is a coordinate system that is either fixed or moving at a constant speed along a straight-line path, relative to a fixed point. Imagine that the stars are fixed in space so that the distance between one star and another never changes. (They are not fixed. The stars are moving relative to each other.) Now imagine that you create a Cartesian coordinate system; a set of three mutually orthogonal axes that you label x, y, and z. Your Cartesian coordinate system is a reference frame. Now as long as your reference frame is not rotating, and is either fixed or moving at a constant velocity relative to the (fictitious) fixed stars, then your reference frame is an *inertial reference frame*. Note that velocity has both magnitude and direction and when we stipulate that the velocity of your reference frame must be constant in order for it to be an inertial reference frame, we aren't just saying that the magnitude has to be constant but that the direction has to be constant as well. The magnitude of the velocity is the speed. So, for the magnitude of the velocity to be constant, the speed must be constant. For the direction to be constant, the reference frame must move along a straight-line path. As you may have surmised, most of the time we assume reference frames like in our lab to be a good approximation of an inertial frame, even though we are on a planet rotating on its axis, orbiting around the sun, and moving away from the center of the universe.

8.6 ANSWER TO CHAPTER QUESTION

An Aristotelian would claim that when an arrow, which has been shot from a bow, flies through the air, the air must provide a force on the arrow to propel it forward, since the natural state of the arrow is at rest (Figure 8.11).

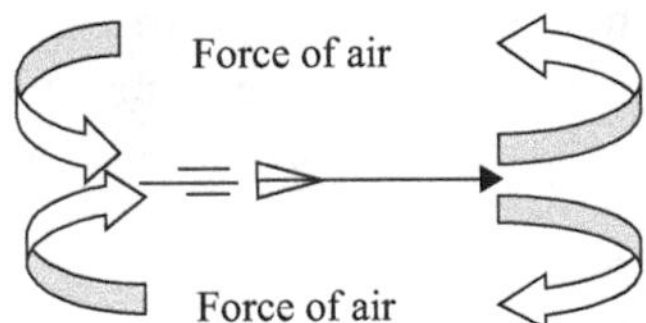

FIGURE 8.11 The Aristotelian view of the arrow flying through the air.

Aristotle believed that motion that was contrary to the natural state of an object, such as the arrow flying through the air. This was known as *violent motion*. The Aristotelian explanation of the arrow's flight begins with the arrow cutting through the air, which produces a small vacuum, and since nature abhors a vacuum the air rushes back into its place propelling the arrow forward. This explanation makes common sense, but it does not provide a way to analyze the situation.

A Galilean philosopher, like Newton, would claim that the arrow was initially propelled forward by a force, provided by the bow string, and after that, the arrow moved ahead due to its own inertia.

In fact, the arrow will probably be slowing down a bit during its flight, because of air resistance (Figure 8.12).

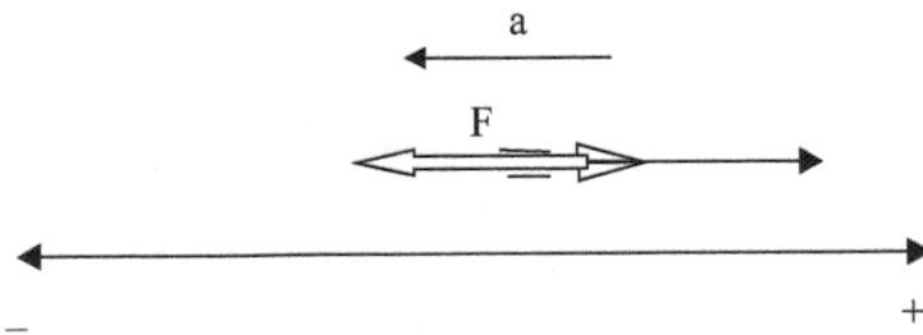

FIGURE 8.12 The Galilean view of the arrow flying through the air.

In this case, the velocity will still be in the (positive) direction of motion, but the force will be in the opposite direction (negative), so by Newton's Second Law the acceleration will also be in the negative direction, so the arrow is slowing down.

8.7 QUESTIONS AND PROBLEMS

8.7.1 Multiple-Choice Questions

1. A 500 kg horse is connected to a 600 kg sleigh, with a horizontal rope of negligible mass. The sleigh is located on a level section of packed snow, which has negligible friction with the sleigh rails. The horse begins to pull and the sleigh accelerates at 3 m/s^2 across the snow. Which of the following statements correctly describes why the sleigh accelerates forward?
   A. The horse pulls with a force that is greater than the weight of the sleigh.
   B. The net force on the sleigh is in the forward direction.
   C. The horse pulls forward on the sleigh with a greater force than the force with which the sleigh pulls back on the horse.

2. A person is pulling a cart of mass of 10 kg forward with a force of 30 N forward. If the cart is moving at a constant velocity of 2 m/s forward, the magnitude of the net force on the cart is:
   A. 0 N B. 10 N C. 20 N D. 30 N

3. A 500 kg horse is connected to a 600 kg sleigh, with a horizontal rope of negligible mass. The sleigh is located on a level section of packed snow, which has negligible friction with the sleigh rails. The horse begins to pull and the sleigh accelerates at 3 m/s^2 across the snow. The tension in the rope must be:
   A. 0 N B. 500 N C. 1000 N D. 1500 N E. 1800 N F. 3300 N

4. In Figure 8.13, a person is pulling a cart, with a mass of 100 kg, with a force of 300 N. If the cart is moving forward with a constant velocity of 2 m/s forward, what is the magnitude of the net force on the cart?
   A. 0 N B. 100 N C. 200 N D. 300 N

5. In Figure 8.13, a person is pulling a cart, with a mass of 100 kg, with a force of 300 N. If the cart is moving forward with a constant acceleration of 2 m/s^2 forward, what is the magnitude of the net force on the cart?
   A. 0 N B. 100 N C. 200 N D. 300

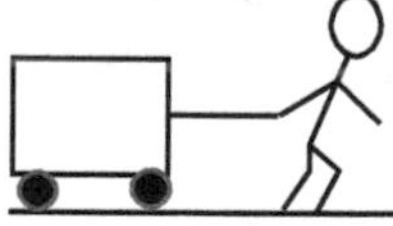

FIGURE 8.13 Multiple-choice questions 4 and 5 and problem 1.

(6–8) In Figure 8.14, a person is pulling on a horizontal rope attached to cart 1, which is attached to cart 2 by another horizontal rope.

The horizontal force applied by the person causes the two carts to accelerate forward at 2 m/s^2. Friction between each cart and the track is negligible and each cart has a mass of 100 kg.

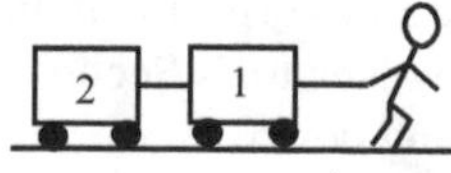

FIGURE 8.14 Multiple-choice questions 6–8.

6. The tension in the rope between the person and cart 1 is greater than, less than, or equal to the tension between cart 1 and cart 2?
   A. greater than B. less than C. equal to

7. The tension in the rope between the carts must have a magnitude of which of the following?
   A. 0 N B. 100 N C. 200 N D. 400 N

8. The tension in the rope between the person and cart 1 must have a magnitude of which of the following?
   A. 0 N B. 100 N C. 200 N D. 400 N

9. A spaceship is accelerating through empty space powered by its rocket engines. When the rocket engines are suddenly turned off, the spaceship will:
   A. suddenly come to a stop.
   B. slow down at a constant rate and come to a stop.
   C. continue to move at a constant velocity.
   D. continue to accelerate.

10. An elevator car with a person included weighs 2.0×10^4 N. If the car and person are accelerating upward during the first few seconds of its ascent, is the tension in the elevator cable during this initial part of the ascent greater than, less than, or equal to 2.0×10^4 N? (As shown in Figure 8.15, the elevator cable is attached to the top of the elevator and it is used to raise and lower the elevator.)
    A. greater than B. less than C. equal to

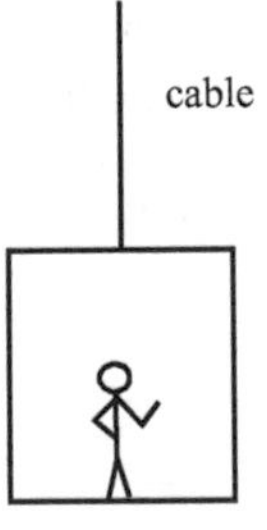

FIGURE 8.15 Multiple-choice question 10 and problem 2.

8.7.2 Problems

1. As shown in Figure 8.13, a person pulls on a horizontal rope attached to a 200 kg wagon that moves across the floor with negligible friction. Compute the acceleration of the wagon if the net force in the horizontal direction on the crate is 100 N.
2. Shown in Figure 8.15, an elevator car with a person included weighs 2.0×10^4 N. If the car and person have an acceleration of $0.25\,m/s^2$ upward during the first few seconds of its ascent, what is the tension in the elevator cable during this initial part of the ascent? (The elevator cable is attached to the top of the elevator and it is used to raise and lower the elevator.)
3. A 210 kg motorcycle accelerates from 0 to 55 mph (24.6 m/s) in 6.0 seconds. What is the magnitude of the net force causing the acceleration?

4. A stationary hockey player makes a pass of the 0.2 kg puck, with her stick. Assume the ice is perfectly horizontal and the puck stays on the ice during the entire contact with the stick. To make the pass, the stick is in contact with the puck over a horizontal distance 0.4 m and during this contact, the puck accelerates from rest to a speed of 5 m/s in the horizontal direction. Compute the net force on the puck during the time the stick is in contact with the puck.
5. A rocket has a total mass of 5.0×10^4 kg. How large is the force produced by the engine (the thrust) when the rocket is accelerating upward at 20 m/s^2?
6. An old piece of furniture, with a mass of 80 kg, is tossed out of a window of a building being prepared for demolition. If the average force of the ground on the piece of furniture is 2×10^5 N to bring it to rest, compute the minimum acceleration (deceleration) of the piece of furniture.
7. A tiny spring-loaded toy with a mass of 0.0123 g leaves the ground with a vertical speed of 0.40 m/s, when it pops. Assuming a constant upward acceleration, if the base of the toy is in contact with the ground for 1.0 ms during the time it exerts a downward force on the ground, compute the force that the toy exerts on the ground during this pop.
8. A large block of wood is nailed to structure so that it cannot move. A bullet, with a mass of 1.8 g and traveling at 350 m/s in a perfectly horizontal direction, is shot into the large block of wood. The bullet comes to a stop after traveling 13 cm straight into the block of wood. Compute the average force the wood exerts on the bullet to bring it to rest.
9. A 2,000 kg car starting from rest is positioned at the starting line of a 300-m long, straight, and horizontal track. The car accelerates at a constant rate from rest to 36 m/s (about 80 mph) in 100 m of traveling straight down the track. Compute the net force on the car during the acceleration of the car from 0 to 36 m/s in the first 100 m of the car's run down the track.
10. Two blocks are located on a long, horizontal, frictionless surface, as shown in Figure 8.16. Block-1, which has a mass of 5 kg, is located at rest at the origin and Block_2, which has a mass of 15 kg, is also at rest and positioned directly to the right of and in contact with Block-1. A horizontal force (F) of 4 N is applied on Block-1 in the +x-direction, until the blocks are moving together at a speed of 1.6 m/s. How far from the origin does Block-1 travel in the time in which the force is applied?

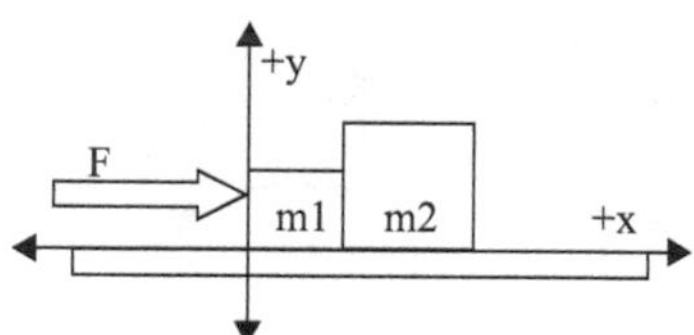

FIGURE 8.16 Problem 10.

9 Dynamics 2

9.1 INTRODUCTION

With the techniques of dynamics and kinematics established in the previous chapters, this chapter is focused on the application of these concepts to the analysis of more complicated systems. The force of friction and the techniques used to integrate it into dynamics-based analysis are formally introduced in the chapter. Some of the subtleties associated with normal force are discussed and worked with into several examples. This chapter is an extension of Chapter 8 focused on applying the concepts and techniques of dynamics to study systems with multiple forces each possibly with more than one component.

> **Chapter question:** If you are trying to move a large piece of furniture across a floor, is it better to try pushing it or pulling it? This question will be answered at the end of the chapter by applying the techniques developed around the normal force and the force of friction presented in the chapter.

9.2 FRICTION

The force of *Friction*, F_f, is any force or collection of forces that resists the motion of an object and cannot otherwise be attributed to a specific force or forces. In this way, friction is like a syndrome in medicine, since it is the resistance to the motion of an object, without a fundamental explanation. For example, if you push a large crate and it doesn't move, the common explanation is that the crate does not move because the force of friction between the floor and the crate is too large. This frictional force is a collection of forces, fundamentally electromagnetic forces between surface atoms and forces of surfaces pressing against each other. We don't need to explain every individual interaction for the analysis, just their collective effect that is summarized as friction.

Since the crate in Figure 9.1 remains at rest, while a force is applied, the static frictional force is equal and opposite of the applied force.

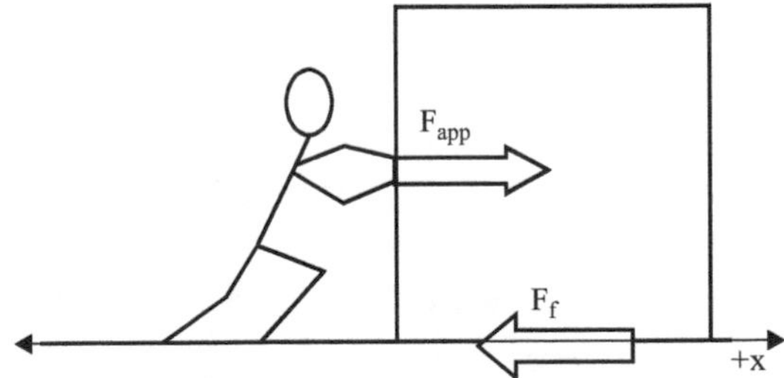

FIGURE 9.1 Applied and frictional forces on a crate.

As long as the crate remains at rest, the magnitude of the frictional force is equal to the applied force. If an object is moving while friction is acting on the object, the force of friction depends on the normal force on the object and the relative roughness of the surface. This surface roughness factor is known as the *coefficient of friction*, μ. The value of μ depends on the surface roughness of the object and the surface it is on. The coefficient of friction is maximum when an object is at

DOI: 10.1201/9781003308065-9

rest and decreases when the object is in motion. Thus, to distinguish these situations, there are two coefficients of friction, static and kinetic, μ_s and μ_k, respectively, for each object. The coefficient of friction depends on the exact objects involved, so for each problem, it is either given or the unknown that you are asked to find. The coefficient of friction values given in Table 9.1 are provided as examples and are not the same for all these interacting surfaces.

TABLE 9.1
Coefficients of Friction for a Few Material Pairs

Material 1	Material 2	Surface Conditions	μ_s	μ_k
Wood – waxed	Wet Snow	Clean and dry	0.14	0.1
Rubber	Asphalt	Clean and dry	0.9	0.5–0.8

To compute the magnitude of the maximum frictional force, F_f, on a stationary object or the magnitude of the frictional force on an object sliding across a surface, the appropriate coefficient of friction, μ_s or μ_k, is multiplied by the normal force, F_N, on the object given by equation (9.1) as:

$$F_f = \mu F_N, \tag{9.1}$$

Note that this is only the magnitude of the frictional force; the direction of the force will need to be determined by the physical situation.

Since friction is just another force, it is integrated into the analysis of dynamics in the same way as the other forces. Following are some examples of applying the concepts and techniques of dynamics with frictional forces included. In addition, situations in which the normal changes will be introduced as part of this analysis.

Example

A box with a mass of 2 kg is pulled with a horizontal applied force of 10 N and the object accelerates at 3 m/s². Find the coefficient of friction for this box and its surface.

Step 1: Generate a free-body diagram of the box that is presented in Figure 9.2.

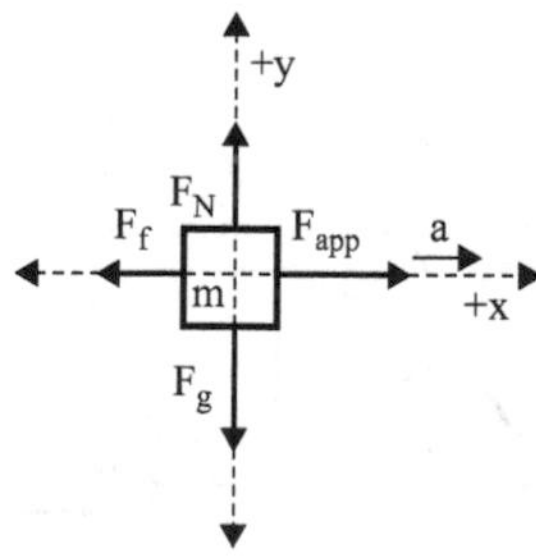

FIGURE 9.2 Free-body diagram of a box on a surface.

Step 2: Find the components of the force and acceleration vectors. In this case, because the forces are only in one direction, the components can be found by inspection.

$$F_{appx} = F_{app} \quad \text{and} \quad F_{appy} = 0$$

$$F_{Nx} = 0 \quad \text{and} \quad F_{Ny} = +F_N$$

$$F_{\text{gx}} = 0 \quad \text{and} \quad F_{\text{gy}} = -F_{\text{g}}$$

$$F_{\text{fx}} = -F_{\text{f}} \quad \text{and} \quad F_{\text{f}} = 0$$

Step 3: Write Newton's Second Law for the x-components of the forces as:

$$\Sigma F_{\text{x}} = ma_{\text{x}}$$

$$F_{\text{appx}} + F_{\text{Nx}} + F_{\text{gx}} + F_{\text{fx}} = ma_{\text{x}}$$

$$F_{\text{app}} + 0 + 0 + -F_{\text{f}} = ma$$

$$F_{\text{app}} - F_{\text{f}} = ma$$

Write Newton's Second Law for the y-components as:

$$\Sigma F_{\text{y}} = ma_{\text{y}}$$

$$F_{\text{appy}} + F_{\text{Ny}} + F_{\text{gy}} + F_{\text{fy}} = ma_{\text{y}}$$

$$0 + F_{\text{N}} - F_{\text{g}} + 0 = 0$$

Step 4: Solve these expressions to find the normal force and the force of friction. First, find the force of friction with the expression from the x-components,

$$F_{\text{f}} = F_{\text{app}} - ma = 10\ \text{N} - (2\ \text{kg})\left(3\ \frac{\text{m}}{\text{s}^2}\right) = 10\text{N} - 6\text{N} = 4\ \text{N}$$

Next, find the normal force with the expression from the y-component,

$$F_{\text{N}} = F_{\text{g}} = mg = (2\ \text{kg})\left(9.8\ \frac{\text{N}}{\text{kg}}\right) = 19.6\ \text{N}$$

To find the coefficient of friction solve equation (9.1) for μ,

$$\mu = \frac{F_{\text{f}}}{F_{\text{N}}} = \frac{(4\ \text{N})}{(19.6\ \text{N})} = 0.204$$

This coefficient will be different for different boxes and surfaces, even if they are made from the same material.

Since the force of friction is just the normal force multiplied by a coefficient, often the most difficult part of understanding frictional force problems is finding the normal force. As mentioned in the previous chapter, *normal forces* are the forces between the object being analyzed and surfaces in contact with the object. Normal Forces are an important part of applying Newton's Laws, since objects on the earth's surface have a force of gravity on them and they are not accelerating through the surfaces upon which they are located. In this way, a normal force is required to maintain equilibrium with the weight of the object. Remember, the normal force gets its name from the mathematical word for perpendicular—there are not abnormal forces. The other important application of the normal force is the computation of the magnitude of the force of friction on an object. As discussed and analyzed in several examples in Chapter 8 and the first example of dynamics with friction in this chapter, for objects at rest or with only unbalanced forces in the horizontal direction on flat horizontal surfaces, the normal force exerted by the horizontal surface on the object is equal and opposite to the weight of the object. There are situations in which the magnitude of the normal force is NOT equal to the weight of an object. The next few examples illustrate two common arrangements in which this is the case.

9.3 INCLINED PLANES

For objects on inclined planes, the normal force is *not* equal to the weight of the object. For example, as shown in Figure 9.3, a block with a mass of 2 kg is being pushed up a flat ramp that makes an angle $\theta = 20°$ with the horizontal by an applied force of 10 N that is in a direction parallel to the ramp.

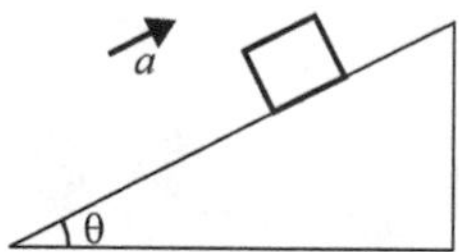

FIGURE 9.3 A block on an inclined plane.

Find the acceleration of the block up the ramp if the coefficient of friction between the ramp and the block is 0.15.

Solution:

Step 1: Create a free-body diagram of the block, as demonstrated in Figure 9.4.

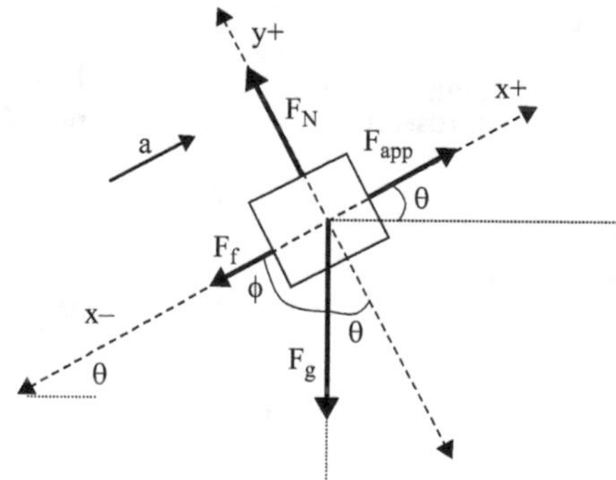

FIGURE 9.4 Free-body diagram of a box accelerated up an inclined plane.

Here, the major change in the free-body diagram is the rotation of the coordinate system so that the +x-axis is parallel to the surface of the inclined plane and thus the +y-axis is perpendicular to the surface. Even though rotating the axis may seem like a big change, it is well worth it in the end, since it greatly simplifies the analysis. If the coordinate system was left horizontal and vertical the applied force, the normal force, and the directions of motion would all have two components (x and y), but with the rotation, the weight is the only force with x- and y-components.

Step 2: Find the components of the force and acceleration vectors. In this case, because the axes were rotated, the only vector with two components is the force of gravity, the rest are only in one direction. This is the advantage of rotating the axes.

$$F_{appx} = F_{app} = 10\text{ N} \quad \text{and} \quad F_{appy} = 0$$

$$F_{Nx} = 0 \quad \text{and} \quad F_{Ny} = +F_N$$

$$F_{gx} = F_g\cos(270° - \theta) \quad \text{and} \quad F_{gy} = F_g\sin(270° - \theta)$$

$$F_{gx} = 19.6\text{ N}\cos(270° - 20°) \quad \text{and} \quad F_{gy} = 19.6\text{ N}\sin(270° - 20°)$$

$$F_{gx} = -6.70\text{ N} \quad \text{and} \quad F_{gy} = -18.42\text{ N}$$

The angle of (270° – θ) is determined by starting at the rotated +x-axis and moving counter clockwise until the F_g vector is reached. It can also be explained that since the –y-axis is rotated counter clockwise away from the weight vector by an angle equal to the angle of the inclined plane θ from the horizontal.

$$a_x = a \quad \text{and} \quad a_y = 0$$

Step 3: Write Newton's Second Law for the x-components of the forces as:

$$\Sigma F_x = ma_x$$

$$F_{appx} + F_{gx} + F_{fx} = ma_x$$

$$F_{app} - F_g - F_f = ma$$

Write Newton's Second Law for the y-components as:

$$\Sigma F_y = ma_y$$

$$F_{Ny} + F_{gy} = ma_y$$

$$F_N + -18.42\text{ N} = 0$$

Step 4: Solve these expressions to find the normal force, then the frictional force, and then the acceleration. First, find the normal force with the expression from the y-components.

$$F_N = 18.42\text{ N}$$

Next, use the normal force to find the force of friction

$$F_f = \mu\ F_N = (0.1)(18.42\text{ N}) = 1.842\text{ N}$$

Next, find the acceleration with the expression from the x-components

$$a = \frac{10\text{ N} - 6.70\text{ N} - 1.842\text{ N}}{2\text{ kg}} = 0.73\ \frac{\text{m}}{\text{s}^2}$$

9.4 APPLIED FORCE AT AN ANGLE RELATIVE TO THE HORIZONTAL

Another situation which alters the magnitude of the normal force and thus the frictional force is when a force is applied to an object at an angle other than parallel to a surface. For example, as demonstrated in Figure 9.5, if a force is applied through the tension in a rope that makes an angle of θ to a box dragged across a horizontal surface, the y-component of the force slightly lifts the box off the surface thus reducing the contact force between the surface and the box.

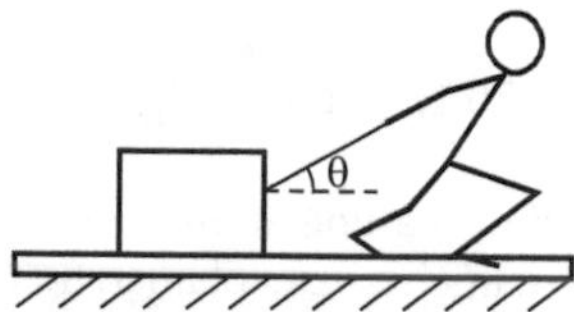

FIGURE 9.5 Object on a flat horizontal surface upon which a force is applied at an angle θ to the horizontal.

Putting some simple numbers into the situation described in Figure 9.5 provides an example for computing the coefficient of friction between the box and the surface. Assume the applied force is 100 N, the angle of the rope is $\theta = 30°$, the mass of the box is 40 kg, and the acceleration of the box is 0.5 m/s^2 while the tension is applied.

Solution:

Step 1: Generate a free-body diagram of the box, as shown in Figure 9.6.

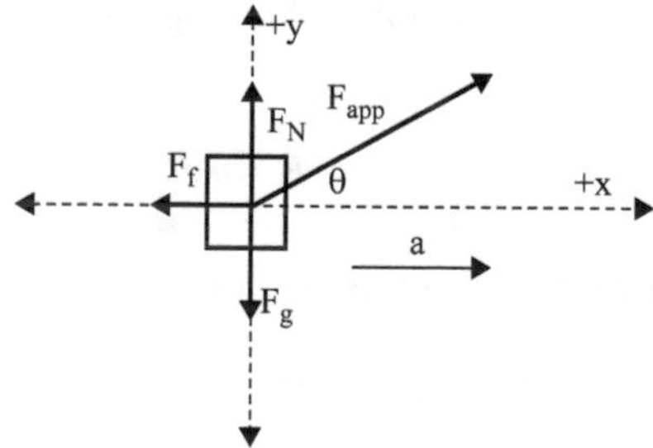

FIGURE 9.6 Free-body diagram of an object on a flat horizontal surface upon which a force is applied at an angle θ to the horizontal.

Step 2: Find the components of the force and acceleration vectors.

$$F_{\text{appx}} = F_{\text{app}}\cos(\theta) \quad \text{and} \quad F_{\text{appy}} = F_{\text{app}}\cos(\theta)$$

$$F_{\text{Nx}} = 0 \quad \text{and} \quad F_{\text{Ny}} = +F_{\text{N}}$$

$$F_{\text{gx}} = 0 \quad \text{and} \quad F_{\text{gy}} = -F_{\text{g}}$$

$$F_{\text{fx}} = -F_{\text{f}} \quad \text{and} \quad F_{\text{fy}} = 0$$

$$a_x = a \quad \text{and} \quad a_y = 0$$

Step 3: Write Newton's Second Law for the x-components of the forces as:

$$\Sigma F_x = ma_x$$

$$F_{appx} + F_{gx} + F_{Nx} + F_{fx} = ma_x$$

$$F_{app}\cos(\theta) - F_f = ma$$

Write Newton's Second Law for the y-components as:

$$\Sigma F_y = ma_y$$

$$F_{appy} + F_{gy} + F_{Ny} = ma_y$$

$$F_{app}\sin(\theta) - F_g + F_N = 0$$

Step 4: Solve these expressions to find expressions for the acceleration and the normal force. First, solve the y-components to find the normal force as:

$$F_N = F_g - F_{app}\sin(\theta)$$

$$F_N = mg - F_{app}\sin(\theta)$$

$$F_N = 409.8\text{ N} - 100\text{ N}\sin(30°) = 359.8\text{ N}$$

Next use the x-components to find the force of friction

$$F_f = F_{app}\cos(\theta) - ma$$

$$F_f = 100\text{ N}\cos(30°) - (40\text{ kg})\left(0.5\ \frac{\text{m}}{\text{s}^2}\right) = 66.6\text{ N}$$

To find the coefficient of friction solve equation 9.1 for μ,

$$\mu = \frac{F_f}{F_N} = \frac{(66.6\text{ N})}{(359.8\text{ N})} = 0.185$$

There is an upper limit for the frictional and normal forces. If other forces make it so that the normal force or the static frictional force would have to be greater than its maximum possible value, something must happen. In the case of the normal force, the surface breaks. In the case of the static frictional force, the object breaks free and starts sliding. The moment the applied force is large enough to overcome the static friction force the block starts sliding. Given that, up until that point, the static frictional force is what is keeping the object from sliding, the static frictional force is always in the direction opposite the direction the molecules in contact with the surface would slide if there were no static friction. Regarding that upper limit, the maximum possible static frictional force is just the product of the coefficient of static friction and the normal force. This is how this coefficient is defined.

9.5 LOCOMOTION

In all the cases presented thus far in this chapter, friction has always been in a direction opposite the direction of motion, but this is not always the case. Friction can be in the direction of motion in some important situations. One important case is all types of locomotion: walking, running, and even wheel-driven devices like bikes, cars, or trains.

As you run, you push against the ground and the ground pushes you forward. The force the ground pushing back on you is friction. You can prove this to yourself by trying to run on ice. For objects that use wheels to propel you forward, the wheels with drive use friction to push you forward. This is the case of the back wheel of a bicycle. As you peddle the bike, the chain translates the force to a torque, which spins the wheel, and the back wheel pushes back on the road and the friction of the road pushes back to propel the bike, and you, forward. The front wheel does not push back on the road, so friction pushes it back so it is opposite the way of the motion (Figure 9.7).

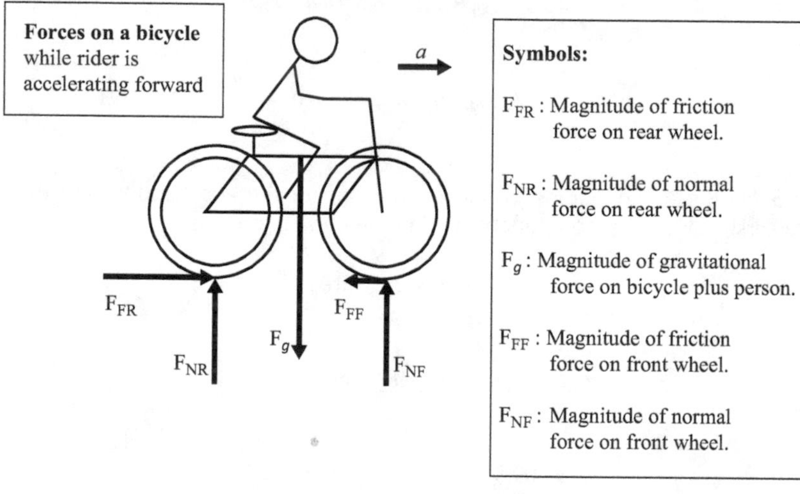

FIGURE 9.7 Free-body diagram of a bike.

9.6 ANSWER TO CHAPTER QUESTION

It is often easier to pull a large piece of furniture across a floor instead of pushing it even if the coefficients of friction are the same in both situations. Assume you place a moving cloth under the piece of furniture; when you push the piece of furniture the normal force is approximately the same as the weight of the object, but when you pull up on the cloth, you decrease the normal force and thus decrease the frictional force (Figures 9.8 and 9.9).

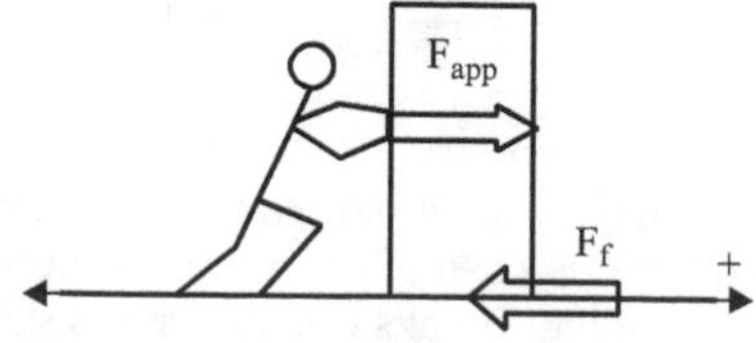

FIGURE 9.8 Pushing the furniture across the floor.

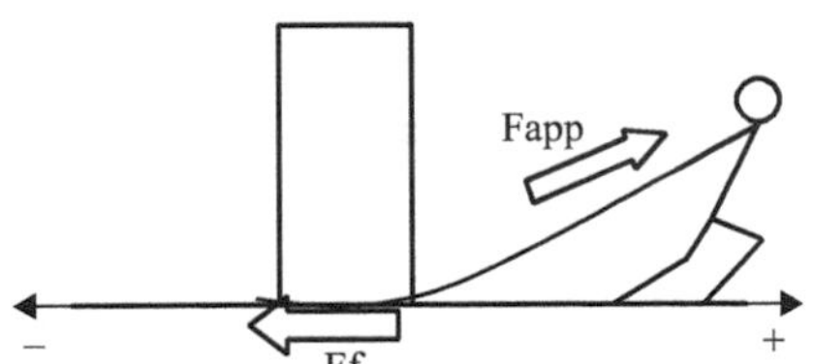

FIGURE 9.9 Pulling on a moving blanket placed under the furniture.

9.7 QUESTIONS AND PROBLEMS

9.7.1 Multiple-Choice Questions

1 & 2. In Figure 9.10, the crate has a mass m and there is a constant tension in the rope of F_T and the crate is moving to the right at a constant speed.

1. Is the magnitude of the force of friction on the crate greater than, less than, or equal to the force of tension, F_T?
   A. greater than B. less than C. equal to

2. Is the magnitude of the acceleration of the crate after 2 s of the person pulling on the crate with this force positive, negative, or zero?
   A. positive B. negative C. zero

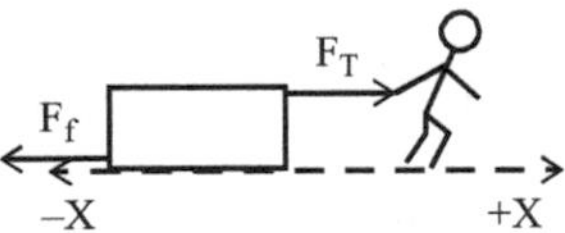

FIGURE 9.10 Multiple-choice questions 1–4 and problems 1–4.

3 & 4. In Figure 9.10, the crate has a mass m and there is a constant tension in the rope of F_T and the crate is moving to the right with a constant acceleration.

3. Is the magnitude of the force of friction on the crate greater than, less than, or equal to the force of tension, F_T?
   A. greater than B. less than C. equal to

4. Is the magnitude of the acceleration of the crate after 2 seconds of the person pulling on the crate with this force positive, negative, or zero?
   A. positive B. negative C. zero

5 & 6. In Figure 9.11, the pulley is frictionless and massless but the track is *not* frictionless.

5. If the mass of object 2 in Figure 9.11 was increased, will the magnitude of the frictional force on object 1 increase, decrease, or remain the same?
   A. increase. B. decrease. C. remain the same.

6. If block 2, in the Figure 9.11, is accelerating downward, how does the tension F_T in the string compare with the magnitude F_{g2} of the gravitational force on object 2?
   A. $F_T < F_{g2}$ B. $F_T = F_{g2}$ C. $F_T > F_{g2}$

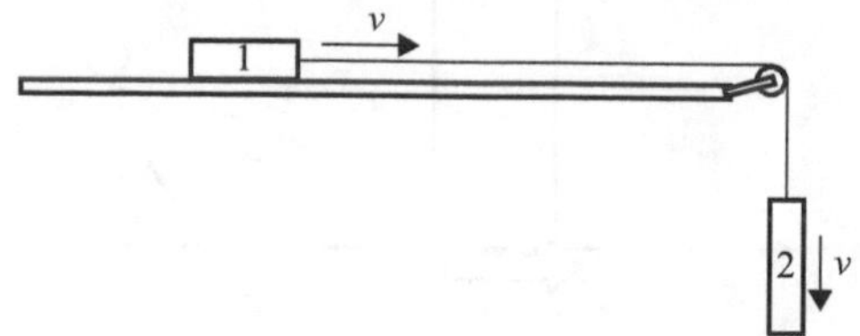

FIGURE 9.11 Multiple-choice questions 5 and 6.

Questions 7–10 refer to the situation depicted in Figure 9.12, in which the two crates with equal mass are attached together with rope-2 and a man standing at the top of the inclined plane, which makes an angle θ with the horizontal, holds onto rope-1, which is attached to crate (1). At the instant depicted, rope-1 is sliding through the man's hands as he pulls back on the rope slowing the acceleration of the crate down the inclined plane to a constant magnitude. There is friction between the crate and the inclined plane.

7. Is the net force on mass-1 greater than, less than, or equal to the net force on mass-2?
   A. greater than B. less than C. equal to

8. Is the tension in rope-1 greater than, less than, or equal to the tension in rope-2?
   A. greater than B. less than C. equal to

9. Is the force of friction on crate-1 greater than, less than, or equal to force of friction on crate-2?
   A. greater than B. less than C. equal to

10. Is the acceleration of crate-1 greater than, less than, or equal to the acceleration of crate-2?
    A. greater than B. less than C. equal to

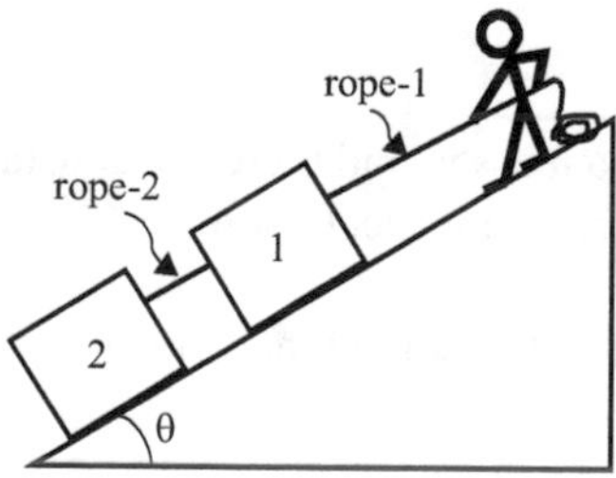

FIGURE 9.12 Multiple-choice questions 7–9 and problem 8.

9.7.2 Problems

1. In Figure 9.10, the crate has a mass of 10 kg and the person has a mass of 20 kg. The tension in the rope is 30 N and the crate is moving to the right at a constant speed of 2 m/s. Compute the magnitude of the force of friction on the crate.
2. In Figure 9.10, the crate has a mass of 10 kg and the person has a mass of 20 kg. The tension in the rope is 30 N and the crate is moving to the right at a constant speed of 2 m/s. Compute the coefficient of friction between the crate and the surface over which it is dragged.

3. In Figure 9.10, the crate has a mass of 10 kg and the person has a mass of 20 kg. If the tension in the rope is 30 N and the crate is moving to the right with a constant acceleration of 2 m/s². Compute the magnitude of the net force on the crate.
4. In Figure 9.10, the crate has a mass of 10 kg and the person has a mass of 20 kg. The tension in the rope is 30 N and the crate is moving to the right at a constant acceleration of 2 m/s². Compute the coefficient of friction between the crate and the surface over which it is dragged.
5. A tow truck, with a mass of 5,000.0 kg, tries to move an illegally parked four-wheel drive truck, with a mass of 3,000.0 kg. Since the four-wheel drive truck has its emergency brakes on, and is in gear, the wheels will not spin. This does not deter the determined tow truck operator from trying to drag the truck along the level street. The cable on the tow truck makes an angle of 50° from the horizontal when attached to the four-wheel drive truck, with the higher end of the cable being the end attached to the tow truck. The coefficient of kinetic friction between the tires and the road is 0.5. Assume the tow truck operator is able to get the four-wheel drive truck to slide along the pavement. Once it is sliding, at an instant when the tension in the cable of the tow truck is 1.5×10^4 N, what is the acceleration of the four-wheel drive truck?
6. A person is pushing a crate of mass 60.0 kg along a flat horizontal floor at a constant velocity. The force of the person on the crate is forward and downward at an angle of 14.0° below the horizontal and has a magnitude of 400.0 N. Find the coefficient of kinetic friction for the crate/floor interaction.
7. A block of mass 1.60 kg is sliding downward the flat surface of a ramp that makes an angle of 34.0° with the horizontal. The coefficient of friction governing the interaction between the surface of the ramp and the bottom of the block is 120. Find the acceleration of the block.
8. The two crates in Figure 9.12 are attached together with rope-2 and a man standing at the top of the incline holds onto rope-1, which is attached to crate (1). At the instant depicted, the man pulls back on the rope slowing the acceleration of the crate down the inclined plane to a constant magnitude of 1.5 m/s². It is important to understand that at this instant, the rope is sliding through the man's hands he is pulling back so that rope-1 has a significant tension in it but the crates are accelerating down the incline. The incline makes an angle $\theta=40°$ with the horizontal, the mass of crate 1 is 50 kg and the mass of crate 2 is 150 kg. The coefficient of friction between each crate and the plane is 0.3. Compute the tension in rope-1 at the instant depicted. *(Hint: Combine mass 1 and mass 2 into one mass with a total mass of $(m_1 + m_2)$ to solve the problem.)*
9. Two crates are originally at rest on an inclined plane, which makes an angle of $\theta=20°$ with the horizontal. The coefficient of friction between the surface and the crates and the incline is 0.15. The crates have a mass of $M_1=75$ kg and $M_2=110$ kg. A guy trying to move the crates up the incline pushes on the crates with a constant force of 1,800 N in a direction directly up the incline $\theta_F=20°$. Compute the displacement of the crates in 2 seconds, while the guy is pushing. *(Hint: Combine M_1 and M_2 into one mass with a total mass of $(m_1 + m_2)$ to solve the problem.)* (Figure 9.13)

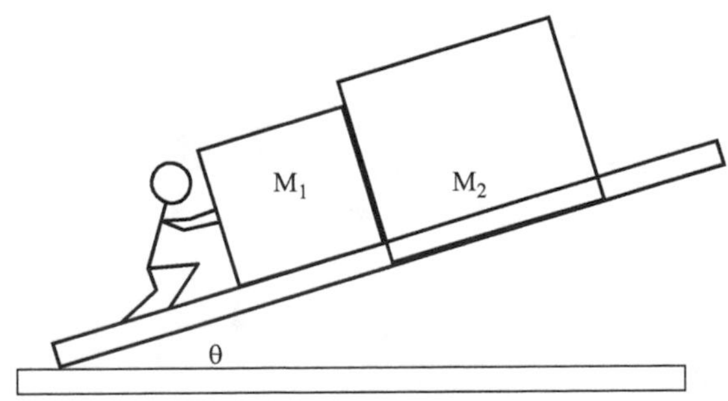

FIGURE 9.13 Problem 9.

10. A crate is originally at rest on a smooth, flat, and horizontal surface. The coefficient of friction between the surface and the crate is 0.21 and the crate has a mass of $M = 120\,\text{kg}$. A guy trying to move the crates across the surface pulls on a rope attached to the bottom of the crate with a constant force of 500 N at an angle of $\theta_F = 20°$, relative to the horizontal. Compute the displacement of the crates in the 2.0 seconds, while the guy is pulling (Figure 9.14).

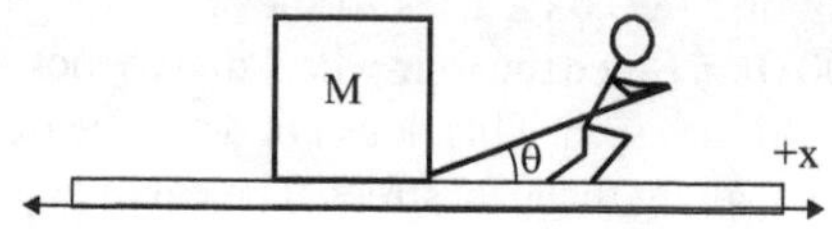

FIGURE 9.14 Problem 10.

10 Static Fluids

10.1 INTRODUCTION

In this chapter, the concepts central to the application of forces associated with an object at rest in a fluid are presented. After defining terms, such as density and volume, the concept of pressure is presented and the expressions needed to compute the pressure in different scenarios are provided. The buoyant force is defined in the context of Archimedes' Principle and several examples applying this force with a dynamics-based analysis are provided.

Chapter question: The buoyant force is responsible for keeping objects, like a boat, afloat in water. The question is, is there a buoyant force on a fish at rest in the center of the water in an aquarium? The buoyant force is one of the critical concepts discussed in the chapter and this question will be answered at the end of the chapter.

10.2 PRESSURE, VOLUME, AND DENSITY

10.2.1 Pressure

The understanding of forces on objects in fluids is an important step in understanding the analysis needed for many biological systems. Since fluids, which include both liquids and gases, are a substance that have no fixed shape, they conform to the shape of the container in which it they are located. Thus, almost all living organisms experience the forces associated with fluids. Where a fluid is in contact with a surface, that surface exerts a normal force on the fluid. By Newton's third law, the fluid is exerting a force, of the same magnitude but opposite direction, on the surface. The force being exerted by the fluid on the surface is referred to as the force due to fluid pressure. Pressure itself is the magnitude of force (F) per area (A); therefore, pressure is a scalar that is represented in equation (10.1).

$$P = \frac{F}{A} \tag{10.1}$$

From this definition of pressure, the SI unit for pressure is $\left(\frac{\text{N}}{\text{m}^2}\right)$, which is given the name Pascal, abbreviated as Pa. In general, the pressure has different values at different points in a fluid. To find the force being exerted on a surface with which a fluid is in contact, the value of the pressure at that point in the fluid is multiplied by the area of that surface and the force vector is assigned a direction perpendicular to and toward the area element.

For a flat surface in contact with a fluid in which the pressure is the same at all points in the fluid across that surface, the magnitude of the force is given by solving equation (10.1) for force to give: $F = P\ A$ and the direction of the force is perpendicular and into the surface as demonstrated in Figure 10.1.

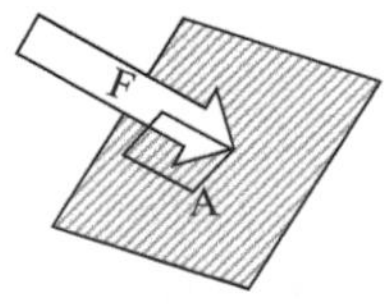

FIGURE 10.1 The pressure force is perpendicular to the area.

DOI: 10.1201/9781003308065-10

The air which we breathe is a fluid, and at the surface of the planet, the pressure due to the atmosphere at a particular time and place is considered a constant. While the pressure in the atmosphere varies significantly with height above the surface and from day to day, at heights having only small differences compared to the thickness of the atmosphere and at a particular instant of time, it is a good approximation to use an average value of:

$$P_{atm} = 1.013 \times 10^5 \text{ Pa}$$

It does change with altitude in that for the first 1,000 m it drops off by about 1% of that value with every 100 m increase in altitude.

Example 10.1

Find the magnitude of the force due to air pressure on the writing surface of the whiteboard at the front of your classroom. The width of the board is $w=3.5825$ m and its height is $h=1.132$ m.

$$A = w\ h = (3.5825 \text{ m})(1.132 \text{ m}) = 4.05539 \text{ m}^2$$

$$F = P\ A = \left(1.013 \times 10^5 \text{ Pa}\right)\left(4.05539 \text{ m}^2\right) = 410{,}811 \text{ N}$$

Considering that 1 m² is approximately the surface area of the average child, the force is even greater for an adult. This is a huge amount of force on our bodies due to the atmosphere. It is amazing that we all live in a fluid that pushes on us this hard every moment of every day.

10.2.2 VOLUME

The volume V of an object or a sample of a fluid is the amount of three-dimensional space that the object or sample occupies. For a right rectangular cuboid of length L, width w, and height h, as depicted in Figure 10.2, the volume is given by $V = (L\ w\ h)$.

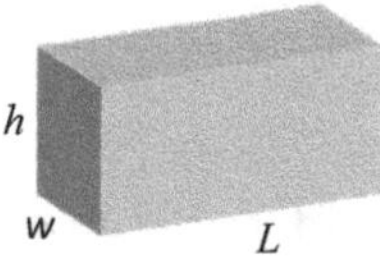

FIGURE 10.2 Volume of a right rectangular cuboid.

A right rectangular cuboid is an example of a prismatic column.

A prismatic column is one for which every cross section has the same two-dimensional shape. A cross section is either one of the two end faces that would be formed by cutting the column at right angles to the direction in which it extends. For the right rectangular cuboid, the two-dimensional shape that every cross section along the entire length of the column has is a rectangle of width w and height h. The volume of any prismatic column is the area A of its cross section (which is equivalent to the area of one end face) times the length of the column given in equation (10.2) as:

$$V = AL \tag{10.2}$$

In the case of the right rectangular cuboid, the area of a cross section is the product of the width and the height of the cuboid, $A=w\ h$, which if substitute for A in equation (10.2) produces the well-known expression for volume as the product of the length, the width, and the height, $V=(L\ w\ h)$. Another common example is the right circular cylinder, sketched in Figure 10.3.

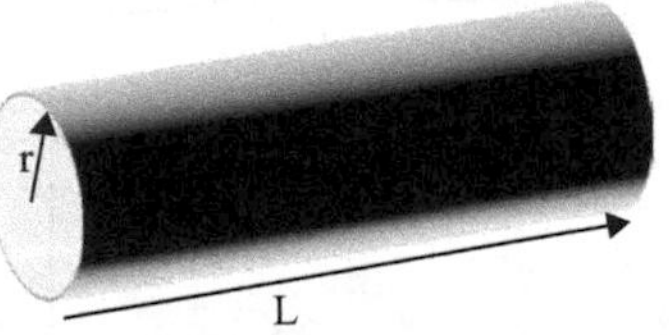

FIGURE 10.3 Volume of a cylinder.

In this case, the area (A) of the end face is $A = \pi r^2$ and the length of the cylinder (L), so the product of these two quantities together can be used in equation (10.2) to give a volume of $V = \pi r^2 L$.

10.2.3 Density

The mass density of an object is the mass-per-volume of the object. In this chapter, the mass density will be referred to as simply the density and will be symbolized by the Greek letter rho (ρ). Density is a property of the material of which an object consists. For an object made of a material that has uniform density, the density of the material of which the object is made is related to the mass (m) and volume (V) of the object by equation (10.3) as:

$$\rho = \frac{m}{V} \tag{10.3}$$

To emphasize that density is a property of the material of which an object is made, whereas mass and volumes are properties of the object itself, note that if an object that is made of a single material is cut in half and half is discarded, the half which remains has half the mass of the original object and it has half the volume of the original object, but it still has the same density. Thus, the density of the material is known as an *intensive variable* since it does not depend on the quantity of the material present. Pressure is another intensive variable since it is measured per area and another intensive variable like temperature will be presented throughout this text, when they are important to the understanding of the topic.

Two objects having the same shape but each made of a material that has a different density than the other, will have different masses. For instance, a ball of solid lead with a diameter of 10 cm will have a much greater mass than the Styrofoam ball with the same diameter. Some materials, like a Styrofoam ball, can be compressed. In compressing an object, it occupies less space, that is, its volume is reduced, but its mass remains constant. Hence, by compressing an object, like the Styrofoam ball, its density is increased. The Styrofoam ball may be compressed to the point that it may have the same density as the lead ball, but its volume will no longer be the same as the lead ball. Unlike solids, a gas is easily compressed by increasing its pressure. Most liquids, such as water, experience a negligible amount of compression, even with a significant increase in pressure. As such, most liquids are referred to as incompressible fluids.

10.3 DEPENDENCE OF PRESSURE ON DEPTH

When swimming, if you dive down deep in the water, even just a meter or two, you may have sensed a pressure difference in the water. For fluids near the surface of the earth, the greater the depth in the fluid, the greater the pressure. For incompressible fluids, the density is approximately the same throughout the fluid and there is a fairly simple relation between the pressure in the fluid and depth. Consider a column of incompressible fluid, within a larger body of that fluid, as depicted in Figure 10.4.

The body of fluid is continuous but considers one part of the fluid that has the shape of a vertical right rectangular cuboid of cross-sectional area A. There is a force due to the fluid pressure downward on the top surface of the column. There are forces due to the fluid pressure on the sides of the column, but the net leftward force on the right side of the column exactly cancels the net rightward

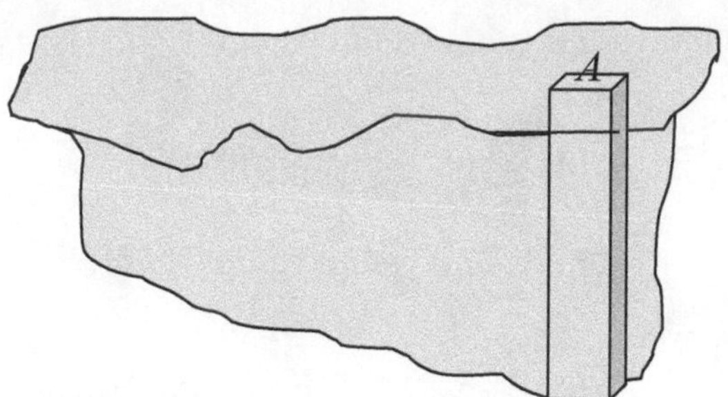

FIGURE 10.4 A column of fluid within a body of incompressible fluid.

force on the left side of the column, and the net backward force on the front side of the column exactly cancels the net forward force on the back of the column.

For simplicity, these opposing forces can be omitted from the free-body diagram of the forces on the column of water in Figure 10.5. There is a force due to the pressure of the water acting upward on the bottom of the column. The coordinate system is established at the bottom of the column where h_{BOT} is defined as the height of the bottom of the column and h_{TOP} is defined as the height of the top of the column.

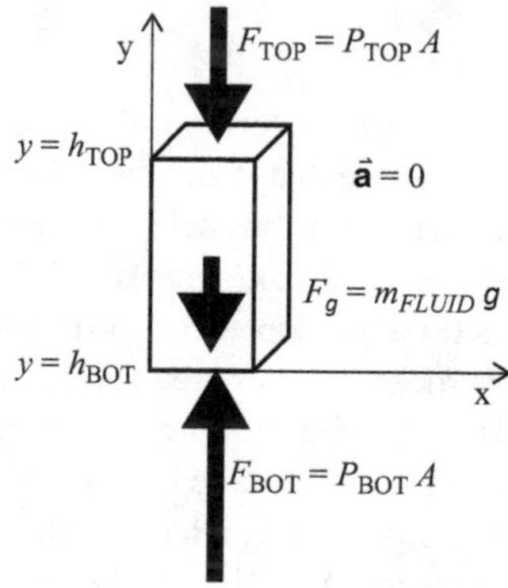

FIGURE 10.5 Free-body diagram of the unbalanced forces on the column of water.

Writing the y-components of the vectors in the free-body diagram yields:

$$F_{TOPy} = -F_{TOP} \quad F_{BOTy} = +F_{BOT} \quad F_{gy} = -F_g$$

Writing out the static expression in the y-direction yields:

$$\Sigma F_y = 0$$

$$F_{TOPy} + F_{gy} + F_{BOTy} = 0$$

$$-F_{TOP} - F_{gy} + F_{BOT} = 0$$

$$-F_{TOP} - m_{FLUID}\mathbf{g} + F_{BOT} = 0$$

Given that the density of the fluid is the mass per volume, $\rho_{FLUID} = \dfrac{m_{FLUID}}{V_{FLUID}}$, the mass of the fluid is the product of the volume and the density of the fluid, $m_{FLUID} = V_{FLUID}\rho_{FLUID}$. Since V_{FLUID} =

$V = A(h_{TOP} - h_{BOT})$, the mass of the fluid is $m_{FLUID} = \rho_{FLUID} A(h_{TOP} - h_{BOT})$. This is where the incompressibility of the fluid is important. If the fluid was compressible, the density would be greater at the bottom than it is at the top and $\rho_{FLUID} = \frac{m_{FLUID}}{V_{FLUID}}$ would not apply. This is the case with the atmosphere, but not with water. So, for water, the force analysis continues as:

$$-P_{TOP}A + \left[-\rho A(h_{TOP} - h_{BOT})g\right] + P_{BOT}A = 0$$

Dividing both sides by A and solving for P_{BOT} results in equation (10.4) as:

$$P_{BOT} = P_{TOP} + \rho \mathbf{g}(h_{TOP} - h_{BOT}) \tag{10.4}$$

This specifies that the pressure at depth $(h_{TOP} - h_{BOT})$ relative to a given point in a fluid is the pressure at that given point plus $\rho \mathbf{g}(h_{TOP} - h_{BOT})$.

Example 10.2

Find the pressure at a depth of 4 m below the surface of a freshwater lake.

Step 1: Sketch a diagram for the example in Figure 10.6.

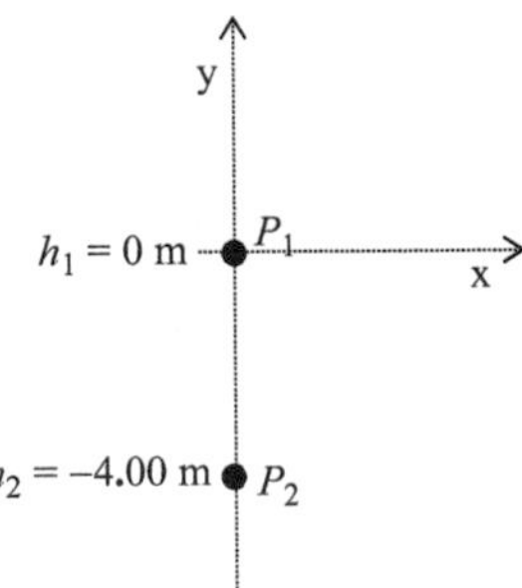

FIGURE 10.6 Diagram for Example 10.2.

Step 2: Gather the important quantities and values. The density of fresh water is $1{,}000 \frac{\text{kg}}{\text{m}^3}$.

The surface of a lake is in contact with the atmosphere so the pressure at the top is atmospheric pressure,

$$P_1 = 1.013 \times 10^5 \text{Pa}.$$

Step 3: Plug into equation (10.4):

$$P_{BOT} = P_{TOP} + \rho g(h_{TOP} - h_{BOT})$$

$$P_2 = P_1 + \rho g(h_1 - h_2)$$

$$P_2 = 1.013 \times 10^5 \text{Pa} + \left(1{,}000 \frac{\text{kg}}{\text{m}^3}\right) 9.8 \frac{\text{m}}{\text{s}^2}\left[0 \text{ m} - (-4 \text{ m})\right]$$

$$P_2 = 1.405 \times 10^5 \text{Pa}$$

Note that the units of the second term work out to N/m², which is Pa. Also, note that this process is slightly different than the other examples since pressure is not a vector so a free-body diagram is not required.

10.4 ARCHIMEDES' PRINCIPLE

Archimedes' Principle states that for objects in a fluid, there is an upward force on the object due to the fluid that has a magnitude equal to the weight of the fluid displaced by the object. This upward force is called the *buoyant force* and is responsible for objects floating in water or hot air balloons floating in air. For an object immersed completely in a fluid, the net pressure-times-area is *the static fluid force* on an object submerged in a fluid. The vector sum of the forces on all the infinite number of infinitesimal surface area elements making up the surface of an object is *upward* because the pressure increases with depth and the lateral pressures on the left, right, back, and front sides are equal and opposite.

Referring to Figure 10.7, the upward pressure-times-area force on the bottom of an object (F_{BOT}) is greater than the downward pressure-times-area force on the top of the object (F_{TOP}). The result is the buoyant force, which is a net upward force on any object that is either partially or totally submerged in a fluid. Therefore, the expression for the buoyant force is found by first finding the difference in pressure on the bottom and the top of the object. So, starting with equation (10.4),

$$(P_{BOT} - P_{TOP}) = \rho \mathbf{g}(h_{TOP} - h_{BOT})$$

Referring back to Figure 10.7, set $\Delta P=(P_{BOT}-P_{TOP})$ and set $h=(h_{TOP}-h_{BOT})$ since it is the height of the object, so

$$\Delta \mathrm{P} = \rho \mathbf{g} h$$

The difference in pressure is the net upward force per area on the object, where the area is the cross-sectional area A of the object. Thus, the net upward force on the submerged object in Figure 10.7 is:

$$F_{net} = (\Delta \mathrm{P})A = (\rho \mathbf{g} h)A.$$

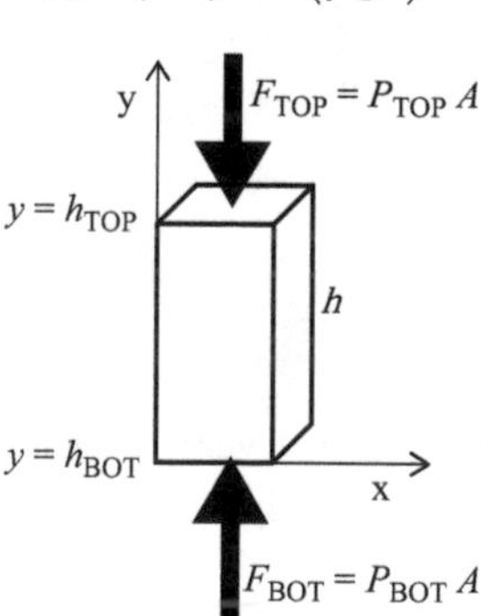

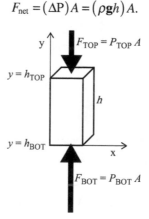

FIGURE 10.7 Free-body diagram of the unbalanced forces on an object submerged in an incompressible fluid like water.

Since the height of the object times its cross-sectional area is the volume (V) of the object, the net upward force on an object in a fluid is given in equation (10.5) as:

$$F_B = \rho \mathbf{g} V. \tag{10.5}$$

This is the magnitude of the buoyant force, F_B, which is directed upward and is the weight of the fluid displaced. Since the volume and density of the fluid are just the mass of the fluid displaced by the object and the product of the mass, and the gravitational field strength (g) is the weight of the fluid displaced, this force is equal in magnitude to the weight of the fluid displaced by the object.

In practice, submerge the object in still water, and release the object from rest, one of three things will happen. The object will experience an upward acceleration and bob to the surface, the object will remain at rest, or the object will experience a downward acceleration and sink. Remember that the buoyant force is always upward, so why do objects sink? The reason is that, after you release an object in a fluid like water, the buoyant force is not the only force acting on the object. The gravitational force still acts on the object when the object is submerged. Recall that the earth's gravitational field permeates everything. For an object that is touching nothing of the substance but the fluid it is in, the free-body diagram for this object looks like the one in Figure 10.8.

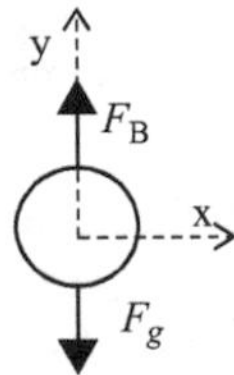

FIGURE 10.8 Free-body diagram of an object that is only in contact with a fluid.

The only difference is the relative lengths of the arrows, signifying the difference in magnitude between the buoyant force and the force of gravity. For the three scenarios discussed at the beginning of this paragraph, if the object accelerates up to the surface of the water $F_B > F_g$, if the object remains at rest $F_B = F_g$, and if the object is sinking $F_B < F_g$.

10.5 BUOYANT FORCE EXAMPLES

Example 10.1

An object of mass .580 kg and volume $2.035 \times 10^{-4}\,\text{m}^3$ is suspended at rest by a string in water. Find the tension in the string (Figure 10.9).

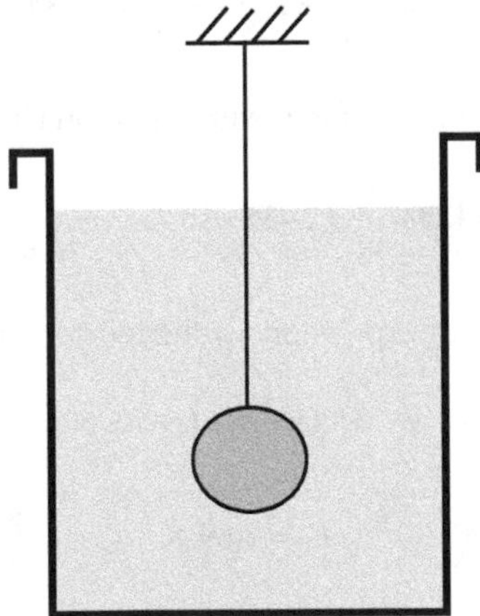

FIGURE 10.9 Diagram for buoyant force Example 10.1.

Solution:

Step 1: Sketch the free-body diagram of the object as demonstrated in Figure 10.10.

Notice that there are two upward forces, the tension in the sting and the buoyant force. They are sketched side by side so that they can be distinguished from each other; there is no actual offset of the forces.

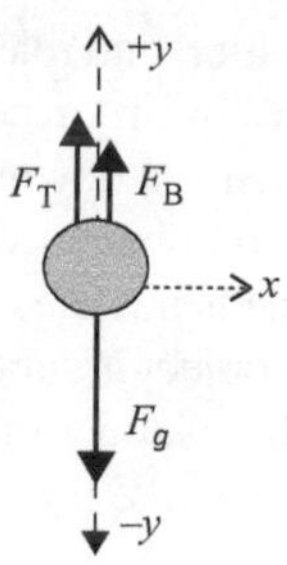

FIGURE 10.10 Free-body diagram of the object in Example 10.1.

Step 2: Write down the y-components of the vectors in the free-body diagram.

$$F_{Ty} = F_T \quad F_{By} = F_B \quad F_{gy} = -F_g$$

Step 3: Apply Newton's Second Law for the y-components of the vectors for a static stone.

$$\Sigma F_y = 0$$

$$F_{Ty} + F_{By} + F_{gy} = 0$$

$$F_T + F_B + (-F_g) = (m)0$$

$$F_T = F_g - F_B$$

Step 4: Solve the individual forces and then plug back into the expression we found in Step 3. Find the magnitude of the force of gravity:

$$F_g = m\mathbf{g} = (.58\text{ kg})\,9.8\frac{\text{N}}{\text{kg}} = 5.684\text{ N}$$

Find the magnitude of the buoyant force with equation (10.5).

$$F_B = \rho_w V\mathbf{g} = 1{,}000\frac{\text{kg}}{\text{m}^3}\left(2.035\times10^{-4}\text{m}^3\right)9.8\frac{\text{N}}{\text{kg}} = 1.9943\text{ N}$$

Plugging these values into our expression for the tension yields

$$F_T = 5.684\text{ N} - 1.9943\text{ N}$$

$$F_T = 3.69\text{ N}$$

Example 10.2

A boat weighs 2×10^6N. Calculate the volume of water displaced by the boat when it is at rest in fresh water.

Step 1: Sketch the free-body diagram of the boat in Figure 10.11

Step 2: Write the y-components of the vectors in the free-body diagram:

$$F_{gy} = -F_g \quad F_{By} = F_B$$

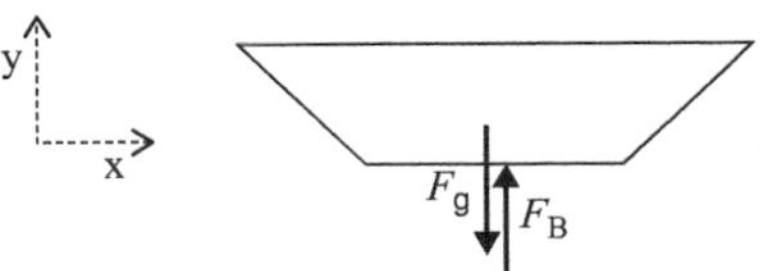

FIGURE 10.11 Free-body diagram of the boat in Example 10.2.

Step 3: Applying Newton's Second Law for the y-components of the vectors of a static situation yields:

$$\Sigma F_y = 0$$

$$F_{gy} + F_{By} = 0$$

$$-F_g + F_B = 0$$

$$F_B = F_g$$

$$F_B = 2 \times 10^6 \,\text{N}$$

Step 4: Solve the expressions:
Solve Archimedes' Principle: $F_B = \rho_w V_{DF} \mathbf{g}$ for the volume displaced V_{DF} yields:

$$V_{DF} = \frac{F_B}{\rho_w \mathbf{g}} = \frac{2{,}000{,}000 \ \cancel{\text{N}}}{\left(1{,}000 \ \frac{\cancel{\text{kg}}}{\text{m}^3}\right)\left(9.8 \ \frac{\text{N}}{\cancel{\text{kg}}}\right)} = 204.1 \ \text{N}$$

10.6 ANSWER TO THE CHAPTER QUESTION

If a fish is at rest in the center of an aquarium filled with water, there is a buoyant force on the fish that has a magnitude that is equal to the weight of the fish. As sketched in the free-body diagram of the fish in Figure 10.12, the forces on the fish include the gravitational force (F_g) and the buoyant force (F_B). The magnitude of these two forces is the same, so the fish is in a static situation. Fish have a swim bladder that they can fill with either oxygen, extracted from the water by their gills, or water to regulate the buoyant force on them by controlling the amount of water that they displace and thus their buoyant force.

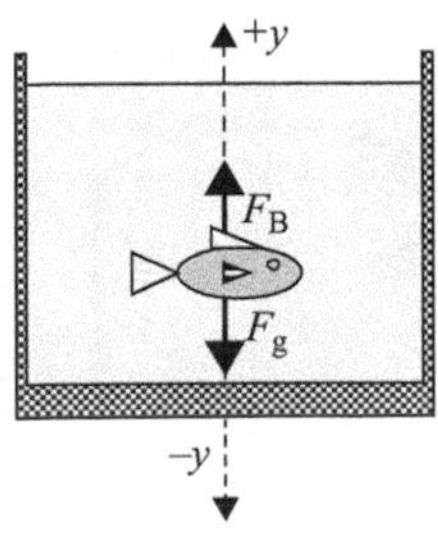

FIGURE 10.12 Free-body diagram of a fish in a container of water.

10.7 QUESTIONS AND PROBLEMS

10.7.1 Multiple-Choice Questions

1. Pressure is defined as the force per length, area, or volume?
   A. length B. area C. volume

2. Is the pressure on an object at a depth in a fluid proportional to the density of the fluid or the density of the object, or both?
   A. density of the fluid B. density of the object C. both

3. According to Archimedes' Principle, is the buoyant force proportional to the density of the fluid or the density of the object?
   A. density of the fluid B. density of the object

4. As an object that starts fully submerged sinks in a fluid, like water, does the buoyant force increase, decrease, or remain the same as the object moves downward through the fluid?
   A. increase B. decrease C. remain the same

5. As shown in Figure 10.13, a block is suspended from a string and lowered into a container so that it is completely submerged in fresh water, but not touching the bottom of the container. Is the force of tension in the string greater than, less than, or equal to the weight of the block?
   A. Greater than B. Less than C. Equal to

6. As shown in Figure 10.13, a block is suspended from a string and lowered into a container so that it is completely submerged in fresh water, but not touching the bottom of the container. Is the buoyant force on the block greater than, less than, or equal to the weight of the block?
   A. Greater than B. Less than C. Equal to

7. If the original cube in Figure 10.13 was substituted for another cube with the same dimensions but made out of more dense material, the buoyant force due to the fresh water on the denser cube will be greater than, less than, or equal to the buoyant force on the original cube?
   A. greater than B. less than C. equal to

8. If the original cube in Figure 10.13 was substituted for another cube made of the same material, with a smaller mass, the buoyant force due to the fresh water on the smaller cube will be greater than, less than, or equal to the buoyant force on the original cube?
   A. greater than B. less than C. equal to

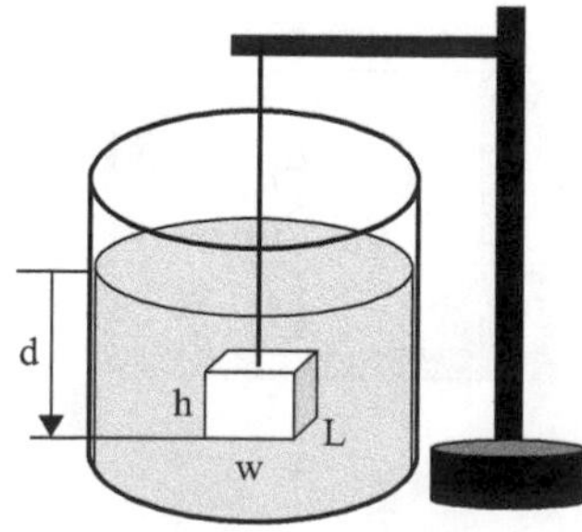

FIGURE 10.13 Multiple-choice questions 5–10 and problem 5.

9. If the fresh water in Figure 10.13 is substituted with glycerin, which has a greater density than water, will the buoyant force on the cube be greater than, less than, or equal to the buoyant force on the cube in fresh water of the original situation?
   A. greater than B. less than C. equal to

10. If the string is cut in Figure 10.13 and the cube sinks to the bottom of the container, will the buoyant force on the cube due to the freshwater be greater than, less than, or equal to the buoyant force on the cube in the original arrangement?
   A. greater than B. less than C. equal to

10.7.2 Problems

1. Compute the pressure at a depth of 5.00 m in fresh water, given the atmospheric pressure of $P_1 = 1.013 \times 10^5\,\text{Pa}$ and a density of fresh water of $1{,}000\,\frac{\text{kg}}{\text{m}^3}$.
2. Compute the pressure at a depth of 50 m in sea water, given the atmospheric pressure of $P_1 = 1.010 \times 10^5\,\text{Pa}$ and a density of sea water of 1,015 kg/m^3.
3. Compute the magnitude of the force being exerted, by the water, on the flat horizontal bottom of an above ground swimming pool filled to a depth of 2.00 m with fresh water and having a circular bottom with a radius of 3 m. Compare this to the weight of the water in the swimming pool.
4. A cube of material weighs 100 N and is floating with 30% of the cube above the water line, what is the volume of the cube?
5. As shown in Figure 10.13, a cube of solid material with a density of 2,700 kg/m^3 and a mass of 21.6 g is hung from a piece of string and lowered into a container of fresh water, which has a density of $\rho_{\text{water}} = 1{,}000\,\text{kg/m}^3$. When the cube is totally submerged and at rest in the water, what is the tension in the string?
6. An object that weighs 10 N displaces a volume of water that weighs 6 N when it is totally submerged in a container of water. The object is completely submerged and resting on the bottom of a container full of water. Compute the magnitude of the normal force acting on the object.
7. A ping-pong ball of mass 2.70 g and diameter 40.0 mm is tied by a string to the bottom of a container of saltwater having density 1,027 kg/m^3. The ball is totally submerged in the saltwater and the length of the string segment connecting the ball to the bottom of the container is 4 cm. Find the tension in the string.
8. A spherical balloon that is filled with helium has a diameter of 30.0 cm and is attached to the upper end of a 125.0 cm segment of string. The lower end of the string segment is tied to the back of a chair. The balloon is at rest. The tension in the string is.0100 N. The density of the air surrounding the balloon is 1.225 kg/m^3. Find the mass of the balloon (including the helium with which it is filled). Neglect the mass of the string.
9. A 360-kg polar bear is sitting on an iceberg that is floating in calm seawater with 90% of its volume below the water level. What is the mass of the iceberg? Assume that the density of sea ice and sea water is 900 and 1,020 kg/m^3, respectively.
10. A cube of material, with 10 cm long sides, weighs 39.2 N. The object is tied to a string and lowered into a container of fresh water, as shown in Figure 10.13. Compute the tension in the string when the object is completely submerged.

11 Fluid Dynamics

11.1 INTRODUCTION

This chapter is an introduction to many of the important topics in fluid dynamics, especially as applied to biological systems. In many naturally occurring situations, the viscosity or fluid friction is an important factor in understanding the motion of the organism of interest. To be able to apply the correct form of the viscous drag force on objects moving through a fluid, the Reynold's number, a dimensional-parameter, is also discussed. With these tools, the analysis of objects moving through viscous fluids is presented in several examples. As an extension of the analysis, the concepts and techniques are applied to viscous fluid flowing through a system. This analysis leads to the expressions that can be applied to study the movement of blood flowing through the circulatory system.

Chapter question: How does the erythrocyte sedimentation rate (ESR) of red blood cells falling through blood indicate inflammation? This question will be answered at the end of the chapter after the topics of fluid dynamics are studied.

11.2 VISCOSITY

Viscosity (μ) is a measure of the friction associated with moving fluids. This friction can be between layers of fluid moving at different speeds or the friction on an object moving through a fluid. Viscosity is defined by the situation described in Figure 11.1, in which two plates, each with an area (A) have a layer of fluid, with a thickness (d), between them.

A force (F_{App}) is applied to the top plate so that it moves with constant speed (v_c) and the bottom plate remains fixed at rest ($v = 0$ m/s). The thin layer adjacent to the top plate moves at the same constant speed as the plate and the thin layer adjacent to the bottom layer is at rest. The layers between the plates have speeds that range from v_c to zero in a quadratic fit. Please see the curve of velocity vectors in Figure 11.1, for a depiction of the velocity distribution across the vertical layers of fluid between the plates. For this arrangement, the constant of proportionality between the applied force (F), the area of the plates (A), thickness of the fluid (d), and the constant velocity (v_c) of the top plate is the viscosity (μ) given in equation (11.1) as:

$$\mu = \frac{(F)(d)}{(A)(v_c)} \qquad (11.1).$$

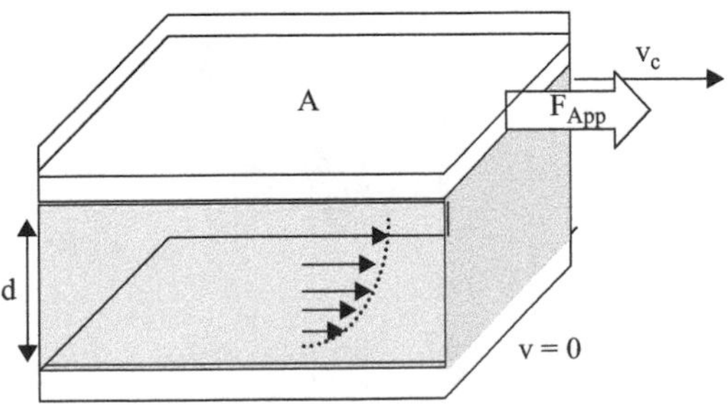

FIGURE 11.1 Arrangement used to define viscosity.

DOI: 10.1201/9781003308065-11

Plugging in the units for all the quantities in equation (11.1) results in the units of "Pascals seconds" for viscosity:

$$\frac{(\mathrm{N\,m})}{\mathrm{m}^2\frac{\mathrm{m}}{\mathrm{s}}}=\left(\frac{\mathrm{N}}{\mathrm{m}^2}\right)\mathrm{s}=(\mathbf{Pa})(\mathbf{s})$$

Plugging in the equivalent units for a Newton results in another set of units, [kilograms/((meters) (second))], that are often used when studying the motion of an object in a fluid.

$$\frac{(\mathrm{N\,m})}{\mathrm{m}^2\frac{\mathrm{m}}{\mathrm{s}}}=\frac{\left(\frac{\mathrm{kg}\,\cancel{\mathrm{m}}}{\mathrm{s}^{\cancel{2}}}\,\cancel{\mathrm{m}}\right)}{\cancel{\mathrm{m}}^2\frac{\mathrm{m}}{\cancel{\mathrm{s}}}}=\frac{\mathbf{kg}}{\mathbf{m\,s}}.$$

Table 11.1 provides values of density and viscosity for some common fluids:

TABLE 11.1
Density and Viscosity of Common Fluids

Fluid (at 20°C)	Density [kg/m³]	Viscosity Pa(s) = [kg/(m s)]
Water	999.74	0.001005
Olive oil	910	0.09150
Engine oil	890	0.3500
Air	1.177	1.846×10^{-5}

It is interesting to note that the density of liquids such as olive oil and water are not very different but their viscosities are very different.

11.3 VISCOUS DRAG FORCE AND THE REYNOLDS NUMBER

As an object moves through a fluid, there is a resistive drag force. Depending on the speed at which the object moves through a fluid and the viscosity and density of the fluid, the object will move through the fluid with either a laminar (steam line) flow shown in Figure 11.2, a turbulent flow demonstrated in Figure 11.3, or a combination of these types of flow.

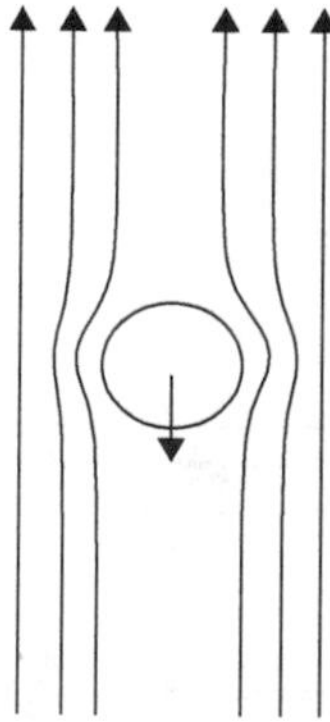

FIGURE 11.2 Laminar (stream line) flow.

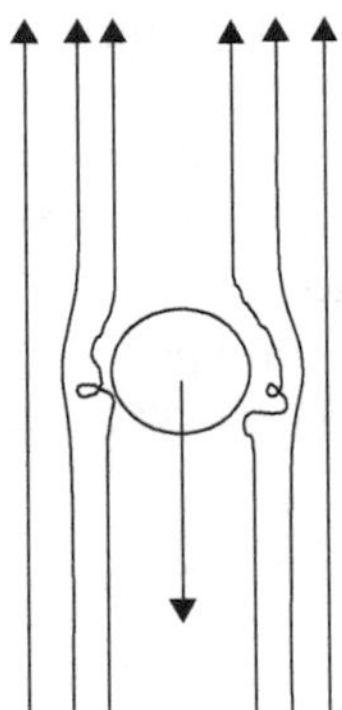

FIGURE 11.3 Turbulent flow.

In some cases, the drag force associated with the viscosity of the fluid is negligible and in other cases it is significant and even when it is significant there are two different expressions that can be used to compute the magnitude of the force.

Fortunately, there is a way to tell which expression to use to find the drag force. The key is the value of a parameter called the Reynolds number, Re, of the system given in equation (11.2) as:

$$\mathrm{Re} = \frac{(\rho v d)}{\mu} \tag{11.2}$$

where ρ is the density (kg/m^3) of the fluid, v is the magnitude of the velocity of the object moving through the fluid (m/s), μ is the viscosity (N·s/m^2), and d is the characteristic length (m) of the object. Notice that the Re is unitless. If the units for the variables are input into equation (11.2), they all cancel out.

$$\mathrm{Re}\ (\text{units}) = \frac{\left(\frac{\text{kg}}{\text{m}^3}\ \frac{\text{m}}{\text{s}}\ \text{m}\right)}{\frac{(\text{kg})}{\text{m s}}} = 1$$

For an object with a circular cross section, the radius of the object is the characteristic length and:

a. the fluid flow is streamlined if Re<2,000.
b. the fluid flow is a combination of laminar flow and turbulent flow, if Re is between 2,000 and 3,000.
c. the fluid flow is turbulent, if Re>3,000.

At low Re, viscous drag prevails and the magnitude of the drag force, given in equation (11.3), is proportional to the magnitude v of the velocity of the object as it moves through the fluid:

$$F_\mathrm{d} = bv \tag{11.3}$$

where b is the linear drag coefficient. This drag coefficient is based upon the geometry of the object and the properties of the fluid. At high Re, turbulent drag prevails and the magnitude of the drag force is proportional to the square of the velocity given in equation (11.4) as:

$$F_\mathrm{d} = Kv^2 \tag{11.4}$$

where K is the quadratic drag coefficient. Between these values of the Re, the drag force is a combination of the two and a specific equation for the situation must be found by experiment.

Equations (11.3) and (11.4) can be used to find the acceleration of an object in the presence of a drag force, but since the magnitude of the force changes as the velocity changes, it creates a situation in which calculus must be used to find the acceleration. Therefore, the analysis carried out in the following section will be restricted to the point after the object has accelerated to a speed at which the forces balance and the object no longer accelerates. Fortunately, this is a common situation especially in many biological systems, so it is worth pursuing the analysis of objects upon which a drag force is present and a constant velocity has been reached.

11.3.1 Analysis at a Low Re

For situations in which the Re is low, the drag force is proportional to the velocity of the object as shown in equation (11.3), $F_d = bv$. For a spherical object moving through a fluid at a low speed, the drag force coefficient depends on the properties of the object and the fluid. Based on the geometry of the situation, the drag coefficient for a spherical object moving at low speeds through a viscous fluid is: $b = 6\pi\mu r$. This relationship was derived by George Stokes in 1851, from the Navier–Stokes equations, which are the fundamental partial differential equations of fluid dynamics. Therefore, the low Re viscous drag force on a sphere falling through a liquid is known as *Stokes law*, given in equation (11.5) as:

$$F_d = 6\pi\mu rv \tag{11.5}$$

with r being the radius of the sphere. For different shapes, the coefficients in front of the equation change so the analysis done in this text will be restricted to spherical objects.

As the equation indicates, the drag force increases linearly as the speed of the sphere increases. Thus, the magnitude of the drag force will increase as the speed of the sphere increases due to the force of gravity on the sphere. Eventually, this drag force and other upward forces on the falling sphere will grow to a point at which the net force on the sphere will be zero. At this point, the sphere will continue to move downward through the fluid at a constant speed, which is known as the terminal velocity, v_t. Remember that at this point, the sphere must be moving downward to have a drag force but it will no longer accelerate.

Example

As depicted in Figure 11.4, a small metal sphere with a radius of 3.0 mm and a mass of 1.05 g is dropped into a large graduated cylinder filled with shampoo, which has a density of 1,250 kg/m^3.

If the metal sphere quickly reaches a terminal velocity of 0.1 m/s, what is the viscosity of the shampoo?

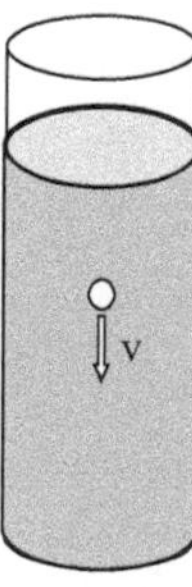

FIGURE 11.4 A ball falling through a fluid at a constant speed.

Step 1: Generate a free-body diagram of the metal sphere moving at a terminal velocity through the shampoo, as shown in Figure 11.5.

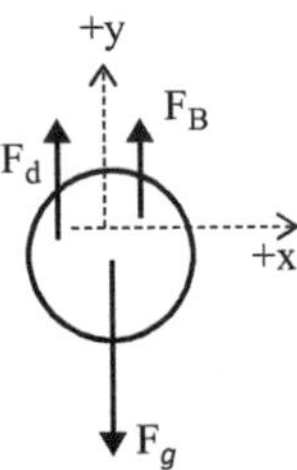

FIGURE 11.5 Free-body diagram of a sphere falling through a liquid.

Step 2: Generate the expressions of the y-components of the force vectors in Figure 11.5.

$$F_{dy} = F_d, \qquad F_{By} = F_B, \qquad F_{gy} = -F_g$$

Step 3: Since the sphere is moving at its terminal velocity, the sphere is in a static situation, so the application of Newton's Second Law is that the net force is set to zero.

$$\Sigma F_y = 0$$

$$F_{dy} + F_{By} + F_{gy} = 0$$

$$F_d + F_B - F_g = 0$$

$$F_d = F_g - F_B$$

Step 4: Substitute Stoke's Law, equation (11.5), for the drag force F_d in the previous expression.

$$6\pi\mu r v_t = F_g - F_B$$

Solving for the viscosity yields:

$$\mu = \frac{F_g - F_B}{6\pi r v_t}$$

Then, substituting expressions for the force of gravity and buoyant force with subscripts to distinguish the mass of the sphere (m_s) from the density (ρ_f) of the fluid results in equation (11.6) as

$$\mu = \frac{m_s g - \rho_f V g}{6\pi r v_t} \tag{11.6}$$

Converting all values to base SI units, inserting the expression of the volume of a sphere is $V = \frac{4}{3}\pi r^3$ and plugging in the numbers from the problem into equation (11.6) gives,

$$\mu = \frac{(0.00105\ \text{kg})\left(9.8\ \frac{\text{N}}{\text{kg}}\right) - \left(1{,}250\frac{\text{kg}}{\text{m}^3}\right)\left(\frac{4}{3}\pi\left(3\times10^{-3}\,\text{m}\right)^3\right)\left(9.8\ \frac{\text{N}}{\text{kg}}\right)}{6\pi\ \left(3\times10^{-3}\text{m}\right)\left(0.1\ \frac{\text{m}}{\text{s}}\right)}$$

Since a Pascal is $\text{Pa} = \frac{\text{N}}{\text{m}^2}$, the final answer is:

$$\mu = \frac{(0.01029\text{ N})-(0.00139\text{ N})}{0.00565\ \frac{\text{m}^2}{\text{s}}} = 1.58\ \frac{\text{Ns}}{\text{m}^2} = 1.58\ (\text{Pa})\text{s}$$

A quick calculation of the Re for this problem indicates the choice of equation (11.3) for the drag force was the correct choice.

$$\text{Re} = \frac{\rho v d}{\mu} = \frac{\left(1{,}250\frac{\text{kg}}{\text{m}^3}\right)\left(0.1\ \frac{\text{m}}{\text{s}}\right)\left(3\times10^{-3}\text{m}\right)}{1.58\ \frac{\text{Ns}}{\text{m}^2}} = 0.237$$

This example illustrates that equation (11.6) provides a straightforward way to experimentally determine the viscosity of liquids. You may have seen a device at a local gas station in which a series of glass cylinders are filled with different brands of motor oils and each has a steel sphere in the oil. When you flip the device over you can see the difference in viscosity of the oils, since the spheres fall at different rates. It is important to note that equation (11.6) is only valid for spherical objects moving at small velocities through a fluid with significant viscosity. For example, this expression works for very small metal balls moving through a fluid such as shampoo, pancake syrup, or motor oil, but it does not work for the same objects moving through water or air.

11.3.2 Analysis at a High Re

For a situation in which the Re is high, turbulence is significant and the drag force is proportional to the square of the velocity, as shown in equation (11.4), $F_d = Kv^2$. Because the viscosity and density of air is so low, air friction is the most common situation in which the quadratic drag force is applied. This is a high Re situation in which turbulence dominates. For air resistance, this K coefficient is proportional to the cross-sectional area A of the object, the density ρ of the fluid (air in this case), and the (unitless) drag coefficient C that indicates the roughness of the surface in a way similar to the frictional coefficient so K is:

$$K = \frac{1}{2}C\rho A$$

This gives a final expression for the magnitude of the drag force for air resistance in equation (11.7) as:

$$F_d = \frac{1}{2}C\rho A v^2 \qquad (11.7)$$

Example

Find the terminal velocity of a baseball with a mass of 149 g, a drag coefficient of 0.3, and a radius of 3.7 cm falling through the air, which has a density 1.204 kg/m³ and a viscosity of $1.82\times10^{-5}\ \frac{\text{Ns}}{\text{m}^2}$.

Step 1: Create a free-body diagram for the baseball falling at its terminal velocity, as shown in Figure 11.6.

Step 2: Generate the expressions of the y-components of the force vectors in Figure 11.6.

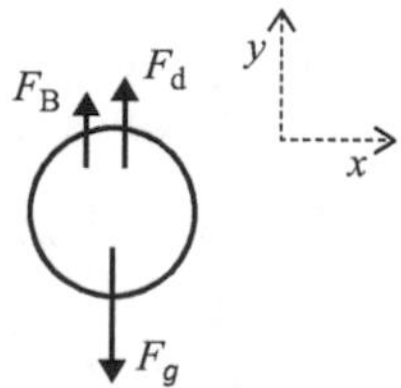

FIGURE 11.6 Free-body diagram of the baseball falling through air at a terminal velocity.

$$F_{dy} = F_d, \quad F_{By} = F_B, \qquad F_{gy} = -F_g$$

Step 3: Since the baseball is moving at its terminal velocity, the baseball is in a static situation so the application of Newton's Second Law is that the net force in the y-direction is set to zero.

$$\Sigma F_y = 0$$

$$F_{dy} + F_{By} + F_{gy} = 0$$

$$F_d + F_B - F_g = 0$$

$$F_d = F_g - F_B$$

Step 4: Solve for the terminal velocity by first substituting equation (11.7), for the drag force F_d in the previous expression:

$$\frac{1}{2} C\rho A v^2 = F_g - F_B$$

Solve for the terminal velocity results in

$$v_t = \sqrt{\frac{2(F_g - F_B)}{C\rho A}}$$

The weight of the baseball is: $F_g = \text{mg} = (0.149 \text{ kg})\left(9.8 \ \frac{\text{N}}{\text{kg}}\right) = 1.46 \text{ N}.$

Given the radius of the baseball, the volume of the ball and the cross-sectional area of half a sphere can be found:

$$V = \frac{4}{3}\pi\left(3.7\times10^{-2}\text{m}\right)^3 = 2.12\times10^{-4}\text{m}^3$$

$$A = \frac{1}{2}4\pi\left(3.7\times10^{-2}\text{m}\right)^2 = 8.6\times10^{-3}\text{m}^2$$

The reason that half the area of the sphere is used as the cross-sectional area is that the bottom half of the baseball's area is "cutting" through the air as it falls. The volume is needed to show that the buoyant force, although present, is so small that it can be neglected for air resistance problems.

$$F_B = \rho V g = \left(1.204 \frac{\text{kg}}{\text{m}^3}\right)\left(2.12\times10^{-4}\text{m}^3\right)\left(9.8 \ \frac{\text{N}}{\text{kg}}\right) = 0.0025 \text{ N}$$

It is clear from this quick calculation that the buoyant force for air on this baseball is much smaller than the weight of the object, so air resistance problems are usually negligible.

$$v_t = \sqrt{\frac{2(1.46 \text{ N} - 0.0025 \text{ N})}{(0.3)\left(1.204 \ \frac{\text{kg}}{\text{m}^3}\right)\left(8.6\times10^{-3}\text{m}^2\right)}} = 30.6 \ \frac{\text{m}}{\text{s}}$$

If the buoyant force is dropped from this analysis, the terminal velocity would still be 30.6 m/s since the density of air is so small compared to the density of objects falling through air at high Re. This is commonly the case for air resistance, so the common expression for the terminal velocity for an object falling in air does not include the buoyant force and is given in equation (11.8) as:

$$v_t = \sqrt{\frac{2mg}{C\rho A}} \tag{11.8}$$

A quick calculation of the Re for this problem indicates the choice of equation (11.4) for the drag force was correct.

$$\mathrm{Re} = \frac{\rho v d}{\mu} = \frac{\left(1.204\frac{\mathrm{kg}}{\mathrm{m}^3}\right)\left(30.6\frac{\mathrm{m}}{\mathrm{s}}\right)\left(3.7\times10^{-2}\,\mathrm{m}\right)}{1.81\times10^{-5}\,\frac{\mathrm{Ns}}{\mathrm{m}^2}} = 75{,}313.2$$

11.4 FLUID FLOW THROUGH A SYSTEM

Much of the same analysis developed to study the motion of objects moving through a fluid can be employed to study the flow of fluids through a system. This analysis is particularly important for studies in which viscosity is significant such as in the study of blood flow through the body. To make the transition to this analysis, a few terms must be defined. The first term is the *Volume flow rate* (Q), which is the volume of fluid (ΔV) flowing through a region in a given time (Δt).

Referring to the fluid signified by the grayed-out section in Figure 11.7, the volume of this section is the product of the cross-sectional area and the length x. If the cross-sectional area is a constant, the length, x, is the only quantity changing so the area, A, can be pulled out of the change leaving, $\frac{\Delta x}{\Delta t}$, which is just the magnitude of the velocity (v) or the speed of the fluid, so the volume flow rate is just the product of the cross-sectional area and the velocity of the fluid.

$$Q = \frac{\Delta V}{\Delta t} = \frac{\Delta(Ax)}{\Delta t} = A\frac{\Delta x}{\Delta t} = Av \tag{11.9}$$

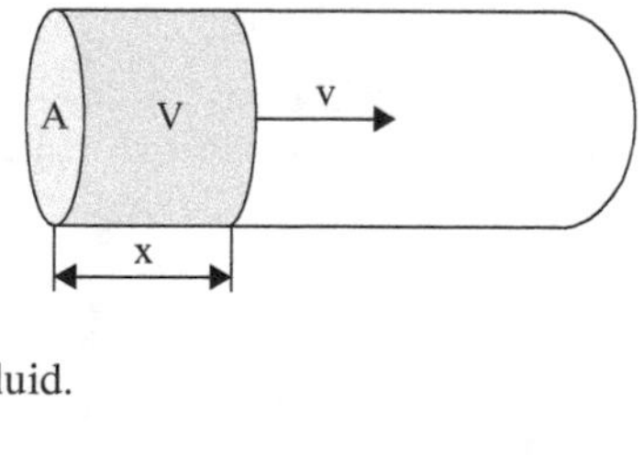

FIGURE 11.7 Displacement of a fluid.

The next term is the pressure gradient, which is denoted by the symbol (ΔP), and is just the pressure difference between the two ends of the tube

$$\Delta P = P_1 - P_2 \tag{11.10}$$

Note that the pressure gradient results in a force applied to the fluid from left to right, in Figure 11.8, and the fluid in the diagram is moving at a constant speed in that direction, thus there must be an equal and opposite force pointing from right to left.

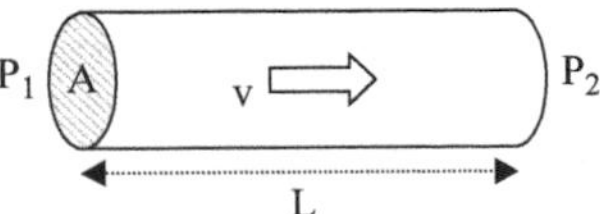

FIGURE 11.8 Pressure gradient.

11.4.1 Continuity Principle

With these terms defined, the analysis of fluid flowing through a system begins with the continuity principle, which is a statement of common sense. It states that for any section of a single pipe, filled with an incompressible fluid, through which the fluid is flowing, the amount of fluid that goes in one end in any specified amount of time is equal to the amount that comes out at the other end in the same amount of time. Referring to Figure 11.9, the continuity principle is given here as equation (11.11) as:

$$Q_1 = Q_2 \tag{11.11}$$

An interesting consequence of the continuity principle is the fact that, in order for the volume flow rate to be the same in a wide part of the pipe as it is in a narrow part of the pipe, the speed of the fluid must be greater in the narrow part of the pipe. That is why water comes out of a hose with a nozzle at a higher speed then without the nozzle on the hose.

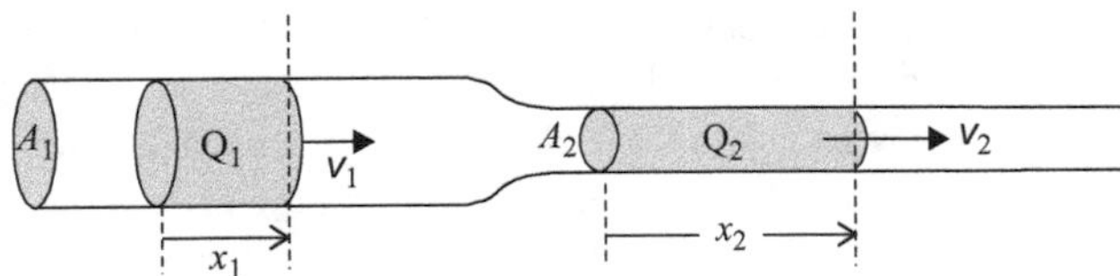

FIGURE 11.9 Continuity principle.

A quantitative relation between the speed at position 1 and the speed at position 2 can be derived starting with equal volumes of the fluid in Sections 11.1 and 11.2. As depicted in Figure 11.9, the continuity principle can be expressed in terms of the area and velocity of the fluid in each section in equation (11.12) as:

$$A_1 v_1 = A_2 v_2 \tag{11.12}$$

11.4.2 Laminar Flow

As with the flow of a fluid around an object as it falls through a fluid, there are limitations on the analysis for the fluid flow through a system of pipes or veins. The two different kinds of flow are laminar flow and turbulent flow. In steady-state laminar flow through a pipe, the position of particles of the fluid in any particular cross section of the pipe uniquely determines the path it will take through the pipe. Laminar flow can be depicted by means of a streamlined diagram, such as the one in Figure 11.10, where each particle moves through the system in a forward direction without turbulence.

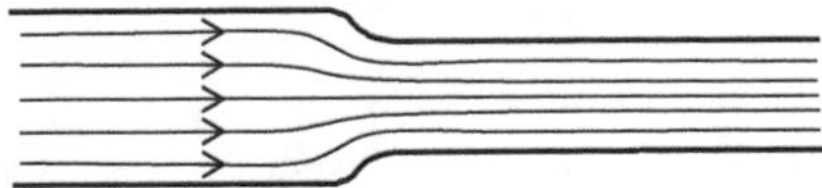

FIGURE 11.10 Streamlines of laminar flow.

The magnitude of the velocity is represented by the spacing of the lines: where the lines are closer together, the magnitude of the fluid velocity is greater. In a turbulent flow, the fluid churns and the path that a particle takes is unpredictable. The following analysis assumes laminar flow, since the analysis required for turbulent flow is beyond the scope of this textbook. It is interesting to note that when flow through a system becomes turbulent, the analysis required changes and other changes often result in the overall system.

For a system like the one described in Figure 11.9, the characteristic length for the Re calculation, equation (11.2), is the diameter of the pipe. The restrictions on the analysis are similar to that for the objects falling through the liquid in that:

a. The fluid flow is laminar, if Re<2,000.
b. The fluid flow is a combination of laminar flow and turbulent flow, if Re is between 2,000 and 3,000.
c. The fluid flow is turbulent, if Re>3,000.

Thus, the following analysis is only legitimate for systems with Re below 2,000.

11.4.3 Hagen–Poiseuille

For a cylindrical pipe, as shown in Figure 11.8, the pressure force $F_P = \Delta P\, A$ is applied across a system to a fluid. Since the fluid is moving at a constant speed, the applied pressure force is equal and opposite to the drag force on the fluid as it moves through the system. There are viscous forces between the fluid and the walls of the pipes and between the layers of fluid moving through the pipe. The middle of the fluid moves faster than the layers of fluid closest to the walls of the pipe. Together, these drag forces result in an expression similar to Stokes' law but with different coefficients because of the difference in the geometry of a sphere and a cylinder and the effects of the layers of fluid. The drag force on a viscous fluid moving through a cylindrical pipe is given by equation (11.13) as:

$$F_d = 8\pi\mu L v \tag{11.13}$$

where m is the viscosity of the fluid, L is the length of the pipe, and v is the speed of the fluid in the pipe. Notice that equation (11.13) looks like equation (11.5) for the spherical object falling through a fluid. The differences are that there is an 8 in equation (11.13) instead of a 6 and the length L of the pipe is used instead of the radius of the sphere.

Since the fluid is moving at a constant speed, the pressure force is equal and opposite to the drag force, so the pressure force is equal in magnitude to the drag force:

$$(\Delta P)A = 8\pi\mu L v$$

This expression is normally rearranged to give: $(\Delta P) = \dfrac{8\pi\mu L v}{A}$.

Then, another A is multiplied to the numerator and denominator of the right-hand side to give $(\Delta P) = \dfrac{8\pi\mu L(Av)}{A^2}$, and it is observed that the volume flow rate (Q) is the product of area and velocity of the fluid to give the Hagen–Poiseuille Equation as equation (11.14) as:

$$(\Delta P) = \frac{8\pi\mu L(Q)}{A^2} \tag{11.14}$$

with a pipe or a vein as the system a circular cross section the area is $A = \pi r^2$, giving the Hagen–Poiseuille equation for a pipe or vein as equation (11.15) as:

$$(\Delta P) = \frac{8\pi\mu L(Q)}{\pi^2 r^4} \tag{11.15}$$

It is clear from equation (11.15) that the most important factor relating blood flow and pressure difference is the radius of the blood vessels. It varies like the radius to the fourth power. Since the pressure difference is related to the inverse of the radius to the fourth power, slight changes in radius can cause large changes in blood flow. In addition, since the viscosity of the blood and length of the blood vessels are fairly constant, it is changes in radius that generally affects blood flow and blood pressure. For example, a doubling of the radius of a blood vessel creates a flow of 16 times the original flow at the same pressure. In the body, however, flow does not conform exactly to this relationship because this relationship assumes long, straight tubes (blood vessels), a Newtonian fluid (e.g., water, not blood, which is non-Newtonian), and steady, laminar flow conditions. Nevertheless, the relationship clearly shows the dominant influence of vessel radius on blood flow and therefore serves as an important concept to understand how physiological (e.g., vascular tone) and pathological (e.g., vascular stenosis) changes in vessel radius affect pressure and flow, and how changes in heart valve orifice size (e.g., in valvular stenosis) affect flow and pressure gradients across heart valves.

Example

Given that the density of blood is 1,060 kg/m^3, the average speed of blood across the cross section of the aorta is 0.3 m/s, the diameter of the aorta is 2 cm, and the viscosity of blood is 4 mPa·s, determine whether the blood flow through the aorta is laminar, turbulent, or somewhere in between.

Solution

First, to make sure the units work out, we change the units of the viscosity:

$$\mu = 4\times10^{-3}\,\frac{\text{N}\cdot\text{s}}{\text{m}^2} = 4\times10^{-3}\,\frac{\text{kg}\cdot\text{m}}{\text{s}^2}\,\frac{\text{s}}{\text{m}^2} = 4\times10^{-3}\,\frac{\text{kg}}{\text{s}\cdot\text{m}}$$

Next, we substitute all the values into the expression for the Re.

$$\text{Re} = \frac{\rho v d}{\mu} = \frac{1{,}060\,\dfrac{\text{kg}}{\text{m}^3}\left(.3\,\dfrac{\text{m}}{\text{s}}\right)(.02\text{ m})}{4\times10^{-3}\,\dfrac{\text{kg}}{\text{s}\cdot\text{m}}}$$

$$\text{Re} = 1{,}590\,(\text{with no units})$$

The Re turns out to be less than 2,000, indicating that the flow of blood through the aorta is laminar, but not to a great extent.

11.5 ANSWERS TO THE CHAPTER QUESTIONS

How does the erythrocyte sedimentation rate (ESR) of red blood cells falling through blood indicate inflammation? To answer this question, we need to know that when the body is experiencing inflammation, red blood cells clump together. Also, experimentally, a clump of red blood cells falls at a speed on the order of a centimeter per hour, so slowly that we are clearly dealing with viscous drag, the drag that is proportional to v itself rather than v^2. Taking our expression for the viscosity of a fluid in terms of the terminal velocity of a spherical object falling through it, equation (11.16), and solving for the terminal velocity, we obtain:

$$v_t = \frac{m_s g - \rho_f V g}{6\pi r \mu} \tag{11.16}$$

with m_s the mass of a blood cell, ρ_s the density of the blood cells, and ρ_f as the density of the liquid blood, the terminal velocity of the blood cells is

$$v_t = \frac{\rho V g - \rho_{DF} V g}{6\pi r \mu} = \frac{(\rho - \rho_{DF}) V g}{6\pi r \mu} = \frac{(\rho - \rho_{DF}) \frac{4}{3}\pi r^3 g}{6\pi r \mu} = \frac{2(\rho - \rho_{DF}) g}{9\mu} r^2$$

where we have approximated both a red blood cell and a clump of red blood cells as spheres by replacing the volume with $\frac{4}{3}\pi r^3$. We can think of a clump of red blood cells as an object having the same density as a red blood cell but a larger radius. In our final expression for the terminal velocity we see that the larger the radius, the greater the terminal velocity, so clumps of red blood cells sink faster than individual red blood cells. In carrying out the ESR test, one puts a sample of blood in a vial and waits. As the red blood cells fall, a layer of clear liquid forms at the top of the vial. The thickness of that layer increases with time. After a fixed amount of time, on the order of an hour, one measures the thickness of the layer. If the thickness is greater than that for the blood of a person with no inflammation, one knows the blood cells were falling faster than they do when they are not clumped together, indicating that they are indeed clumped together, thus indicating inflammation in the body of the person from which the blood was drawn.

11.6 QUESTIONS AND PROBLEMS

11.6.1 Multiple-Choice Questions

1. Spherical beads are dropped into liquids with different densities. All the liquids have the same viscosity. Will the beads have the greatest terminal velocity while moving through the liquid with the highest density or the liquid with the lowest density?
   A. highest B. lowest
   C. The density of the fluid doesn't influence the terminal velocity.

2. Spherical beads with the same radius but with different densities are dropped into a liquid. Will the bead falling with the greatest terminal velocity have the highest density or the lowest density?
   A. highest B. lowest
   C. The density of the bead doesn't influence the terminal velocity.

3. Spherical beads are dropped into liquids with different viscosities. All liquids have the same density. Will the beads have the greatest terminal velocity when they are moving through the liquid with the highest density or the liquid with the lowest viscosity?
   A. highest B. lowest
   C. The viscosity of the fluid doesn't influence the terminal velocity.

4. The terminal velocity of a baseball falling through air can be computed without considering the buoyant force of the air on the ball because the buoyant force is tiny compared to the gravitational force on the ball. If the buoyant force of air were considered, would the calculated terminal velocity of the ball be slightly greater than, slightly less than, or exactly equal to the terminal velocity of the baseball computed without considering the buoyant force?
   A. slightly greater than B. slightly less than C. exactly equal to

5. For fluid in motion, the continuity principle is a statement of the fact that:
   A. There are no gaps in a fluid.
   B. The center-to-center distance between molecules making up a fluid is the same throughout the fluid.
   C. The rate at which fluid flows out of a pipe segment at one end is equal to the rate at which fluid flows into that pipe segment at the other end.
   D. The greater the velocity of a fluid, the greater the flow rate, independent of the cross-sectional area of the pipe through which the fluid is flowing.
   E. There are no jump discontinuities in the velocity of a fluid as a function of time.

6. For the case of fluid of density 900.0 kg/m^3 flowing through a pipe of diameter 4 inches at a rate of 35 gallons per minute, the greater the viscosity of the fluid, the
   A. smaller the Re B. greater the Re

7. A small sphere is dropped into a wide, deep container filled with a fluid with a significant viscosity and density, such as glycerin. The sphere reaches its terminal velocity very quickly after it begins to fall through the fluid, therefore its speed increases at a greater rate, increases at a reduced rate, or remains constant for the rest of its trip to the bottom of the container.
   A. greater rate B. reduced rate C. remains constant

8. A small sphere is dropped into a wide, deep container filled with a fluid with a significant viscosity and density, such as glycerin. The sphere reaches its terminal velocity very quickly after it begins to fall through the fluid, therefore at terminal velocity the buoyant force on the sphere is greater than, less than, or equal to the weight of the sphere.
   A. greater than B. less than C. equal to

9. A small sphere is dropped into a wide, deep container filled with a fluid with a significant viscosity and density, such as glycerin. The sphere reaches its terminal velocity very quickly after it begins to fall through the fluid, therefore at terminal velocity the drag force on the sphere is greater than, less than, or equal to the weight of the sphere.
   A. greater than B. less than C. equal to

10. A small sphere is dropped into a wide, deep container filled with a fluid with a significant viscosity and density, such as glycerin. The sphere reaches its terminal velocity very quickly after it begins to fall through the fluid, therefore at terminal velocity the sum of the buoyant force and the drag force on the sphere is greater than, less than, or equal to the weight of the sphere.
    A. greater than B. less than C. equal to

11.6.2 Problems

Terminal velocity of objects falling through a fluid:

1. A steel sphere, with a density of 7,850 kg/m^3 and a radius of ½ mm, is dropped into a graduated cylinder filled with glycerol, which has a density of 1,261 kg/m^3 and a viscosity of 1.412 Pa s. Assuming a low Re, compute the terminal velocity of this steel sphere falling through glycerol.
2. Compute the Re for the steel sphere falling through glycerol in problem 1, given the terminal velocity of the steel sphere is 0.0025406 m/s.

3. A ping-pong ball has a mass of 2.7 g and a radius of 2 cm and is falling through air at a terminal velocity of 8 m/s. Assuming the buoyant force is insignificant, find the drag coefficient for this ping-pong ball.
4. A large and deep container is filled with a liquid, which has a viscosity of 1.108 Pa s and a density of 2.900 kg/m^3. A brass sphere with a mass of 0.35 g and a radius of 0.215 cm is held above the surface of the liquid. The sphere is released into the liquid, from rest, and it accelerates downward until it reaches a terminal velocity. Compute the terminal velocity of the sphere as it moves through the fluid in this low Re situation. Note that the density of brass is 8,400 kg/m^3.
5. A large and deep container is filled with a liquid, which has a density of 1,240 kg/m^3 and a viscosity of 1.50 Pa s. An aluminum sphere with a diameter of 1 cm, a density of 2,700 kg/m^3, and a mass of 1.413 g is held above the surface of the liquid. The sphere is released into the liquid, from rest, and it accelerates downward until it quickly reaches a terminal velocity. The sphere continues to fall through the liquid at its terminal velocity until it reaches the bottom of the container. Compute the terminal velocity of this sphere in this liquid assuming the buoyant force is significant and the Re is low.
6. A large and deep container is filled with a liquid, which has a density of 1.000 kg/m^3. A sphere with a mass of 0.2 kg and a radius of 0.03 m is held above the surface of the liquid. The sphere is released into the liquid, from rest, and it accelerates downward until it reaches a terminal velocity of 1.2 m/s. Compute the viscosity of this liquid.
7. Compute the Re for the sphere falling through the liquid in problem 6, given the terminal velocity of the sphere is 1.2 m/s and the viscosity is 1.256 Pa s.

Fluid flow through a system:

8. Water is flowing through a hose with an inner diameter of 2 cm at a rate of 2.0 gallons per minute. What is the average speed (in meters per second) of the water the cross section of the hose? Note that 1 inch = 2.54 cm and 1 m^3 = 264.172 gallons. Assume the hose to be completely full of water at all times.
9. A pipe with an inner radius of 2.0 cm is connected to a pipe with an inner radius of 1.00 cm. Both pipes are completely full of water and the speed of the water in the smaller part of the pipe is 4.0 m/s. Find the speed of the water in the larger pipe.
10. For problem 10, calculate the Re for the water flowing in the smaller part of the pipe and state whether the water flow in that pipe is turbulent, laminar, or a combination of the two. Use the density and viscosity of water of 1,000 kg/m^3 and .001 Pa s, respectively.

APPENDIX: GEL ELECTROPHORESIS (SYNTHESIS OPPORTUNITY 1)

An example of using dynamics in the study of a complex system with biological applications.

Gel electrophoresis is used in the study of DNA, RNA, and proteins, as seen in the gel in Figure 11.11.

The goal of this technique is to separate molecules of interest by their size to categorize them and/or to learn about the molecules. In a gel electrophoresis device, the key is that the terminal velocity v_t of a denatured protein with negatively charged sodium dodecyl sulfate (SDS) molecules attached, moving through a polyacrylamide gel under the influence of an electric field, is inversely proportional to their mass. The goal of this appendix is to identify why this is the case and to further discuss the physics of SDS PAGE (sodium dodecyl sulfate polyacrylamide gel electrophoresis).

The starting point is related to establishing rough estimates of some of the physical characteristics of the proteins used in a common system to investigate what forces need to be included in the derivation. According to (Viney & Fenton1998), the denatured proteins are rod shaped and all have

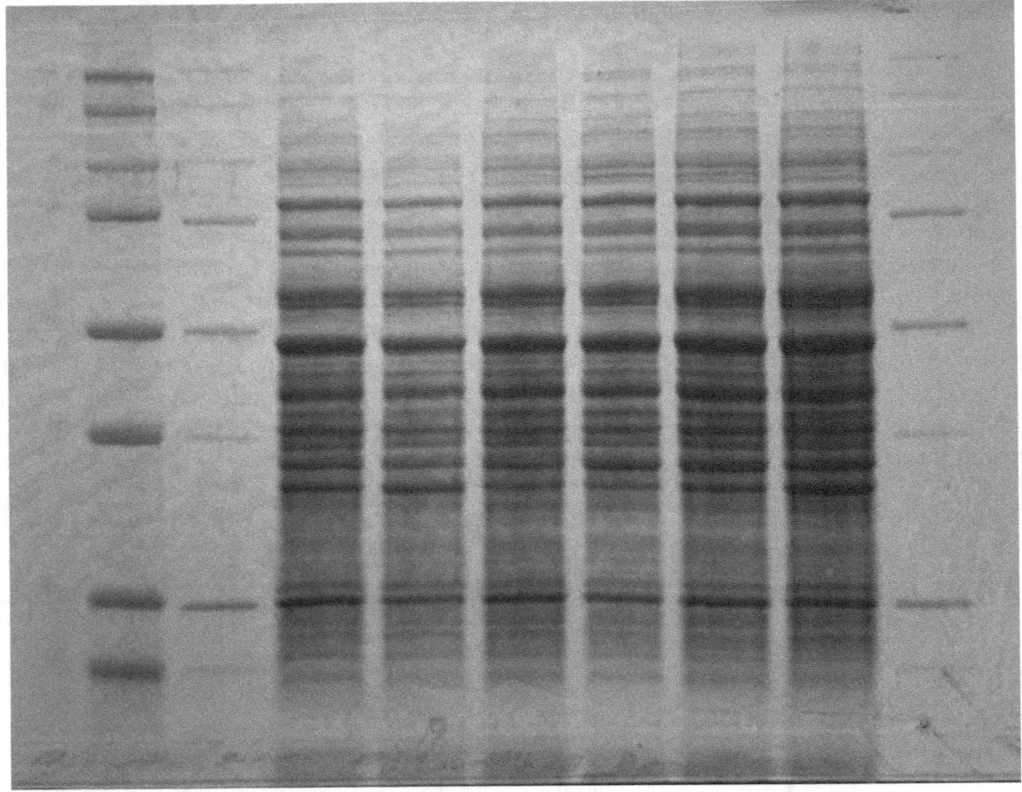

FIGURE 11.11 Gel of proteins made using gel electrophoresis.

the same diameter meaning that the masses are directly proportional to their lengths. The linear mass density, λ, of the proteins is the mass, m, per length, L, of the protein. Therefore, the mass of the protein is related to the length, simply as the product of the length and mass density, $m=\lambda L$. The mass density of a typical protein is $\rho=1.41\text{g}/(\text{cm})^3$ (Fischer et al., 2004). The units of this density can be converted to from g/(cm)³ to kDa/(nm)³ as follows:

$$\rho = 1.41\frac{\boldsymbol{g}}{(\text{cm})^3}\frac{1\,\text{Da}}{1.66054\times10^{-24}\,\boldsymbol{g}}\left(\frac{1\,\text{cm}}{1\times10^7\,\text{nm}}\right)^3 = 849\frac{\text{Da}}{(\text{nm})^3}.$$

The mass density is mass per volume and the volume of a cylinder, as shown in Figure 11.8, is the product of the cross-sectional area, A, and the length, L, of the cylinder:

$$\rho = \frac{m}{V} \text{ for this configuration } \rho = \frac{m}{AL}$$

$$\frac{m}{L} = \rho A, \text{ but since } A = \pi r^2, \text{ then } \frac{m}{L} = \rho\pi r^2, \text{ and } \mathrm{r} = (D/2) \text{ so } \frac{m}{L} = \rho\pi\left(\frac{D}{2}\right)^2$$

$$\text{Resulting in a mass per length of}: \lambda = \rho\pi\left(\frac{D}{2}\right)^2 = 849\frac{\text{Da}}{(\text{nm})^3}\pi\left(\frac{1\,\text{nm}}{2}\right)^2.$$

Given the nominal diameter of an amino acid is 1 nm, from (Marko & Ansari, 2013) as an estimate for the diameter D of the rod-shaped proteins used to make the gel, the linear mass density is

$\lambda = \rho\pi\left(\frac{D}{2}\right)^2$, where the nominal diameter of an amino acid is 1 nm, from (Marko & Ansari, 2013) as an estimate for the diameter D of the rod-shaped proteins used to make the gel The PubChem article, referenced in the bibliography (PubChem), shows the neutral SDS molecule as including one Na^+ ion, which leads us to conclude that in solution, the SDS anion has a charge of −1e (where the e is the charge of a single proton, 1.6022×10^{-19} C). According to (Viney & Fenton, 1998), the SDS anion attaches to proteins at the rate of one SDS anion to two amino acids. Note that the charge q on a protein of length L is given by $q = \chi L$, where χ is the linear charge density.

As depicted in Figure 11.12, the gel is subjected to an upward-directed applied electric field.

Being negatively charged, the proteins in the gel experience an electric force in the downward direction. Thus, the proteins move downward. In the model of (Viney & Fenton, 1998), the proteins are subjected to an upward-directed drag force, which has two contributions: a drag force due to the interaction of the proteins with the gel itself and a drag force due to the interaction of the proteins with positively charged ions moving upward, in the direction opposite that of the negatively charged proteins themselves.

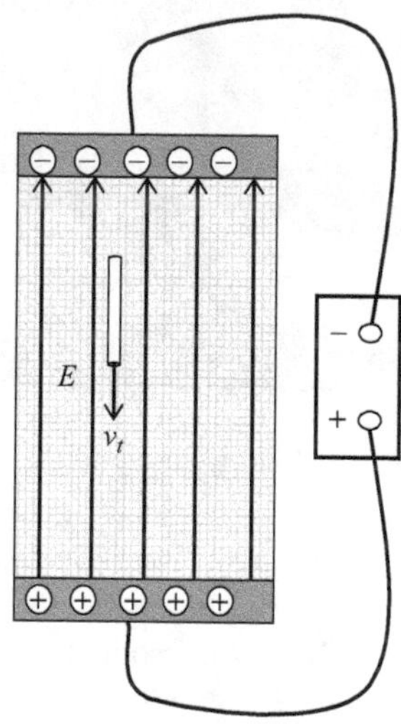

FIGURE 11.12 Gel electrophoresis device.

Regarding the electric field: The bottom of the gel rests in a small tub of an aqueous buffer solution. The top of the gel forms the bottom surface of another tub of buffer solution. A wire extends along the bottom of the upper tub, just above the gel. Another wire extends along the bottom of the lower tub, inside the lower tub. Both wires are bare and both wires are submerged in the buffer solution. The buffer solution is full of both positive and negative ions and hence is a good conductor. The two wires are called electrodes. The buffer solution in the upper tub maintains electrical contact between the upper electrode and the top of the gel, and the buffer solution in the lower tub maintains electrical contact between the lower electrode and the bottom of the gel. A power supply connected to the electrodes is used to create an upward-directed electric field in the gel.

The electric field produced by the power supply is referred to as the applied electric field. When an electric field is applied to a region of space occupied by a medium (such as the polyacrylamide gel in this case), the applied electric field causes the charged particles in the medium to rearrange themselves in a manner that enables them to create their own electric field in the medium, one that is in the direction opposite that of the applied electric field. More specifically, to create an upward-directed field, the power supply maintains a negative charge on the electrode (metal wire) that is above and parallel to the upper edge of the gel, and it maintains a positive charge on the electrode that is below and parallel to the lower edge of the gel. This creates a redistribution of charge in the gel, which puts, on average, positively charged particles closer to the negatively charged electrode, and negatively charged particles closer to the positively charged electrode. This charge distribution results in a downward-directed contribution to the total electric field.

In any material other than a perfect conductor, the applied electric field is greater than the electric field due to the charge distribution in medium, and the total electric field is in the direction of the applied electric field, but it is weaker than the applied electric field. The magnitude of the total electric field depends on the actual material making up the medium, in particular, a unitless property of that material known as the dielectric constant represented by the Greek letter κ (kappa). The magnitude E of the total electric field in the medium is given by $E = \frac{E_{\text{applied}}}{\kappa}$ where E_{applied} is the applied electric field, created by the electrodes and the power supply.

Estimates to justify neglecting the gravitational and buoyant forces from the derivation:

To get an estimate for the total electric field E to which the proteins were subjected, refer to (Neelakanta, 1995) where, for a polyacrylamide gel, the dielectric constant is given as 60 (with no units). For example, with an initial applied electric field of 1,211 N/C, the total electric field on the proteins has a magnitude of

$$E = \frac{E_{\text{applied}}}{\kappa} \approx \frac{1{,}211\ \text{N/C}}{60} \approx 20\ \text{N/C}$$

For a protein with a charge of –50e, the charge in Coulombs is: $q = -50\text{e} = -50\left(1.6022 \times 10^{-19}\text{C}\right) = -8.0011 \times 10^{-18}\text{C}$ which yields an electric force of magnitude

$$F_{\text{E}} = |q|\text{E} \approx \left|-8.0011 \times 10^{-18}\text{C}\right| 20\text{N/C}$$

$$F_{\text{E}} \approx 1.6 \times 10^{-16}\text{N}$$

The mass of this protein of 10 kDa is:

$$m = 10{,}000\ \text{Da} = 10{,}000(1.66054 \times 10^{-27}\text{kg}) = 1.66054 \times 10^{-23}\ \text{kg}$$

so the magnitude of the gravitational force on this protein is:

$$F_{\text{g}} = mg = (1.66054 \times 10^{-23}\ \text{kg})\ 9.8\ \text{m/s}^2$$

$$F_{\text{g}} = 1.627 \times 10^{-22}\ \text{N}$$

The magnitude of the gravitational force on the smallest protein is thus roughly $\frac{1.627 \times 10^{-22}\text{N}}{1.6 \times 10^{-16}\text{N}} \times 100\% \approx .0001\%$ of the magnitude of the electrical force on the smallest protein. That percentage applies to all the proteins since, for instance, for the largest protein used in the lab, both the charge and the mass are 7.5 times the corresponding value for the smallest protein, meaning the magnitudes of both forces are 7.5 times of what they are for the smallest protein, leaving the ratio of the magnitudes of the forces the same. So, the gravitational force is negligible compared to the electrical force.

Since the mass and the density of the smallest proteins are known, the volume of the smallest proteins can be found:

$$V = m/\rho = 1.66054 \times 10^{-23}\ \text{kg}/1.41 \times 10^{3}\ \text{kg/m}^3 = 1.177688 \times 10^{-26}\ \text{m}^3.$$

In addition, the density of the gel is known to be $\rho_f = 1.2 \times 10^3 \text{kg/m}^3$, so the magnitude of the buoyant force on the smallest protein is:

$$F_b = \rho_f V g = (1.2 \times 10^3 \text{ kg/m}^3)(1.177688 \times 10^{-26} \text{ m}^3)(9.8 \text{ m/s}^2)$$

$$F_b = 1.38496 \times 10^{-22} \text{ N}$$

The buoyant force has a slightly smaller magnitude than the gravitational force, which agrees with the reality that the proteins would slowly settle to the bottom of the gel if left in the gel for a long time. In addition, the buoyant force on the smallest protein is thus roughly $\frac{1.38496 \times 10^{-22} \text{N}}{1.6 \times 10^{-16} \text{N}} \times 100\% \approx 0.0001\%$ of the magnitude of the electrical force on the smallest protein. That percentage applies to all the proteins since, for instance, for the largest protein used to produce the gel, both the charge and the mass are 7.5 times the corresponding value for the smallest protein, meaning the magnitudes of both forces are 7.5 times of what they are for the smallest protein, leaving the ratio of the magnitudes of the forces the same. So, the buoyant force is also negligible compared to the electrical force.

Given that the buoyant and gravitational forces are negligible, the following is the analysis of the motion of a protein through the gel.

Step 1: Generate a free-body diagram of a protein in the gel with the power supply turned on and thus an electrical downward force, as shown in Figure 11.13.

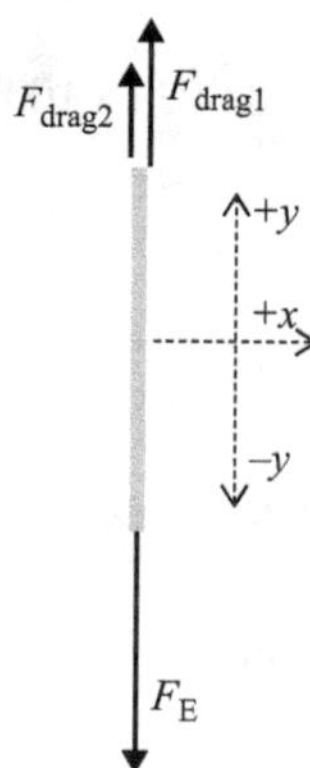

FIGURE 11.13 Free-body diagram of the protein.

As noted in the previous section, the buoyant and gravitational forces can be neglected from this derivation. Note that F_E is the magnitude of the electric force of the total electric field of magnitude E on the protein. $F_E = |q|E$ where q is the charge of the protein. The charge is the charge-per-length χ of $-1\frac{\text{e}}{\text{nm}} = -1.6022 \times 10^{-19} \frac{\text{C}}{\text{nm}}$ times the length of the protein, so $|q|$ can be replaced using $q = \chi$ L. Hence, the electric force on the protein is $F_E = |\chi| EL$.

The magnitude of the drag force on the protein due to the gel by itself is F_{drag1}, this does not include the drag resulting from positive ions moving upward past the protein. The gel is a viscoelastic material, which yields a drag force that is modeled as being viscous drag. The formula for the magnitude of the viscous drag force on an object moving through a fluid is

$$F_{\text{drag1}} = 8\pi\mu_{\text{eff}} L_{\text{characteristic}} v_{\text{t}}$$

where the subscript "eff" standing for *effective* to the μ is inserted to remind the reader that the viscosity of a gel is different from the viscosity of a liquid. The subscript "characteristic" to the symbol L was added to represent the characteristic length of the object since we are using the un-subscripted L to represent the length of the protein, and we have added the subscript t to the symbol v because we are dealing with an object at its *terminal* velocity. This drag force is similar to fluid drag force on a sphere falling through a fluid, *Stokes law*: $F_{\text{d}} = 6\pi\mu rv$. Both equations depend on the viscosity (μ) and the velocity (v), but because of the different geometries of the two situations the 6π is changed to 8π and the radius of the sphere is changed to the length of the cylinder. These values of 6π and 8π are experimentally determined.

F_{drag2} is the magnitude of the drag force on the protein due to positive ions moving upward past the protein. The greater the electric field, the greater the rate at which ions will be moving upward past the protein, so (Viney & Fenton, 1998) judge the magnitude of this force to be proportional to the magnitude E of the total electric field in which the protein finds itself meaning that it can be expressed as:

$$F_{\text{drag2}} = 8\pi\mu_{\text{eff}} L_{\text{characteristic}} k_2 E$$

where k_2 is a constant determined for the system.

Step 2: Find the relevant components of the vectors appearing in our free-body diagram:

$$F_{\text{Ey}} = -F_{\text{E}} \qquad F_{\text{drag1y}} = F_{\text{drag1}} \qquad F_{\text{drag2y}} = F_{\text{drag2}}$$

Step 3: Applying Newton's Second law for the protein moving at a terminal velocity through the gel gives::

$$\sum F_y = ma_y$$

$$F_{Ey} + F_{\text{drag1y}} + F_{\text{drag2y}} = 0$$

$$-F_E + F_{\text{drag1}} + F_{\text{drag2}} = 0$$

$$-|\chi|EL + 8\pi\mu_{\text{eff}} L_{\text{characteristic}} v_{\text{t}} + 8\pi\mu_{\text{eff}} L_{\text{characteristic}} k_2 E = 0$$

(Viney & Fenton, 1998) argue that the characteristic length of the protein has to be somewhere between the diameter of the protein and the length of the protein, and they use $L_{\text{characteristic}} = k_1 L^n$, where k_1 is a constant and the unitless exponent n is to be determined experimentally but it has to be somewhere between 0 and 1 with $n=0$, meaning that k_1 is the diameter of the protein and with $n=1$ meaning that the characteristic length is the actual length L of the protein in which case k_1 would be the number 1. For the in-between values, in SI units, k_1 has to have units of m/m^n for the units of the characteristic length to work out to be meters. Making this substitution yields:

$$-|\chi|EL + 8\pi\mu_{\text{eff}} k_1 L^n v_{\text{t}} + 8\pi\mu_{\text{eff}} k_1 L^n k_2 E = 0$$

This is where the fact that the gel is a viscoelastic material rather than a viscous fluid is considered (Viney & Fenton, 1998). Since the gel consists of long polymers, the polymers can be considered to be in a big tangle and in tension. Consider a small protein moving through the gel will stretch the polymers. Consider a larger protein to consist of a first part equal in size to that of the smaller protein, plus the remainder of the larger protein. When a larger protein moves through the same region, the remainder comes in contact with different sections of the same polymers already stretched by the first part. The fact that they are stretched makes the pressure on the remainder greater than the pressure that would be acting on it if it were the only part, and the remainder further stretches the polymers making the pressure on the first part greater than it would be if there were no remainder. As a result, the longer the protein, the greater the pressure on it. The pressure is providing the normal force for the kinetic frictional force so the greater the pressure, the greater the drag. Note that the pressure also acts on a larger area, but that has already been considered by the characteristic length. Given that the effective viscosity is proportional to L, we make the substitution $\mu_{\text{eff}}=k_3L$ (into our expression for v_t), where k_3 is the constant of proportionality. This results in:

$$-|\chi|EL+8\pi k_3Lk_1L^nv_t+8\pi k_3Lk_1L^nk_2E=0$$

Since $(L)(L^n)=L^{(n+1)}$, the expression becomes equation (11.17)

$$-|\chi|EL+8\pi k_3k_1L^{(n+1)}v_t+8\pi k_3k_1L^{(n+1)}k_2E=0 \tag{11.17}$$

Next, equation (11.17) is be solved for the terminal velocity, v_t. Since $L^{(n+1)}=LL^n$, the L in the denominator can be used to cancel the L in the numerator resulting in the following expression.

$$v_t=\frac{|\lambda|E}{8\pi k_3k_1L^n}-k_2E$$

Substitute $L=m/\lambda$ (length=mass divided by linear density) to get v_t in terms of the mass of the protein:

$$v_t=\frac{|\chi|E\lambda^n}{8\pi k_3k_1}m^{-n}-k_2E$$

Recall that χ is the linear charge density, which is the same for all the proteins (with SDS molecules attached). Furthermore, the applied electric field is adjusted during the run as necessary to keep E constant during the run. Defining the constants

$$c_1=\frac{|\chi|E\lambda^n}{8\pi k_3k_1}\text{ and }c_2=k_2E,$$

results in a simple expression of the terminal velocity of the proteins in a gel electrophoresis experiment in equation (11.18):

$$v_t=c_1m^{-n}-c_2 \tag{11.18}$$

In practice, since all the constants that make up the two coefficients (c_1 and c_2) are difficult to know, a set of proteins with set mass and charge are used to calibrate a gel before it is used to study an unknown sample of proteins. Data are taken and the values of c_1, c_2, and n are found by adjusting the value of n until the data fit a straight line and the slope and y-intercept give c_1 and c_2, respectively.

One final note regarding the statement that the electric field was adjusted during the run as necessary to keep E, the magnitude of the total electric field experienced by the proteins, constant during the run: the effect of the electric field in the gel is to make ions in the gel flow. The electric field exerts a force on the ions that causes them to experience an acceleration that causes the speeds of the ions to increase, which they do until the magnitude of the drag force on the ions equals the magnitude of the electric force, and, from then on, the particles move at constant velocity. The positively charged ions move upward, in the same direction as the electric field, and the negatively charged ions, including the charged proteins under study, move downward. Both the positive ions and the negative ions contribute to the total electric current (charge flow rate) in the circuit (the closed electrical path extending from the positive electrode, out of the positive electrode, down through the buffer in the upper tub, down through the gel, down through the buffer solution in the lower tub, down into the negative electrode, through the power supply, and back out of the power supply to the positive electrode). The total rate of charge flow, called the electric current, is proportional to the total electric field in the gel. So, to maintain a constant total electric field in the gel, all one has to do is to make sure the current is kept constant. There are power supplies that do exactly that, and indeed a constant-current power supply was used to maintain a constant current of.040 A through the gel. To do that, the power supply had to maintain a constant total electric field in the gel, and to do that, the electric field applied by the power supply gradually increased from 1,211 to 2,250 N/C. Note that it is the electric field in the gel that causes the proteins to move, not the electric current; in fact, the moving proteins are *part of* the charge flow whose rate is the electric current.

CONCEPTUAL QUESTION

How does the charge of the 50,000 Da protein (with SDS anions attached) used in gel electrophoresis compare with the charge of the 10,000 Da protein (with SDS anions attached) used in the same device?

Answer:

The charge on one of the proteins used in the lab is directly proportional to its mass, so the 50 kDa protein has five times the charge that the 10 kDa protein has. In the chapter, the mass m was expressed in terms of the linear mass density μ, the length L as $m=\lambda^*L$, the charge q can be expressed in terms of the linear charge density λ, and the length L as $q=\chi^*L$

Solving the former for $L = m/\chi$ and plugging it into the latter yields $q=(\chi/\lambda)$ m explicitly showing q to be proportional to m meaning that if you quintuple the mass, the charge quintuples as well.

EXAMPLE PROBLEM

From values in the table, the charge-per-length (linear charge density = C) of the proteins with SDS anions attached is –1 e/nm = -1.6022×10^{-19} C/nm (Table 11.2). So, $q=CL$.

Expressions:

$$F_E = q \cdot E \qquad F_{D1} = 8\pi k_3 k_1 L^{(n+1)} v_t \qquad F_{D2} = 8\pi k_3 k_1 L^{(n+1)} k_2 E$$

$$F_g = m \cdot g \qquad F_b = \rho_f \cdot V \cdot g \qquad \rho = \frac{m}{V}$$

Problems & Questions: Table 11.2 provides the data for the proteins used in a Gel Electrophoresis (GE) experiment that is referred to in the following problems and questions. For these proteins, preliminary measurements of the gel were made and the value of $k_3k_1=0.7275\times10^{-3}$ Pa*(s/m$^{0.333}$),

TABLE 11.2
Characteristics of Sample Proteins

Mass [Da]	10,000	15,000	20,000	25,000	37,000	50,000	75,000
Estimated length [nm]	50	75	100	125	185	250	375
Estimated charge [e]	–50	–75	–100	–125	–185	–250	–375

Note that: 1 Da = 1.66054×10^{-27} kg and 1e = 1.6022×10^{-19} C.

k_2=9.159×10^{-7} Cm/Ns, and n=0.333 were computed. In addition, it is known that the density of the gel is 1.2×10^{3}kg/m^3 and the density of the proteins is 1.41×10^{3}kg/m^3.

1. Problem: Compute the force of gravity on a 25kDa protein.

$$m = (25{,}000 \text{ Da})\left(1.66054\times10^{-27}\text{kg}\right) = 4.15135\times10^{-23}\text{kg}$$

$$F_g = m * g = \left(4.15135\times10^{-23}\text{kg}\right) * (9.8 \text{ N/kg}) = 4.068323\times10^{-22} \text{ N}$$

2. Problem: Compute the buoyant force on a 25kDa protein in this GE experiment.

$$V = m/\rho = 4.15135\times10^{-23}\text{kg}/1.41\times10^3\text{kg/m}^3 = 2.94421986\times10^{-26}\text{m}^3.$$

$$F_B = \rho_f V g = (1.2\times10^3\text{kg/m}^3)(2.94421986\times10^{-26}\text{m}^3)(9.8 \text{ N/kg}) = 3.46240255\times10^{-22}\text{N}$$

3. Question: Since the proteins will fall through the gel, at a very slow rate, even when the electric field is off, is the force of gravity greater than, less than, or equal to the buoyant force?
   A: The force of gravity is greater than the buoyant force on the protein in the gel.

4. Problem: Compute the Electric force on a 25kDa protein in this GE experiment.

$$F_E = q * E = \left(125 * 1.6022\times10^{-19}\text{C}\right)(20 \text{ N/C}) = 4.0055\times10^{-16}\text{N}$$

5. Problem: Compute the terminal velocity of a 25kDa protein in this GE experiment.
   Going back to the expression of the force equation solved for terminal velocity:

$$v_t = \frac{|\chi|EL}{8\pi k_3 L k_1 L^n} - k_2 E, \text{ since we know } |\chi|EL = F_E = 4.0055\times10^{-16}\text{N}$$

We can plug in all the values we know:

$$v_t = \frac{4.0055\times10^{-16}\text{N}}{8\pi\ 0.7275\times10^{-3}\text{Pa} * \left(\text{s/m}^{1.333}\right)\left(125\times10^{-9}\text{m}\right)^{1.333}} - \left(9.159\times10^{-7}\frac{\text{Cm}}{\text{Ns}}\right)20\frac{\text{N}}{\text{C}}$$

$$v_t = 1.6548\times10^{-5}\text{m/s}$$

6. Question: A protein with a mass of 15kDa will reach a terminal speed, which is greater than, less than, or equal to the terminal speed of the 25kDa protein in the same GE experiment.
   Answer: A smaller protein will reach a higher terminal velocity, because even though the electric force on a smaller protein is less than that on a larger protein the drag forces on a smaller protein decreases by a greater per length percentage than the electric force decreases. This results in a greater per mass difference between the downward electric forces and the upward drag forces on a smaller protein.
7. Question: A protein with a mass of 15kDa will have a net force on it which is greater than, less than, or equal to the net force on the 25kDa protein in the same GE experiment.
   Answer: This is not a trick question, but it is tricky. Remember that both these proteins are moving at a terminal (fixed) velocity so both have a zero acceleration and thus both

have a zero net force. So, the answer is the net force on each is zero, so the net forces are technically equal.

BIBLIOGRAPHY-GEL ELECTROPHORESIS

Arianna Marucco, I. F. (2013). Interaction of fibrinogen and albumin with titanium dioxide. In *Journal of Physics: Conference Series* (Vol. 429, No. 1, p. 012014). IOP Publishing.

Clipper Controls Inc. (2017). Dielectric Constant Values. Retrieved December 2017, from Clipper Controls.

Fischer, H., Polikarpov, I., & Craievich, A. F. (2004). Average protein density is a molecular-weight-dependent function. Protein Sci., *13*(10), 2825–2828.

Marko, J., & Ansari, A. (2013). Basic Physical Scales Relevant to Cells and Molecules. Retrieved December 2017, from Spring 2013 Phys 450/BioE 450: Molecule and Cell Biophysics (University of Illinois at Chicago): http://physicsweb.phy.uic.edu/450/MARKO/N003.html

Milo, R., & Phillips, R. (n.d.). How big is the "average" protein? Retrieved December 2017, from Cell Biology by the Numbers: http://book.bionumbers.org/how-big-is-the-average-protein/

Neelakanta, P. S. (1995). Handbook of Electromagnetic Materials: Monolithic and Composite Versions and Their Applications. Boca Raton, FL: CRC Press.

PubChem. (n.d.). Sodium dodecyl sulfate. Retrieved December 2017, from Pub Chem.

Vallari, R. (2016). Private Communication, Saint Anselm College.

Viney, C., & Fenton, R. A. (1998). Physics and gel electrophoresis: using terminal velocity to characterize molecular weight. Eur. J. Phys., *19*, 575–580.

12 Circular Motion and Centripetal Force

12.1 INTRODUCTION

Until this point in the volume, the motion of the objects studied has been predominantly in one dimension, so the change in the velocity of an object was focused on the change in the magnitude of the velocity vector. In this chapter, the change in velocity of an object in uniform circular motion is all about the change in the direction of the object's velocity. Therefore, this motion, which is common in nature, requires its own analysis. First, by studying the change in the direction of the velocity vectors of an object moving in a circle at a constant speed the centripetal acceleration is derived. From the acceleration the associated net force, known as the centripetal force, is explained. Then, a series of examples employed in these concepts are provided in which the forces of tension, friction, gravity, electrostatics, and magnetism are involved.

Chapter question: A centrifuge is a device that separates solutions, like blood, into its different constituents by spinning the solution at high speeds. The solution is poured into test tubes, loaded into the centrifuge, and spun at a high rate until the constituents of the solution are separated. As a centrifuge spins faster, heavier particles in the solution move away from the center of the circle, toward the bottom of the test tube. In the case of blood, the denser red blood cells move to the outside of the circle with the largest radius r, as shown in Figure 12.1, which is often referred to as the bottom of the tube, the white cells and platelets move to the center of the tube, and the blood plasma moves to the inside, which is the top of the tube.

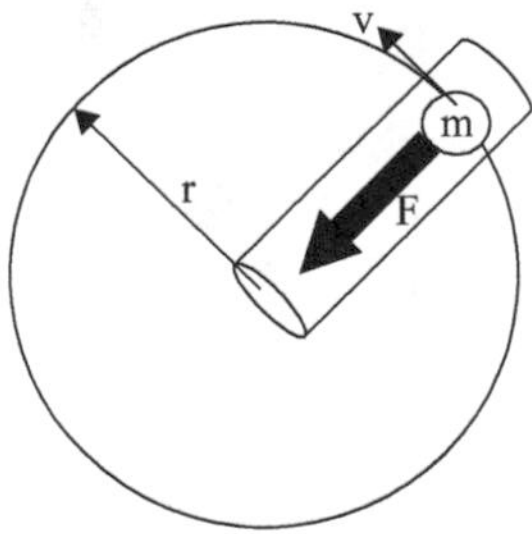

FIGURE 12.1 Chapter question diagram.

The question is, why do red blood cells move to the outside of the circular path that is associated with the bottom of the tube? After studying circular motion, an answer will be provided at the end of the chapter.

12.2 CENTRIPETAL ACCELERATION

An object that is moving in a circle is experiencing acceleration even if the speed of the object is constant, because the direction of the velocity is continually changing. In previous chapters, the motion was restricted to one direction so accelerations inferred a change in speed. In this chapter,

DOI: 10.1201/9781003308065-12

the change in the directional aspect of velocity results in an acceleration even though the speed remains the same.

In Figure 12.2, the location of a car traveling around a circular track at four equally spaced time intervals is shown along with the instantaneous velocity of the car at these four separate moments in time.

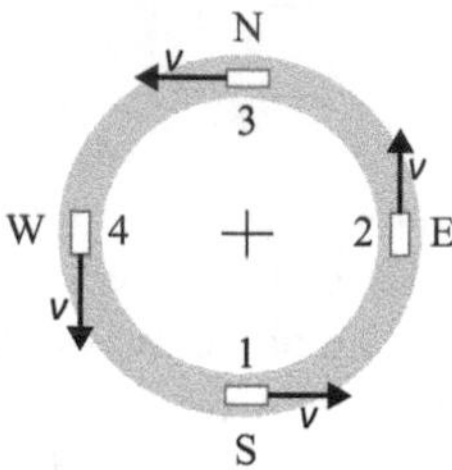

FIGURE 12.2 A car traveling at a constant speed around a circular track.

It is clear that the velocity vector of the car is changing even though the length of the velocity vectors is constant. Because the velocity is changing, the car is experiencing acceleration even though the speed is a constant.

In the sequence of diagrams, as the car goes around the track, its velocity vector is rotating, since it starts pointing East, then North, then West, then South, and back to East, and so on. The velocity vector completes one rotation in the time the car makes it once around the track, so the faster the car is going, the faster the velocity vector is rotating meaning the faster the velocity is changing (in direction) and hence the greater the acceleration. In a similar representation, the location of a ball on the end of a string that is being rotated in a circular path at a constant speed v is presented in Figure 12.3. Here locations 0 and 1 are at two instants of time t_0 and t_1.

The change in the velocity of the ball from locations 0 to 1 is $\overline{\Delta \mathbf{v}} = \vec{\mathbf{v}}_1 - \vec{\mathbf{v}}_0$. To find the acceleration of the ball, divide the change in the velocity by the time interval t_{01} from time t_0 to time t_1 during which that change occurs. This is the average acceleration of the ball

$$\vec{\mathbf{a}}_{c,\,AVG} = \frac{\overline{\Delta \mathbf{v}}}{t_{01}}$$

Keeping the direction of the velocity and position vectors constant for t_1 and switching the direction of the velocity and position vectors by 180° for t_0 to introduce a negative sign to the initial vectors, and sliding the arrows from head to tail to represent the subtraction of these vectors in the diagram, generates two similar triangles in Figure 12.3. It is clear that these triangles share the same angle θ,

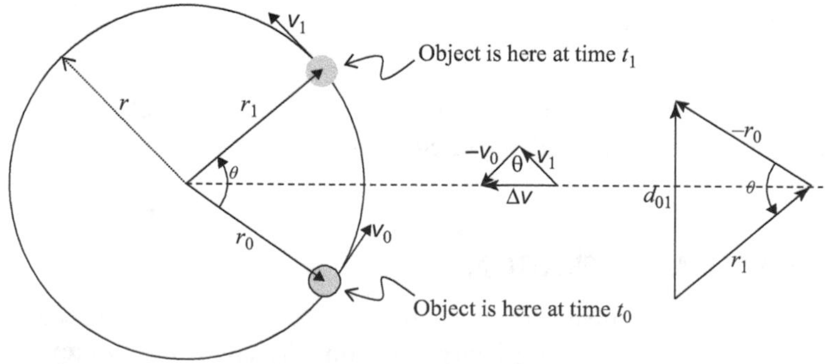

FIGURE 12.3 Geometry of centripetal acceleration.

since the velocity vectors are perpendicular to the position vectors and the position vectors are θ degrees apart. For similar triangles, the ratios of corresponding sides are equal: $\frac{\Delta v}{v_0} = \frac{d_{01}}{r_0}$

Which can be rearranged to give: $\frac{\Delta v}{v} = \frac{d_{01}}{r}$. In both cases, v replaced v_0 and r replaced r_0 because the radius and speed are constants.

Solving for Δv yields: $\Delta v = \frac{v}{r} d_{01}$.

Dividing both sides by t_{01} yields: $\frac{\Delta v}{t_{01}} = \frac{v}{r} \frac{d_{01}}{t_{01}}$.

The left-hand side of the expression is a change in velocity over a change in time, which is what we call the centripetal acceleration a_c and the $\frac{d_{01}}{t_{01}}$ appearing on the right is the magnitude of the average velocity of the particle: $a_{c,\,AVG} = \frac{V}{r} v_{AVG}$. Since the magnitude of the average velocity is the same as the velocity, the right-hand side can be replaced with v^2, resulting in the expression for the magnitude of the *centripetal acceleration*, given in equation (12.1) as:

$$a_c = \frac{v^2}{r}. \tag{12.1}$$

Note the direction of $\overline{\Delta v}$ is directly toward the center of the circle. Thus, the direction of the centripetal acceleration is directed toward the center of the circular motion. This is the case no matter where the object is on the circle. The word centripetal means center-directed. That is why the acceleration associated with the changing direction of the velocity of an object in a circular path is the *centripetal* acceleration.

The result $a_c = \frac{v^2}{r}$ for the centripetal acceleration is good any time an object is moving in a circle. The speed of the object is given as v at the instant for which the centripetal acceleration is $a_c = \frac{v^2}{r}$. If the speed of the object is changing, then the object also has a component of acceleration tangent to the circle that is equal to the rate of change of the speed. That component of acceleration is called the tangential acceleration a_t.

As shown in Figure 12.4, if the speed of the object is increasing, the tangential acceleration, a_t, is in the same direction as the velocity and since it is moving in a circle there is a centripetal acceleration, a_c. If the speed of the object is decreasing, the tangential acceleration is in the direction opposite that of the velocity, but the centripetal acceleration is always still pointing toward the center of the circle. The total acceleration of an object in circular motion is the vector sum of the two component vectors, a_t and a_c, of acceleration. They are always at right angles to each other (whenever they are both nonzero), so they are relatively easy to add. Remember that if an object is moving around in a circle at constant speed, the tangential acceleration a_t is zero, so the only acceleration is the centripetal acceleration a_c pointing toward the center of the circle.

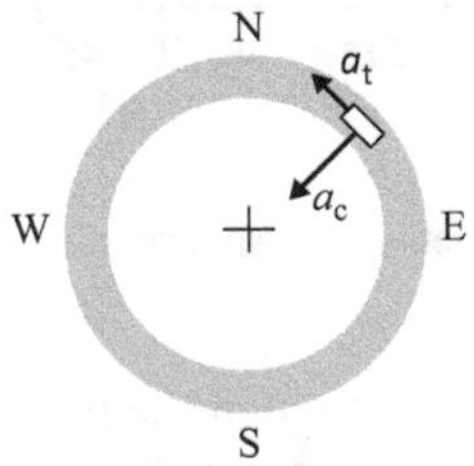

FIGURE 12.4 A car traveling counterclockwise (as viewed from above) around a circular track.

For this chapter, the focus is on circular motion of an object moving at a constant speed, so the tangential will always be zero. This motion of an object moving around in a circle path at constant speed is undergoing *uniform circular motion*, which means the object undergoes only centripetal acceleration, which means the net force on the object is directed toward the center of the circle. In Figure 12.2, note that the compass direction that is toward the center of the circle, and hence the compass direction of the centripetal acceleration, depends on where the car is in the diagram. When the car is at point 1, the centripetal acceleration is Northward. When the car is at point 2, the centripetal acceleration is westward, and so on. The direction of the centripetal acceleration is continually changing but always points toward the center of the circle. This means that the direction of the net force on the object is continually changing but always points toward the center of the circle.

The net force that is causing the centripetal acceleration is sometimes referred to as a centripetal force. It is important to realize that the centripetal force is not a separate force on its own but is the net force on an object that is moving in a circle. Thus, the centripetal force should never appear in a free-body diagram. In contrast, it is correct to include the centripetal acceleration vector in a free-body diagram, since it is the acceleration resulting from the net force.

12.3 HISTORICAL EXAMPLE OF CENTRIPETAL ACCELERATION

In the late 1600s, Isaac Newton studied the motion of the planets and the moon in an effort to produce the best possible star charts for naval navigation. During his studies, he realized that the planets revolve around the sun in orbits that are almost circular. It is true that a few years before Newton's work, Johannes Kepler published three laws of planetary motion that are:

1. The planets orbit the sun in elliptical orbits.
2. A line drawn from the planet to the sun sweeps out equal area in equal time.
3. The square of the period (T) of any planet is proportional to the cube of the planet's mean radius (R). $R^3 = K\,T^2$, where K, which Kepler found to be 3.35×10^{18}, is a constant, which is the same for all planets. A graph of the square of the period of the planet's orbit vs the cube of the period of the planet's orbits gives a line with a slope equal to K, as given in Figure 12.5.

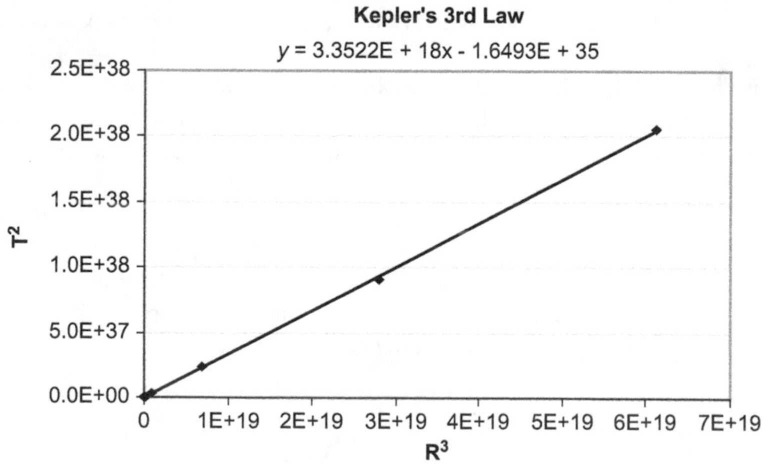

FIGURE 12.5 Plot of T^2 vs R^3 of the planets in the solar system.

What is not commonly known is that even though the planets orbit the sun in elliptical orbits, the eccentricities (a measure of how much an ellipse differs from a circle) of the planet's orbits are so small that that they can be considered circular (Figure 12.6).

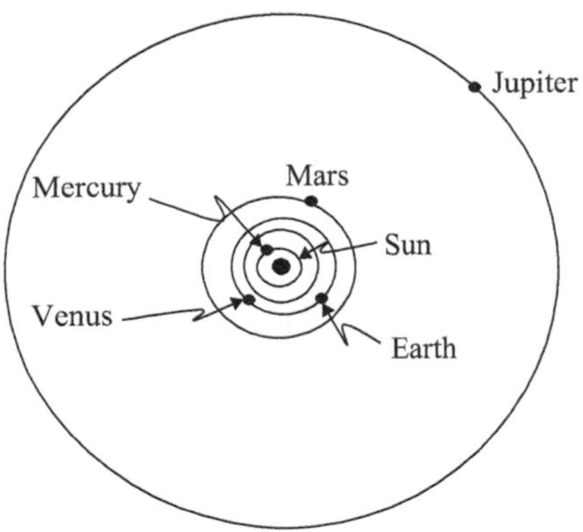

FIGURE 12.6 Orbits of five of the planets in the solar system closest to the sun.

Newton was able to start with his Second Law:

$$\vec{F}_{\text{Net}} = m\vec{a}$$

and set the net force pointing from each planet toward the sun, which is the center of the circle made by the path of the planet's orbit, to his force of gravity, $\frac{\boldsymbol{Gm_s m_p}}{\boldsymbol{r^2}} = ma$. It was Newton that also developed the concept of centripetal acceleration, which he then entered into his Second Law for the planets $\frac{Gm_s m_p}{r^2} = m_p \frac{v^2}{r}$

He canceled the mass of the planets and one of the radii from both sides and solved the equation for v^2: $\frac{Gm_s}{r} = v^2$ since the motion of the planets is circular and at an almost constant speed the velocity is the distance of one circumference of the circle ($2\pi r$) divided by the time it takes to travel around the circle that is called one period (T), so $\boldsymbol{v} = \frac{\boldsymbol{2\pi r}}{\boldsymbol{T}}$. If the velocity is squared and substituted back into the Second Law: $\frac{Gm_s}{r} = \boldsymbol{v}^2$, the result is $\frac{Gm_s}{r} = \frac{\boldsymbol{4\pi^2 r^2}}{\boldsymbol{T^2}}$, which can be rearranged to give:

$$r^3 = \left(\frac{Gm_s}{4\pi^2}\right)T^3$$

which is Kepler's Third Law. If the values of the universal gravitational constant, $G = 6.67 \times 10^{-11}$ Nm^2/kg^2, and the mass of the sun, 1.99×10^{30} kg, are entered into the expression in parentheses of Newton's derivation of Kepler's Third Law, the value of K is found to be 3.36×10^{18} m^3/s^2, which is the same value that Kepler found after 15 years of data analysis. It turns out that the centripetal acceleration analysis is the piece of the puzzle that Newton used to link his work on Gravitation and his work on the Three Force Laws.

12.4 EXAMPLES OF CENTRIPETAL FORCE ANALYSIS

Example 12.1: Gravitational Force as the Centripetal Force

The radius of the moon is 1,737 km, and its mass is 7.346×10^{22} kg. Find the speed of a satellite in a circular orbit having a radius that is 11 km greater than the average lunar radius, so that the satellite can orbit in the closest path to the moon without striking any part of its surface (Figure 12.7).

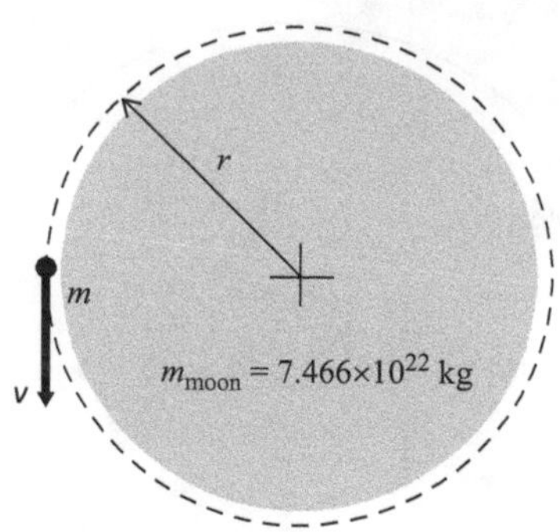

FIGURE 12.7 Diagram for Example 12.1.

Solution:

Gather important information:

$$r = r_{moon} + \text{height above surface}$$

$$r = 1{,}737{,}000 \text{ m} + 11{,}000 \text{ m} = 1{,}748{,}000 \text{ m}$$

Step 1: As shown in Figure 12.8, draw a free-body diagram of the satellite on the left side of the orbit.

The satellite is orbiting in a circle so the acceleration is toward the center of the moon, which is in the +x-direction.

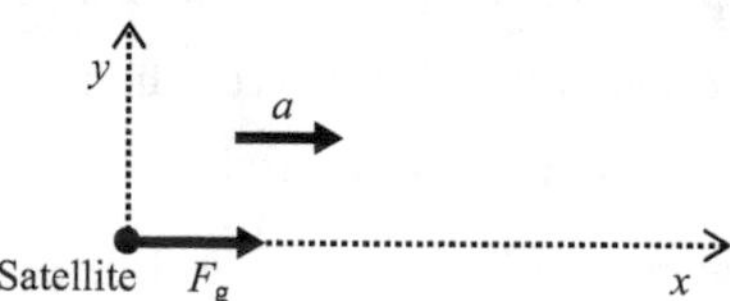

FIGURE 12.8 Free-body diagram for Example 12.1.

Step 2: Find the x-components of the forces on and acceleration of the satellite.

Since the acceleration is the centripetal acceleration: $a_x = a = \dfrac{v^2}{r}$

Since the force is the gravitational force: $F_{gx} = F_g = \dfrac{Gm_{moon}m}{r^2}$

Step 3: Apply Newton's Second Law for the x-components of the vectors:

$$\sum F_x = ma_x$$

$$F_{gx} = ma_x$$

$$\frac{Gm_{moon}m}{r^2} = m\frac{v^2}{r}$$

Step 4: Solve the expression for the velocity of the satellite v. Note that the mass of the satellite cancels out, so our final answer for the speed of the satellite by solving

$$v = \sqrt{\frac{Gm_{moon}}{r}}$$

Plug in the values:

$$v = \sqrt{\frac{6.67 \times 10^{-11} \frac{\text{Nm}^2}{\text{kg}^2}\left(7.346 \times 10^{22} \text{ kg}\right)}{1,748,000 \text{ m}}}$$

to get

$$v = 1,674 \frac{\text{m}}{\text{s}}$$

This is a speed of about 3,745 mph!!

Example 12.2: Electrical Force as the Centripetal Force

In the Bohr model of the hydrogen atom, depicted in Figure 12.9, the centripetal force takes the form of the electrostatic attraction computed with the Coulomb force between the electron and the proton.

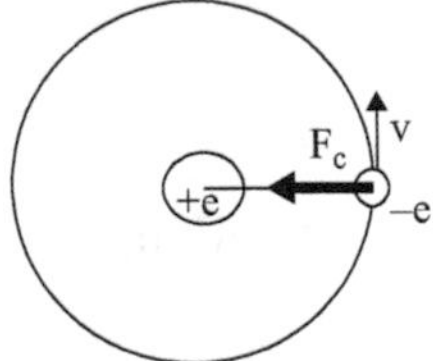

FIGURE 12.9 Diagram of the Bohr model of the hydrogen atom.

Assuming that the radius of the electron orbit is the Bohr radius, which has a value of $r=0.529$ Angstrom$=0.529 \times 10^{-10}$m, find the speed of the electron in its orbit.

Step 1: As shown in Figure 12.10, draw a free-body diagram of the electron on the right side of the orbit.

The electron is orbiting in a circle, so the acceleration is toward the center of the atom, which is the –x-direction.

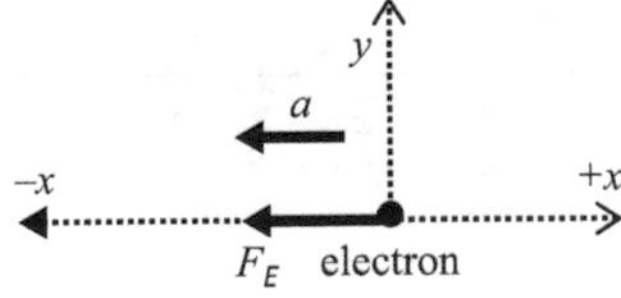

FIGURE 12.10 Free-body diagram for Example 12.2.

Step 2: Find the x-components of the forces on and acceleration of the electron.

Since the acceleration is the centripetal acceleration: $a_x = -a = -\frac{v^2}{r}$.

Since the force is the gravitational force: $F_{Cx} = -F_C = \frac{kq_pq_e}{r^2}$, where q_p and q_e are the magnitudes of the charge on the proton and electron, respectively, which are both equal to the value e = 1.602×10^{19}C.

Step 3: Apply Newton's Second Law for the x-components of the vectors:

$$\sum F_x = ma_x$$

$$F_{Cx} = ma_x$$

$$-\frac{kq_p q_e}{r^2} = -m_e \frac{v^2}{r}$$

Notice that the mass used in Newton's Second Law is the mass of the electron, since it is the charge that is accelerating in a circular path.

Step 4: Solve the expression for the velocity of the satellite v.

$$v = \sqrt{\frac{kq_p q_e}{m_e r}} = \sqrt{\frac{ke^2}{m_e r}}$$

Plug in the values:

$$v = \sqrt{\frac{9\times 10^9 \frac{\text{Nm}^2}{\text{C}^2}\left(1.602\times 10^{-19}\text{ C}\right)^2}{\left(0.529\times 10^{-10}\text{ m}\right)^2}}$$

to get

$$v = 2.186\times 10^6 \frac{\text{m}}{\text{s}}$$

Example 12.3: Magnetic Force as the Centripetal Force

Background: When a charged particle enters a region of space in which there is a magnetic field that exists in a direction perpendicular to the path of the charged particle, the particle will move into a circular path if the magnetic field is strong enough.

Referring to Figure 12.11, if a positively charged particle enters a region of space moving East with a magnetic field pointing into the page, the particle will experience a force Northward, based on the right-hand rule. That force will bend the particle until it is moving Northward. At that point, the force will be westward, and so on. As the force is always pointing toward the center of the circle, it is the centripetal force on the charged particle.

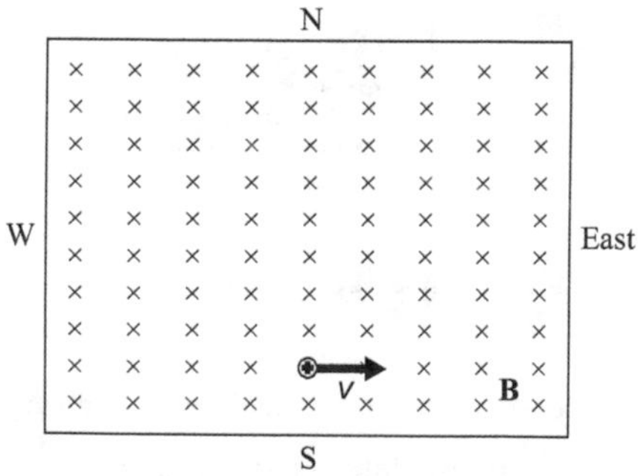

FIGURE 12.11 Diagram of a moving charged particle in a magnetic field.

Example: A proton, in a vacuum, is moving in a circular path due to a perpendicular magnetic field, with a magnitude of .500 T that points straight-downward. At one instant of time, depicted in Figure 12.12, the proton's velocity is 3.00×10^5 m/s due Eastward.

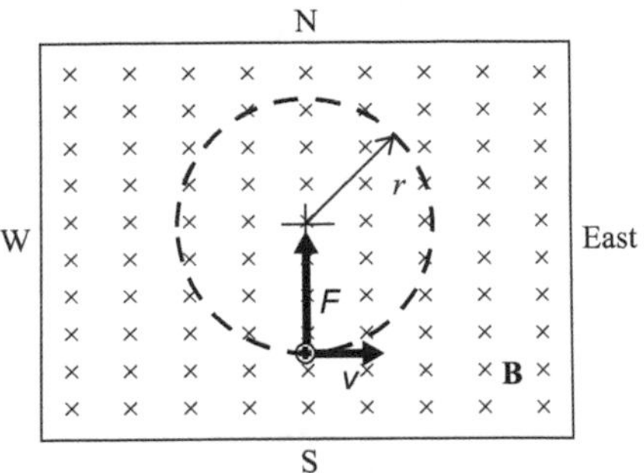

FIGURE 12.12 Diagram of a moving charged particle in a magnetic field and the circular path of the particle.

Find the radius of the path along which the proton is moving. Neglect any gravitational force that might be acting on the proton, since it will be perpendicular to the path of the circle and will be much smaller than the magnetic forces. Note that the charge of a proton is 1.6022×10^{-19}C and its mass is 1.6726×10^{-27}kg.

Solution:

Step 1: As shown in Figure 12.13, draw a free-body diagram of the proton at the point described in Figure 12.12.

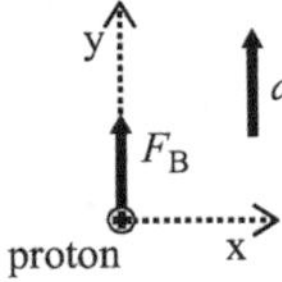

FIGURE 12.13 Free-body diagram for Example 12.3.

The proton is orbiting in a circle so the acceleration is toward the center of the circle, which is the +y-direction.

Step 2: Find the y-components of the forces on and acceleration of the proton.

Since the acceleration is the centripetal acceleration: $a_y = a = -\frac{v^2}{r}$.

Since the force is the magnetic force: $F_{By} = F_B = qvBsin(\theta)$, where q is the charge of the proton, v is the magnitude of the velocity of the proton, B is the magnetic field, and θ is the angle between the velocity of the proton and the direction of the magnetic field. For the proton to travel in a constant velocity circular path, the angle θ is 90°, so $F_B = qvB|\sin 90°| = qvB$.

Step 3: Apply Newton's Second Law for the y-components of the vectors:

$$\sum F_y = ma_y$$

$$F_{By} = ma_y$$

$$qvB = m\frac{v^2}{r}$$

Notice that the mass used in Newton's Second Law is the mass of the proton, since it is the charge that is accelerating in a circular path. Also, notice that there is a v on each side of the equation so one can be canceled from each side.

Step 4: Solve the expression for the radius of the proton's path *r*.

$$r = \frac{mv}{qB}$$

Solving for *r*, substituting values with units, and evaluating leads to our final answer:

$$r = \frac{\left(1.6726 \times 10^{-27}\,\text{kg}\right) 3 \times 10^{5}\,\text{m/s}}{\left(1.6022 \times 10^{-19}\,\text{C}\right).5\text{T}}$$

$$r = .006264\ \text{m}$$

$$r = 6.264\ \text{mm}$$

From these three examples with three different forces of nature, it should be clear that the process of solving problems of circular motion follows the same procedure independent of the forces involved. The same process can be used if the force is something simple like a tension in a rope.

Example 12.4: Tension Force as the Centripetal Force

A rope is tied to a harness on a 500 kg horse and a person holds the rope so that the radius of the circle that the horse makes as it trots around is 10 m. Assuming that the horse is trying to go straight at 4 m/s and the tension is keeping the horse in a circular path, find the tension in the rope (Figure 12.14).

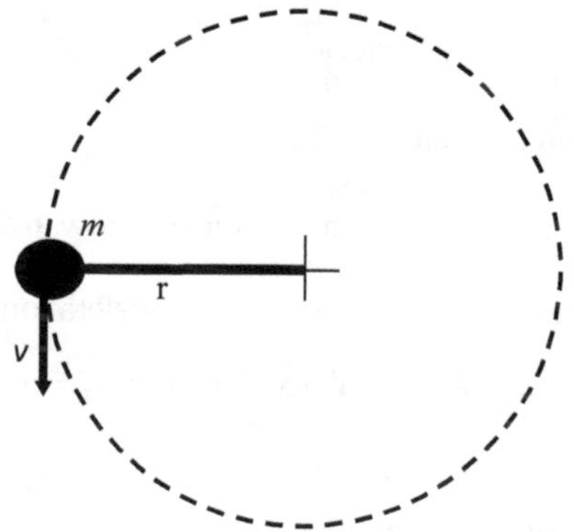

FIGURE 12.14 Diagram for Example 12.4.

Solution:

Step 1: As shown in Figure 12.15, draw a free-body diagram of the horse on the left side of the circle.

The horse is trotting in a circle so the acceleration is toward the center of the circle, which is in the +x-direction.

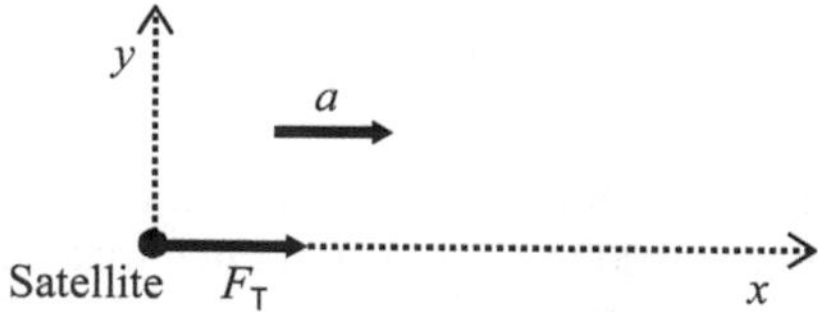

FIGURE 12.15 Free-body diagram for Example 12.4.

Step 2: Find the x-components of the forces on and acceleration of the satellite.

Since the acceleration is the centripetal acceleration: $a_x = a = \frac{v^2}{r}$.

Since the force is the gravitational force: $F_x = F_T$.

Step 3: Apply Newton's Second Law for the x-components of the vectors:

$$\Sigma F_x = ma_x$$

$$F_{Tx} = ma_x$$

$$F_T = m\frac{v^2}{r}$$

Step 4: Solve the expression for the tension in the rope, by plugging in the values in the previous equation.

$$F_T = (500\text{ kg})\frac{\left(4\ \frac{\text{m}}{\text{s}}\right)^2}{10\text{ m}} = 80\text{ N}$$

This is a speed of about 18 lb of force, so not too much for a person to apply.

12.5 CENTRIPETAL ACCELERATION AND ANGULAR VELOCITY

Since the object in uniform circular motion maintains a constant distance away from the center of the circle, its radius, and only the angle around the circle changes with time, it is sometimes easier to describe the motion in terms of quantities such as period, T, frequency, f, and angular velocity, ω.

First these quantities must be defined. The starting point of the definitions is that one cycle is defined as one complete trip around the circle.

Period, T, is the time it takes an object to complete one cycle. It is usually measured in seconds (s).

Frequency, f, is the amount of cycles per second, which is usually measured in cycles/s = Hertz = Hz

Angular velocity, ω, is the angular rate of revolution. It is found in two ways, which are mathematically equivalent.

First, ω can be computed by dividing the tangential velocity by the radius of the motion. $\omega = v/r$

Second, ω can be computed by multiplying the frequency by 2π, $\omega = 2\pi f$.

Since the distance around a circle is $2\pi r$ and the period is the time for one trip around the circle, these two ways of computing ω, given in equation (12.2), are equivalent.

$$\omega = \frac{v}{r} = \frac{(2\pi r/T)}{r} = \frac{2\pi}{T} = 2\pi f \tag{12.2}$$

As expressed in equation (12.3), the centripetal acceleration can be written in terms of the angular velocity of the object,

$$a_C = \frac{v^2}{r} = \omega^2 r \tag{12.3}$$

As shown in equation (12.4), the centripetal force can also be written in terms of the angular velocity,

$$F_c = ma_C = m\frac{v^2}{r} = m\,\omega^2 r \tag{12.4}$$

12.6 ANSWER TO THE CHAPTER QUESTION

12.6.1 The Centrifuge

A centrifuge is a device that separates solutions, like blood, into its different constituents by spinning the solution at high speeds. The solution is poured into test tubes, loaded into the centrifuge, and spun at a high frequency until the constituents of the solution are separated. As a centrifuge spins faster, heavier particles in the solution move away from the center of the circle, toward the bottom of the test tube. In the case of blood, the denser red blood cells move to the bottom of the tube, the white cells and platelets move to the center of the tube, and the blood plasma moves to the top of the tube. The question is, why do the red blood cells move to the bottom of the tube, which is on the outside of the circle of rotation, when the centrifuge is spinning?

Electrostatic forces understood in biology as hydrogen bonds, hold together the water, which makes up 90% of the plasma. The red blood cells, white blood cells, and platelets are suspended in the plasma as they flow through our bodies. If a sample of blood is left in a test tube, the denser red blood cells, which are full of hemoglobin that contains iron, sink to the bottom. So, it is clear that the buoyant force on red blood cells is not large enough to keep them afloat in the blood plasma. In a horizontal test tube at rest, the normal forces between the cells and between the cells and the plasma are the only horizontal forces acting on the cells. As the test tube is rotated, cells bump into each other and the inward pointing normal force between the cells becomes the inward net force, which is the centripetal force.

As the blood is spun in the centrifuge, the rate of rotation (ω), the radius (r), and the mass (m) of the cells dictate the centripetal force needed to keep the cells in a constant circular path. Referring to Figure 12.1, at one particular point in the test tube at a given radius (r) away from the center of rotation all the cells are moving with the same velocity (v). The more massive red blood cells have a greater inertia so they require a greater centripetal force ($F_c = mv^2/r$) to remain at the same radius as the white blood cells or the platelets. Thus, they move to the outer edge of the rotation, which is at the bottom of the tube. When the required centripetal force is greater than what can be provided, the object must move to a path of greater radius. So, the answer to the question is that the red blood cells move to the bottom of the tube, which is the outside of the circle set by the centrifuge, because they require the largest centripetal force to remain in the circular path.

Numerical Example

The terminal velocity of red blood cells in a vertical test tube containing a sample of blood is 10 mm/hour when that test tube is at rest on a test tube rack on a lab bench. What would be the terminal velocity, relative to the test tube, if the test tube were in a centrifuge spinning at 4,000 rpm? Assume the blood cells in question are at a distance of 12 cm from the axis of rotation.

Solution:

$$4{,}000\frac{\text{revolutions}}{\text{minute}} = 4{,}000\frac{\text{revolutions}}{\text{minute}}\frac{\text{minute}}{60\text{ s}} = 66.667\frac{\text{revolutions}}{\text{s}}$$

In one revolution, the location in the test tube at a distance $r = 12$ cm away from the axis of rotation travels a distance $2\pi r = 2\pi(.12\text{ m}) = .75398\text{ m}$. Multiply the number of revolutions per second by the distance per revolution, to get the speed of the tube as it travels along the circle on which it is moving.

$$v_F = 66.667\frac{\text{revolutions}}{\text{s}}\frac{.75398\text{ m}}{\text{revolution}} = 50.266\frac{\text{m}}{\text{s}}.$$

The centripetal acceleration of this location of the tube is:

$$a_c = \frac{v_F^2}{r} = \frac{(50.266 \text{ m/s})^2}{.12 \text{ m}} = 21{,}056 \frac{\text{m}}{\text{s}^2}$$

This is very large compared to the earth's gravitational field strength $\boldsymbol{g} = 9.8\frac{\text{N}}{\text{kg}} = 9.8\frac{\text{m}}{\text{s}^2}$. Measurement of acceleration in a centrifuge is often compared to the gravitational field strength and given in g's. This value can be expressed as:

$$\boldsymbol{g}_F = \frac{\boldsymbol{g}_F}{\boldsymbol{g}}\boldsymbol{g} = \frac{21056 \text{ N/kg}}{9.8 \text{ N/kg}}\boldsymbol{g} = 2{,}148.6\,\boldsymbol{g}$$

Assume the blood cells are subject to *viscous* drag, the terminal velocity in a test tube placed on a lab bench is given by solving equation (12.5) for the terminal velocity, v_t:

$$v_t = \frac{m_{cell}\boldsymbol{g} - (\rho_f V)\boldsymbol{g}}{6\pi r\mu} = \frac{m_{cell} - (m_{fluid})}{6\,\pi\, r_{cell}\,\mu}\boldsymbol{g} \tag{12.5}$$

where m_{DF} is the mass of the fluid displaced by a red blood cell. By inspection, if g is changed to 2,148.6g, then the new terminal velocity is 2,148.6 times the original terminal velocity:

$$v_t' = 2{,}148.6\, v_t = 2{,}148.6\frac{10 \text{ mm}}{\text{h}}\frac{\text{m}}{1{,}000 \text{ mm}}\frac{\text{h}}{3{,}600 \text{ s}} = .0059683\frac{\text{m}}{\text{s}} = 5.9683\frac{\text{mm}}{\text{s}}$$

So, at this speed, the red blood cells move about 6 mm every second, so it doesn't take long to move to the outside of the path, which is the bottom of the tube. It is important to check if the analysis of a viscous force is still acceptable for this much higher speed. Red blood cells are not spherical but disk-shaped, on the order of 7.2 μm in diameter and 2.3 μm thick, but in this calculation, they are approximated as spheres of diameter 5 μm to get an approximate value for the Reynolds number. The density of blood plasma is 1,025 kg/m^3 and the viscosity of blood is approximately 3.5×10^{-3} Pa s. Plugging these values into the equation for the Reynolds number results in:

$$\text{Re} = \frac{\rho_{FLUID}vd}{\mu} = \frac{1{,}025 \text{ kg/m}^3 (.0047744 \text{ m/s})\, 5\times 10^{-6}\text{m}}{3.5\times 10^{-3}\text{Pa s}} = .006991$$

which is well within the values for a laminar flow and a viscous drag.

12.7 QUESTIONS AND PROBLEMS

12.7.1 Multiple-Choice Questions

1. A car goes around a curve of increasing radius at a constant speed. Does the centripetal acceleration of the car as it rounds the curve, increase, decrease, or remain the same?
   A. increases B. decreases C. remains the same

2. A satellite moves in a circular orbit about the center of the earth at a constant distance from the center of the Earth. Thus, the satellite also has a constant speed, acceleration, and net force on it, or all three?
   A. speed B. acceleration C. net force D. all three are constant

3. In calculating the velocity at which a satellite must travel to maintain a specific orbital radius, the mass of the satellite is not needed for which of the following reasons?
   A. The mass of the satellite is extremely small compared to the mass of the Earth.
   B. The gravitational and inertial properties of the satellite's mass cancel each other.
   C. The gravitational force on the satellite is inversely proportional to the square of the orbital radius.

4. The G-7 and the CS satellite have approximately the same mass. The G-7 satellite is in a circular orbit around the Earth with an orbital radius of eight Earth radii, $R_{G7}=8\ R_e$. The CS satellite is in a circular orbit around the Earth, with an orbital radius of two Earth radii, $R_{CS}=2\ R_e$. The tangential speed of the G-7 satellite is greater than, less than, or equal to the speed of the CS satellite?
   A. greater than B. less than C. equal to

5. Consider two satellites in circular orbit about the moon at two different distances from the center of mass of the moon. Which satellite has the greater speed relative to the moon?
   A. The one in the bigger orbit B. The one in the smaller orbit

6. Consider two satellites in circular orbit about the planet Mars at the same distance from the center of the planet. Satellite A has a mass of 30 kg and satellite B has a mass of 60 kg. Which satellite is moving faster?
   A. Satellite A B. Satellite B C. Neither

(7–10) A ball, with a mass of 200 g, is attached to a string and the other end of the string is held at a constant location. The ball is swung around in a vertical circle or radius $r=30$ cm at a constant speed of 2 m/s, as depicted in Figure 12.16.

7. As the ball, in Figure 12.16, continues in its vertical circle when it reaches the bottom, traveling at the same speed, is the tension in the string greater than, less than, or equal to the tension in the string when the ball is at the top of the circle?
   A. greater than B. less than C. equal to

8. For the ball in Figure 12.16, if all other quantities are kept constant and the mass of the ball is increased, will the tension in the string at any point in the path increase, decrease, or remain the same?
   A. increase B. decrease C. remain the same

9. For the ball in Figure 12.16, if all other quantities are kept constant and the speed of the ball is increased, will the tension in the string at any point in the path increase, decrease, or remain the same?
   A. increase B. decrease C. remain the same

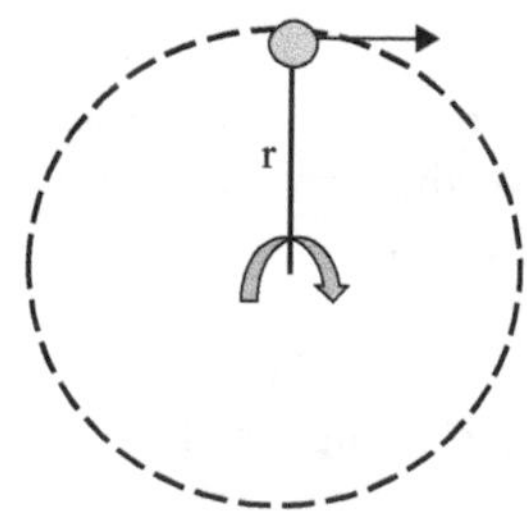

FIGURE 12.16 Diagram for multiple-choice questions 7–10 and problem 7.

10. For the ball in Figure 12.16, if all other quantities are kept constant and the radius of the path of the ball is increased, will the tension in the string at any point in the path increase, decrease, or remain the same?
A. increase B. decrease C. remain the same

12.7.2 Problems

1. A particle with a mass of 12.0 g is moving at a constant speed on a circular path with a radius $r = 5.20$ cm. The period of revolution is $T = 1.5$ seconds. Find the magnitude of the net force on the particle.
2. A space probe is in a circular orbit around Planet X at a constant altitude of 2.5×10^6 m, which has a measured radius of 3.0×10^6 m. If the probe takes 67.5 minutes to make one complete orbit, what is the mass of Planet X?
3. Compute the radius of orbit of a satellite in circular orbit about the earth moving with a speed of 5,000 m/s.
4. For a satellite in geosynchronous orbit about the earth:
   a. Calculate the radius of the orbit.
   b. Calculate the speed (relative to the center of the earth) of the satellite.
5. Find the radius of the orbit of a proton moving with speed 2.56×10^4 m/s in a uniform 1.50 T magnetic field given that the velocity of the proton is perpendicular to the magnetic field.
6. A satellite, with a mass of 500 kg, is in a circular orbit around a planet. The altitude of the satellite above the surface of the planet is a constant value of 2.5×10^6 m. The planet is spherical and has a uniform density. The mass of the planet is 6.00×10^{24} kg, and it has a radius of 3.0×10^6 m. Compute the time for one orbit of the satellite.
7. A ball, with a mass of 200 g, is attached to a string and the other end of the string is held at a constant location. The ball is swung around in a vertical circle at a constant speed of 2 m/s with a constant radius of 30 cm. Compute the tension in the string at the top of the loop.
8. A plane is performing a loop-de-loop, with a radius of 1 km. At the bottom of the loop, the normal force on the 70 kg pilot is directly up and is four times the weight of the pilot, so the pilot is pulling 4 g. Compute the speed of the plane at the bottom of this loop.
9. A singly ionized carbon (C) atom has a charge of $+1.6 \times 10^{-19}$ C and a mass of 19.9×10^{-27} kg. The ion of carbon (C^+) is moving in a circle at a speed of 2.84×10^5 m/s in a magnetic field that is pointing in a direction perpendicular to the circle made by the carbon ion. Find the radius and the period of the carbon ion traveling in a circular path in the magnetic field.
10. A particle of mass 35.0 μg is one of the particles of a suspension in water in a test tube in a centrifuge that is spinning at 475 rpm about a vertical axis, so the test tube is horizontal. The particle is 14.0 cm from the axis of rotation. Find the centripetal force exerted on the particle.

APPENDIX: MASS SPECTROSCOPY (SYNTHESIS OPPORTUNITY 2)

Mass spectroscopy is a device that is used by many scientists, from biochemists to physicists, to measure the mass-to-charge ratio of unknown molecules, including DNA. The basic parts of the device are shown in Figure 12.17.

The molecules to be studied are located in a chamber, named the molecule source in Figure 12.17. In the molecule source, the molecules are ionized by a beam of electrons that collide with the molecules and knock electrons out of the molecules leaving them electrically charged. In this chamber, there is a strong electric field that accelerates the molecules at a speed proportional to their charge and mass, as described in Figure 12.18.

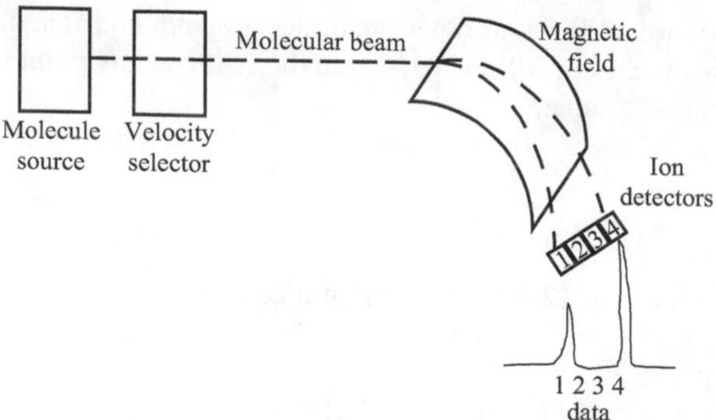

FIGURE 12.17 Diagram of a mass spectrometer.

The force on the molecule is found by multiplying the strength of the electric field (E) by the charge on the ionized molecule, the acceleration of the molecule is just the force on the molecule divided by the mass of the molecule, and the speed of the molecule can be found with knowledge of the acceleration of the molecule and the distance that it is in the area with the electric field.

Example 12.1

Find the acceleration of the molecules in the molecular source for a molecule that has a mass of 6.645×10^{-27} kg and a total charge of 3.204×10^{-19} C in an area with an electric field of 100 N/C and a length of 3.73 mm.

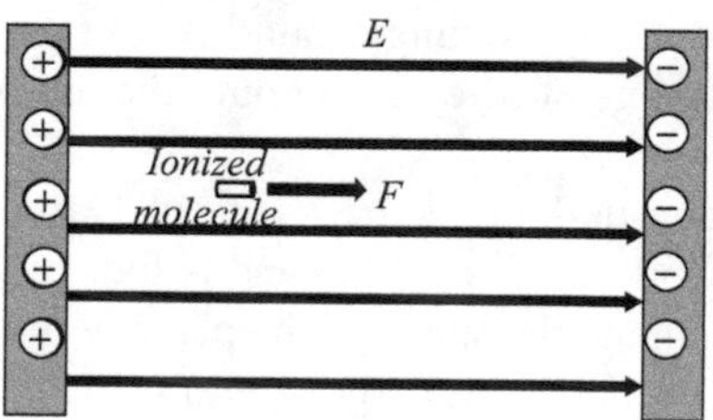

FIGURE 12.18 Molecular source chamber.

Step 1: Draw the free-body diagram of the molecule in the molecular source chamber, as shown in Figure 12.19.
The electric field is in the +x-direction.

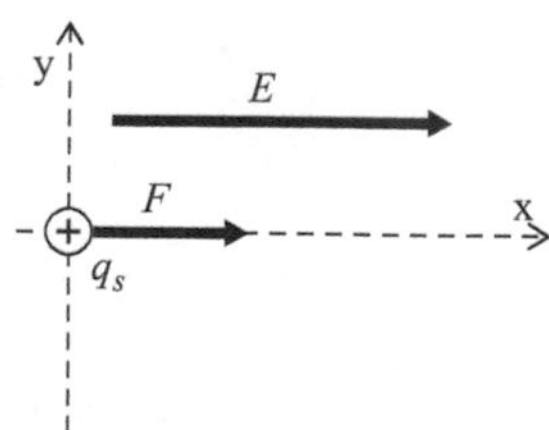

FIGURE 12.19 Free-body diagram of the molecule in the electric field of the molecular source.

Step 2: Compute the magnitude of the electric force on this particle.

$$F = qE = \left(3.204 \times 10^{-19}\text{C}\right)\ (100\ \text{N/C}) = \ 3.204 \times 10^{-17}\ \text{N}$$

Step 3: Compute the acceleration of the molecule.

$$F_x = ma_x$$

$$a_x = (F_x/m) = (3.204 \times 10^{-17}\text{N}/6.645 \times 10^{-27}\text{kg}) = 4.822 \times 10^{9}\text{m/s}$$

Step 4: Compute the velocity of the ion if it travels a distance of 3.73 mm with this acceleration. Using equation (7.7) solved for the velocity, the velocity of the molecules can be found as $v = \sqrt{2\left(4.822 \times 10^{9}\ \text{m/s}\right)\left(6.22 \times 10^{-3}\text{m}\right)} = 6{,}000\ \text{m/s}.$

The speed of the molecules depends on the mass and charge of the molecules along with the electric field, so molecules of different mass will have a different speed. To select out the molecules all traveling at the same speed, a device known as a *velocity selector* is employed. In this example, concepts from previous chapters are put together to explain how only charged particles with a specific velocity are selected for analysis with a mass spectrometer.

Example 12.2

Find the speed and direction in which the molecule in the previous section can cross a region of crossed electric and magnetic fields demonstrated in Figure 12.20.

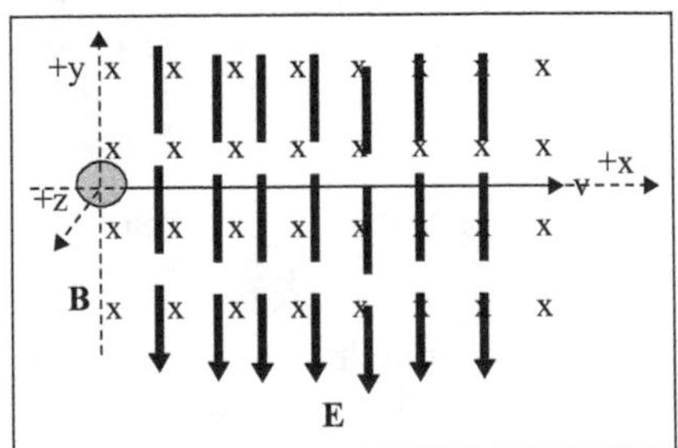

FIGURE 12.20 A molecule moving through crossed electric and magnetic fields.

Given that a region of space is filled with crossed (perpendicular) electric and magnetic fields. For this example, an electric field of 1,200 N/C is directed in the –y-direction as indicated by the vectors pointing toward the bottom of the page and a magnetic field of 0.2 T is directed in the –z-direction, as indicated by the x's in Figure 12.19. Remember the charge on the molecule is +3.204 × 10^{-19}C.

Step 1: Shown in Figure 12.21, a free-body diagram of the molecule in the crossed electric and magnetic fields.

For this particle to be moving at a constant speed, the magnetic force and the electric force must be balanced. Since the molecule has a positive charge and the electric field is in the –y-direction, the electric force on the molecule is also in the –y-direction. With a magnetic field into the page and a velocity of the particle to the right, the magnetic force on the particle is given by the RHR in the +y-direction.

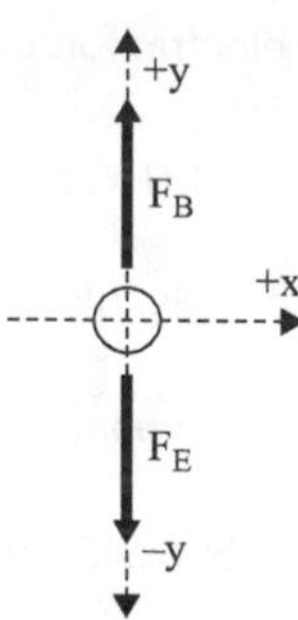

FIGURE 12.21 Free-body diagram of the molecule in the crossed electric and magnetic fields of the velocity selector.

Step 2: Write the Statics in the y-direction for this molecule:

$$F_{B\_y} - F_{E\_y} = 0N$$

Step 3: Since the angle between the magnetic field and the velocity is 90°, the magnetic force is $B\,q\,v$, and the electric force is $q\,E$, the static equations become:

$$B\,q\,v = E\,q$$

Step 4: Since both forces depend on charge, it can be canceled from each side, and it is trivial to solve for v and then plug in the numbers from the example to find the speed of the molecule that can travel through the crossed fields in a straight line since it has a zero net force.

$$v = E/B = (1{,}200\ \text{N/C})/(0.2\ \text{T}) = 6{,}000\ \text{m/s}$$

Notice this is the same speed as the molecule that left the molecule source chamber in step 1 of this process. There will be many other molecules traveling at the same and different speeds, some with the same mass and some with greater or lesser mass depending on where each molecule was in the molecular source, so that the force had a different amount of distance to accelerate each molecule. Therefore, there should be a group of molecules all with different masses exiting the velocity selector all traveling at the same speed.

A beam of molecules with different masses all traveling at the same speed enters the magnetic field of the mass spectrometer. Assume the magnetic field for the mass spectrometer in Figure 12.22 is into the page, and the ions are moving to the right, as shown in Figure 12.22.

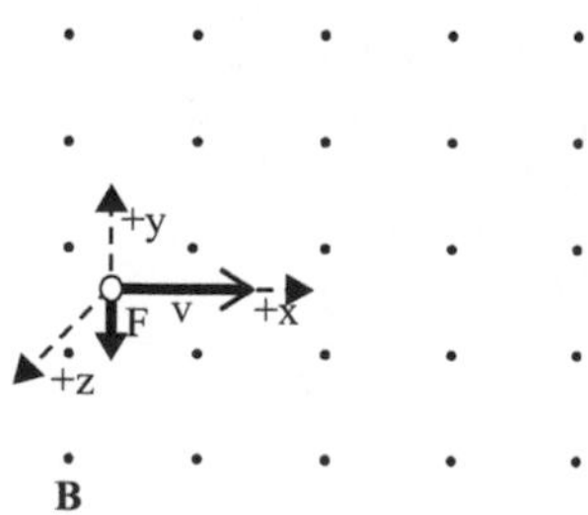

FIGURE 12.22 Diagram of the molecule in the magnetic fields of the mass spectrometer.

Step 2: The expression of the magnitude of the magnetic force on the molecules due to the magnetic field is given in Chapter 6 in equation (6.3)

$$F = B\ q\ v\ \sin(\theta)$$

Notice that the angle θ is 90° so the sin(90°) = 1. Following the steps in Example 3 of this chapter that explains each step of the analysis of the magnetic force as the centripetal force on a charged particle, the results are that the radius r of the path of a charged particle moving a magnetic field is:

$$r = \frac{mv}{qB}$$

The molecule studied in this example has a mass of 6.645×10^{-27} kg, a charge of 3.204×10^{-19} C, and a speed of 6,000 m/s. If the magnetic field is set at 1×10^{-2} T, the radius of curvature is:

$$r = \frac{\left(6.645 \times 10^{-27}\,\text{kg}\right)\left(6{,}000\,\frac{\text{m}}{\text{s}}\right)}{\left(3.204 \times 10^{-19}\,\text{C}\right)\left(1.0 \times 10^{-2}\,\text{T}\right)} = 1.244 \times 10^{-2}\,\text{m} = 1.244\ \text{cm}$$

Assume that this is the radius of the path that corresponds to the particles reaching detector 1 of Figure 12.15, the radius of the path of the particles that reaches detector 4 is larger, so they are the more massive particles. This is how the mass spectrometer separates out particles by their mass.

13 Rotational Motion

13.1 INTRODUCTION

Rotational motion is the motion of an object spinning around an axis that passes through the object itself. This is to be contrasted with translational motion, which is the motion of an object moving through space in a straight or curved path without rotation. As demonstrated in Figure 13.1, a block sliding down an incline moves with only linear motion defined by a displacement, a velocity, and an acceleration. A disk rotating about a fixed axis moves with only rotational motion, but a ball rolling down an incline experiences both rotational and linear motion.

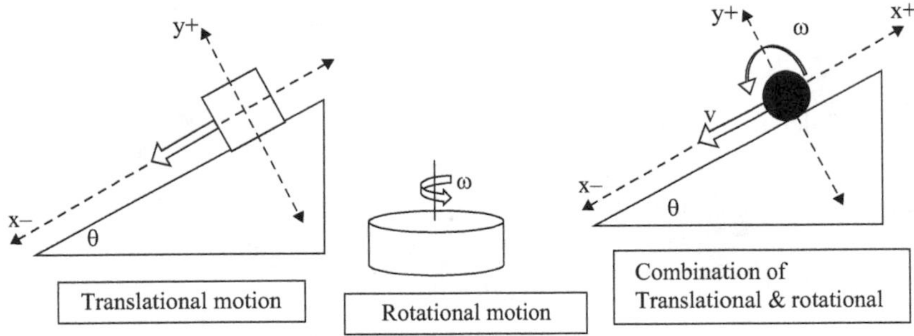

FIGURE 13.1 Examples of different types of motion.

As demonstrated in Chapter 12, when an object is traveling in a circular path, the concepts of translation kinematics are commonly applied to the analysis, but sometimes the concepts of rotational motion, period, frequency, and angular frequency can be applied. So, circular motion provides a transition between the language of translational and rotational motion, which is formalized in this chapter. In addition, rotational dynamics will formalize the connection between the net rotational force, or net torque, on an object and the angular acceleration of the object.

> **Chapter question**: There are bacteria that employ a rotating flagellum, tails that look a bit like a corkscrew, to propel themselves forward. In normal situations, these propulsion systems work well to move these bacteria forward through water. On the other hand, when a drop of water containing these bacteria is placed on a microscope slide, the bacteria begin to move in approximately circular paths at fairly constant speeds (Figure 13.2).

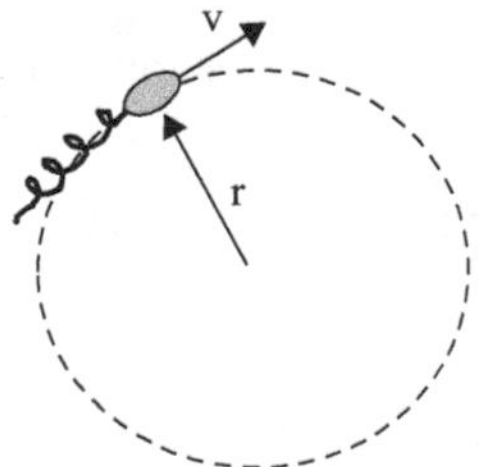

FIGURE 13.2 The path of bacteria moving in a circular path in a water droplet on a microscope slide.

DOI: 10.1201/9781003308065-13

Why does this happen to these bacteria? This question will be answered at the end of this chapter employing the concepts of rotational motion, which is the focus of this chapter.

13.2 ROTATIONAL KINEMATICS

13.2.1 The Radian

The formal analysis of rotational motion is based upon the concept of the angular unit of measurement of radians. As depicted in Figure 13.3, an angle θ, measured in radians, is the ratio of the arc length, s, around the curve to the radius r of the circle.

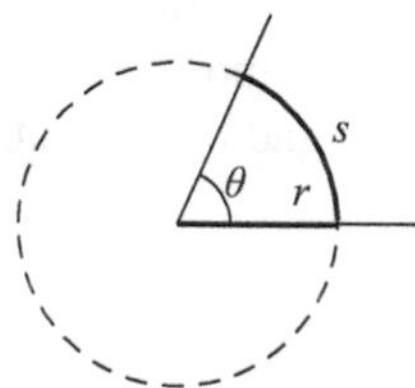

FIGURE 13.3 Definition of the radian.

This is expressed formally in equation (13.1) as:

$$\theta = \frac{s}{r} \tag{13.1}$$

This is best confirmed by plugging into equation (13.1) the arc length all the way around a circle, which is the circumference of the circle ($2\pi r$), and the angle in radians all the way around the circle is the well-known answer of 2π.

$$\theta = \frac{s}{r} = \frac{2\pi r}{r} = 2\pi \text{ Rad}$$

Although it is common to write "Rad" after the number associated with the angle computed in equation (13.1), it is clear that equation (13.1) generates a value of angle that has no units, because it is the ratio of two lengths. So, the radian is not truly a unit; it is just a placeholder that confirms how the angle is measured. Thus, sometimes the Rad is dropped from an expression or part of a computation without explanation.

Example (Radians)

A person walks part of the way around a circle of radius 5.00 m. The angle between a line from the center of the circle to her starting position and a line from the center of the circle to her ending position is 1.00 radians. What is the total distance traveled by the person?

Solution: Having traveled on the circle, the distance traveled is the arc length corresponding to the given angle.

Starting with equation (13.1) for an angle in radians,

$$\theta = \frac{s}{r}$$

solve for the arc length:

$$s = r\theta$$

Substituting values with units yield:

$$s = (5.00\ \text{m})\ 1.00\ \text{radians} = 5.00\ \text{m}$$

Notice that because the radian is an indicator rather than a true unit, it is eliminated in the final answer.

When analyzing a system with rotational kinematics and the angle is given in degrees or some fraction of a rotation, it must be converted to radians. The conversion factor for angles is, 1 rotation = 2π radians = 360°.

13.2.2 Angular Displacement

Consider a rigid body that is free to rotate about a fixed axis. The coordinate system is fixed in space so that it cannot rotate with the object and the z-axis is defined along the axis of rotation. The rigid body is observed from a position that has a positive z-coordinate and is on the z-axis. A reference line is attached to the object for purposes of establishing the angular position of the object at a specific instant of time. The reference line extends from the axis of rotation, perpendicular to the axis of rotation, out to or toward the outer perimeter of the object. The angle that the reference line on the object makes with the +x-direction (measured positive counterclockwise from our viewpoint) is the angular position of the object.

The symbol θ with a single subscript to represent the angular position of the object is used in Figure 13.4 as θ_1 at a time designated t_1.

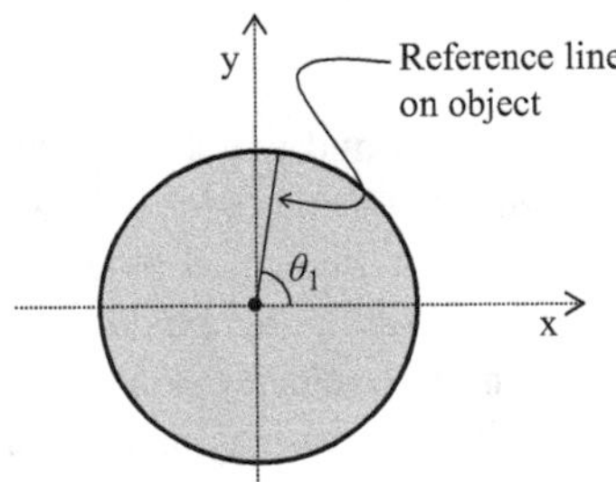

FIGURE 13.4 Position of a rotating rigid body.

The angular displacement that an object undergoes from time t_0 to t_1 is the change in its angular position from time t_0 to time t_1 and symbolized with a double subscript as follows, θ_{01}.

As depicted in Figure 13.5, the angular displacement is from one angular position to another.

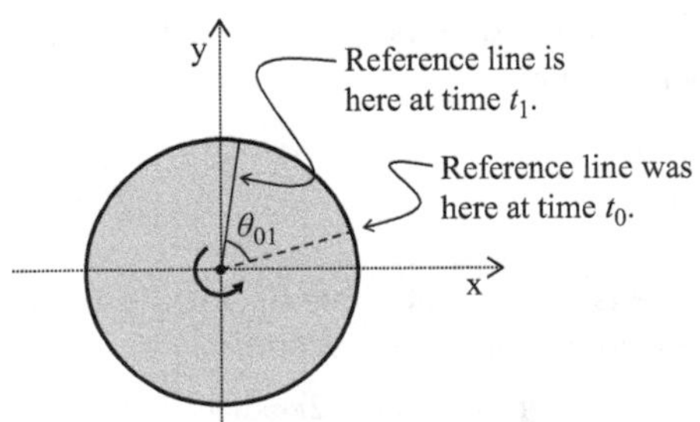

FIGURE 13.5 Displacement of a rotating rigid body.

To represent this vector displacement, a right-hand rule is employed. If you point the thumb of your right hand at your nose, you will be able to wrap the fingers of your right hand in a counterclockwise direction. So a rigid body that is rotating in a counterclockwise direction is said to have a angular displacement in the +z-direction, which is coming out of the page. (There is nothing coming out of the page, this is just a short-hand way of saying counterclockwise as viewed from above.) In the case depicted in Figure 13.5, the angular displacement is in the +z-direction, so the z-component of the displacement is: $\theta_{01z} = +\theta_{01}$.

If the rigid object was rotating in a clockwise direction as observed above, you would need to point the thumb of your right hand at the page and then you could rotate the fingers of your right hand in a clockwise direction; this rotation is in the -z-direction. So, with objects rotating in the x-y plane, the angular displacement vector will point in the –z-direction or the +z-direction.

A total angular displacement from $\vec{\theta}_0$ at time t_0 to an angular displacement of $\vec{\theta}_1$ at time t_1 is given in equation (13.2) as:

$$\vec{\theta}_{01} = \vec{\theta}_1 - \vec{\theta}_0 \tag{13.2}$$

The SI unit for angular displacement is the radian (rad).

Example (Angular Displacement)

What is the magnitude of the angular displacement, in radians, of a bicycle wheel that completes 20 rotations?

Solution: $\theta_{01} = 20 \text{ rotations} = (20 \text{ rotations}) \dfrac{2\pi \text{ radians}}{\text{rotation}} = 125.66 \text{ radians}$

13.2.3 Angular Velocity

The *angular velocity* of an object is how fast and which way the angular position of the object is changing. Equivalently, the *angular velocity* of an object is how fast and which way that object is spinning at a particular instant in time. It is represented by a vector ($\vec{\omega}$) with a magnitude of the spin rate of the object and whose direction is along the axis of rotation that corresponds to the sense of rotation by the right-hand rule. The angular velocity of an object is sometimes referred to as the *instantaneous angular velocity* of that object to distinguish it from the average angular velocity of the object over some time interval. The average value-with-direction of the angular velocity $\vec{\omega}$ over some time interval, say from time t_0 to time t_1, is obtained by dividing the total angular displacement $\vec{\theta}_{01}$ that occurs during that time interval by the duration of the time interval $t_{01} = t_1 - t_0$ and is presented in equation (13.3) as:

$$\vec{\omega}_{\text{Avg}01} = \vec{\theta}_{01} / t_{01} \tag{13.3}$$

The SI unit for angular velocity is the radian per second (rad/s). A common non-SI unit of angular velocity is rpm (revolutions per minute).

Example (Angular Velocity)

The crankshaft of a car engine is spinning at a rate of 2,500 rpm. What is the magnitude of the angular velocity of the crankshaft in radians per second?

Solution: $\omega = 2{,}500 \text{ rpm} = \left(2{,}500 \dfrac{\text{revolutions}}{\text{minute}}\right) \dfrac{2\pi \text{radians}}{\text{revolution}} \left(\dfrac{\text{minute}}{60\text{s}}\right) = 261.8 \dfrac{\text{radians}}{\text{s}}$

13.2.4 Angular Acceleration

If the spin rate of a rotating object is increasing or decreasing, then the object has some angular acceleration. The *angular acceleration* of an object is how fast and which way the angular velocity of the object is changing. It is represented by a vector ($\vec{\alpha}$, the Greek letter alpha) whose magnitude is how fast the angular velocity is changing and whose direction is which way the angular velocity is changing. If the angular velocity of the object is increasing, the angular acceleration is in the same direction as the angular velocity. If the angular velocity of the object is decreasing, the angular acceleration is in the direction opposite that of the angular velocity. The angular acceleration of an object is sometimes referred to as the *instantaneous angular acceleration* of that object to distinguish it from the average angular acceleration of the object over some time interval. The average value-with-direction of the angular acceleration $\vec{\alpha}$ over some time interval, from time t_0 to time t_1, is obtained by dividing the total change in angular velocity $\vec{\omega}_1 - \vec{\omega}_0$ that occurs during that time interval by the duration of the time interval $t_{01} = t_1 - t_0$. The expression of angular acceleration is given in equation (13.4) as:

$$\vec{\alpha}_{\text{Avg01}} = (\vec{\omega}_1 - \vec{\omega}_0)/t_{01} \tag{13.4}$$

The SI unit for angular acceleration is (rad/s)/s, which is usually written rad/s^2.

Given the focus on rotation about a fixed axis, the coordinate system is chosen so that the z-axis is along the axis of rotation so that the only nonzero component of the rotational motion vectors is the z-component.

13.2.5 Rotational Kinematic Equations

Similar to the kinematic equations for translational motion, there are kinematic equations for rotational motion. These rotational kinematic equations follow from the definitions of angular displacement, angular velocity, and angular acceleration. From our definitions for angular velocity and angular acceleration, the following relations can be established. As done in Chapter 7 for the translational kinematic equation, the starting point for an object undergoing constant angular acceleration (α_{01z}) during the time interval from time t_0 to time t_1 in the z-direction is just a rearrangement of equation (13.4) to give equation (13.5) as:

$$\omega_{1z} = \omega_{0z} + \alpha_{01z}\, t_{01} \tag{13.5}$$

Rearranging equation (13.3), the definition of average angular velocity gives an expression for angular displacement.

$$\theta_{01z} = \theta_{0z} + \omega_{\text{ave\_z}}\, t_{01}.$$

The common definition of average $(\omega_{0z} + \omega_{1z})/2$ can be inserted for average angular velocity and then equation (13.5) can be inserted for ω_{1z} to give equation (13.6) for angular displacement from time t_0 to time t_1 as:

$$\theta_{01z} = \theta_{0z} + \omega_{0z} t_{01} + \frac{1}{2}\alpha_{01z} t_{01}^2 \tag{13.6}$$

Equation (13.5) can be solved for the time interval t_{01} and substituted into equation (13.6) to find an expression of the angular velocity in terms of angular displacement and angular acceleration in equation (13.7) as:

$$\omega_{1z}^2 = \omega_{0z}^2 + 2\alpha_{01z}\theta_{01z} \tag{13.7}$$

These expressions, like the ones for translational kinematics, assume that the angular acceleration is a constant in the time interval under consideration. If the angular acceleration is changing, these

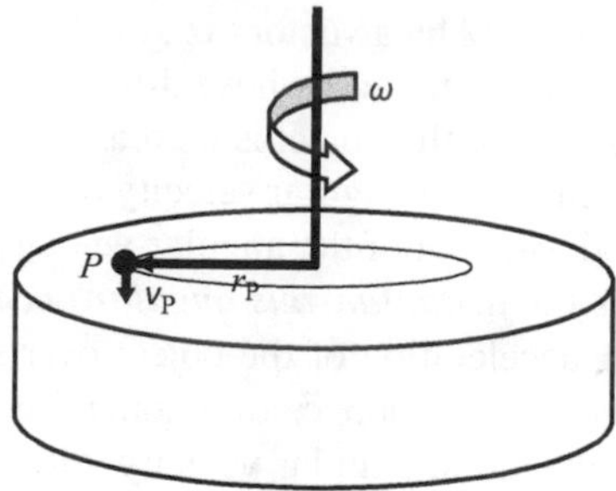

FIGURE 13.6 r_P is the distance point P from the axis of rotation, v_P is the speed of the particle, and ω is the magnitude of the angular velocity of the spinning object.

expressions are not valid and a calculus-based solution must be employed. The best way to understand this relationship is to consider a simple system like the one in Figure 13.6.

When a rigid body is rotating, every particle (except those on the axis of rotation) of that rigid body is going around in a circle. That is, the individual particles of the object are in translational motion. All the particles making up a rigid body in rotation complete one circular orbit about the axis of rotation in one and the same amount of time. However, particles farther from the axis of rotation have farther to go to make it around their circle once, thus, they must be going faster.

Starting with the expression for magnitude of the average angular velocity: $\omega_{Avg01} = \theta_{01}/t_{01}$. For the point P, the spinning disk in Figure 13.6, the symbol r_P represents the distance that the point P is away from the axis of rotation. Multiplying both sides of our expression for the magnitude of the average angular velocity by r_P yields:

$$r_P\ \omega_{Avg01} = r_P\ \theta_{01}/t_{01}$$

From our definition of angular measure of displacement in radians, equation (13.1), the product r_P θ_{01} that appears on the right is equal to the length of the arc that the point P travels along associated with the angular displacement θ_{01}. In other words, $s_{P01} = r_P\,\theta_{01}$. Replacing $r_P\,\theta_{01}$ with s_{P01} in the expression, we have been working on yields:

$$r_P\ \omega_{Avg01} = s_{P01}/t_{01}$$

The expression s_{P01}/t_{01} appearing on the right is the total distance traveled by point P during the time interval t_{01}, divided by the duration of time interval t_{01}, which is the average speed of point P during time interval t_{01},

$$r_P\omega = v_P$$

which is commonly rearranged in equation (13.8) as:

$$v_P = \omega r_P \tag{13.8}$$

Note that the angular velocity is a characteristic of the motion of the rotating rigid body as a whole and hence there is no subscript on the ω in our expression for the speed of a point P of the disk, but both the speed and the distance from the axis of rotation are specific to a particular point (P) of the rigid body and hence the symbols (v_P and r_P) used to represent them do have the subscript P. Furthermore, note that the instantaneous velocity v_P is tangent to the circle on which the point is moving and hence can be considered to be the tangential velocity of the point P. The point P, being part of a rigid body, is a fixed distance from the axis of rotation, meaning it can't have any radial velocity (radial velocity would be a component of the velocity directed straight toward or straight away from the axis of rotation, along the radius of the circle that the particle is moving on), so the tangential velocity is the only component of the velocity of the particle.

There is a similar relation between the tangential acceleration $a_{P,t}$ of a point on a rotating rigid body and the angular acceleration of the rigid body. The relation is given in equation (13.9) as:

$$a_{P,t} = \alpha r_P \tag{13.9}$$

The derivation of equation (13.9) is similar to that for equation (13.8) and is as follows.

Given that at time t_0 and at time t_1 the velocities of point P is $v_{P0} = \omega_0 r_P$ and $v_{P1} = \omega_1 r_P$, respectively.

The change in the velocity in the time interval from t_0 to t_1 is: $v_{P1} - v_{P0} = \omega_1 r_P - \omega_0 r_P$.

The radius can be factored out: $v_{P1} - v_{P0} = (\omega_1 - \omega_0) r_P$

Dividing each side by the time interval t_{01}: $(v_{P1} - v_{P0})/t_{01} = [(\omega_1 - \omega_0/t_{01}] r_P$

The change in velocity over a time interval is the translational acceleration and the change in angular velocity over a time interval is angular acceleration: $a_{P,t} = \alpha r_P$, which is just equation (13.9).

In summary, a disk that is rotating, as shown in Figure 13.7, has rotational variables of (θ, ω, and α) and the linear motion of the point of interest on the edge of the same rotating disk has the translational variables of (s, v, and a).

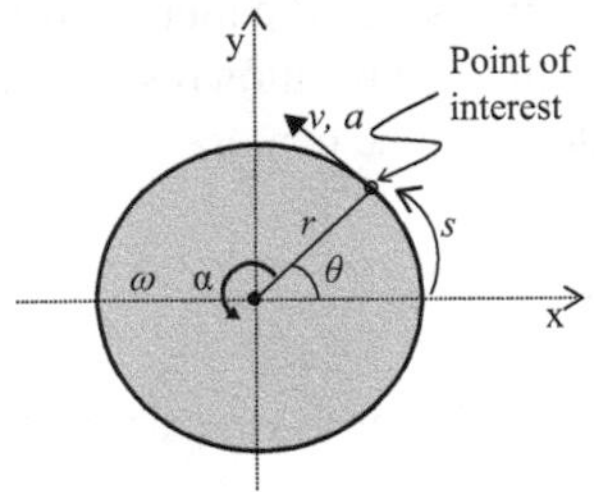

FIGURE 13.7 Rotational kinematics.

The angular displacement of the line extending from the center of disk to the point of interest (θ) is equal to displacement along the arc (s) along the outer edge of the disk divided by the radius of the motion, $\theta = \frac{s}{r}$. The angular velocity (ω) is just the translational velocity (v) divided by the radius of the motion of the point of interest, $\omega = \frac{v}{r}$, and the angular acceleration (α) is just the translational acceleration (a) divided by the radius of the motion of the point of interest, $\alpha = \frac{a}{r}$. As a result, the rotational kinematic equations are generated by writing the displacement in the translational equations as the displacement along the arc length (s) of interest and dividing every term in the translational equations by the radius (r) of the motion. This is summarized in Table 13.1.

TABLE 13.1
Kinematic equations for translational and rotational motion

Translational (Linear)	Rotational
$s = v\,t$	$\theta = \omega\,t$
$s = s_o + v_i\,t + ½\,at^2$	$\theta = \theta_o + \omega_i\,t + ½\,\alpha\,t^2$
$v = \Delta s/\Delta t$	$\omega = \Delta\theta/\Delta t$
$v_f = v_i + at$	$\omega_f = \omega_i + \alpha\,t$
$v_f^2 = v_i^2 + 2\,as$	$\omega_f^2 = \omega_i^2 + 2\alpha\theta$
$a = \Delta v/\Delta t$	$\alpha = \Delta\omega/\Delta t$

The expressions in Table 13.1 help us explain some simple situations. For example, the merry-go-round, in Figure 13.8, is undergoing only rotational motion, but the children are experiencing both linear and rotational motion.

FIGURE 13.8 Two children on a merry-go-round.

The children both have the same rotational motion, since they go around the same amount of times per minute regardless of where they sit on the merry-go-round, but the child (1) furthest away from the rotational axis will have a higher translational speed, since this child travels around a bigger circle than the child (2) sitting closer to the center.

Example (Rotational Kinematics)

The disk in Figure 13.7 has a radius of 0.5 m and is initially spinning with a constant angular velocity of 10 rad/s.

a. Through what angular distance does it turn in 2 seconds?
   Since the disk is spinning counterclockwise the rotational components are in the +z-direction.

$$\theta_z = \theta_{oz} + w_{iz}t + \tfrac{1}{2}a_z t^2 = 0 \text{ rad} + (10 \text{ rad/s})(2 \text{ s}) + \tfrac{1}{2}(0)(2 \text{ s})^2 = 20 \text{ rad}$$

b. Through what translational distance does it turn in 2 seconds?

$$s = \theta r = (20 \text{ rad})(0.5 \text{ m}) = 10 \text{ m}$$

c. If the brakes are applied and the wheel is stopped in 5 seconds, what is its rate of angular acceleration?

$$\alpha_z = \Delta w_z / \Delta t = (0 - 10 \text{ rad/s})/(5 \text{ s}) = -2 \text{ rad/s}^2.$$

d. What is the tangential acceleration of a point on the outer edge of the wheel at the point of interest?

$$a = \alpha_z r = (-2 \text{ rad/s}^2)(0.5 \text{ m}) = -1 \text{ m/s}^2 \text{ along the arc.}$$

e. Through what angular distance did it turn as it stopped (from the point in which the brakes were applied to the point at which it stopped?)

$$\omega_{fz} = \omega_{iz} + \alpha_z t \rightarrow t = (\omega_f - \omega_i)/\alpha_z = (0 \text{ rad/s} - 10 \text{ rad/s})/-2 \text{ rad/s}^2 = 5 \text{ s}$$

$$\theta_z = \theta_{oz} + \omega_{iz}t + \frac{1}{2}\alpha_z t^2 = 0 \text{ rad} + (10 \text{ rad/s})(5 \text{ s}) + \frac{1}{2}\left(-2 \text{ rad/s}^2\right)(5 \text{ s})^2 = 25 \text{ rad}$$

13.3 ANGULAR VELOCITY AND FREQUENCY

Another common way of describing the motion of a rotating object is to describe how often, or frequently, the object rotates per second. The frequency, f, of an object's rotation is the number of complete revolutions the object makes per second. By definition, the frequency is the inverse of period, T, of the rotation, which is the time it takes to make one revolution. This is given in equation (13.10) as:

$$f = \frac{1}{T} \tag{13.10}$$

By definition, the angular velocity, ω, is the change in the angular position per change in time, $\omega = \Delta\theta/\Delta t$.

Since one revolution corresponds to a change in angle, $\Delta\theta$, of 2π radians and the time for one revolution is a period, T, the angular velocity can be related to frequency in equation (13.11) as:

$$\omega = \frac{2\pi}{T} = 2\pi f \tag{13.11}$$

Example (Angular Kinematics)

A wheel, with a radius of 10 cm, spins at 200 rpm.

a. What is its angular velocity in rad/s?
b. What is its linear velocity of a point on the outer edge of the wheel?

Solution:

a. ω = 200 revolution per minute = (200 rev/min)(2π rad/rev)(1 min/60 s)
   ω = 20.9 rad/s = 20.9 Hz
b. v = ω r = (20.9 rad/s)(0.1 m) = 20.9 m/s

13.4 ROTATIONAL DYNAMICS

Just as an unbalanced force is required to change the velocity of an object in translational motion, an unbalanced torque is required to change the angular velocity of an object in rotational motion about an axis of rotation.

For mass m at the end of a light rod of length r that is attached at the origin of an axis if the rod is moving such that the angular velocity about the center is ω, the magnitude of the tangential velocity, v, of the mass will continue until it is acted upon by an unbalanced (net) force.

So, starting with Newton's Second Law

$$\vec{F}_{\text{Net}} = m\,\vec{a}$$

multiply the force, F, by r to produce an expression of torque, assuming the force is applied so that it is perpendicular to the radius, $\tau = F\,r\,\sin(90°) = F\,r$. Next, divide the acceleration, a, by r to produce the expression for angular acceleration $\alpha = \frac{a}{r}$. For the equality of Newton's Second Law not to be violated, the m must be multiplied by r^2 to cancel the other two r's inserted into the equation to produce an expression for torque and angular acceleration.

$$Fr = \left(mr^2\right)\left(\frac{a}{r}\right)$$

The product ($m\ r^2$) is defined as the moment of inertia (I) so that the version of Newton's Second Law for rotational systems can be written in vector form in equation (13.10) as:

$$\vec{\tau} = I\ \vec{\alpha} \tag{13.10}$$

Notice that the restriction put on this derivation is that the applied force is perpendicular to the line from the axis of rotation to the location of the applied force. This restriction does not severely limit the use of equation (13.10), since many systems include rolling wheels or chains on the sprockets of a bicycle have this orientation.

So, for a mass, m, rotating at a distance, r, from the center of rotation, as shown in Figure 13.9, the moment of inertia is (mr^2).

Therefore, the *moment of inertia*, I, of an object is in effect the "rotational mass" of an object. It has units of [(kg)(m^2)] and is always proportional to (mr^2), but for different objects, the moment of inertia has different fractions in the front of the mr^2 that represents the distribution of the mass of the object around the rotational axis. For example, if the mass of the object in Figure 13.9 is distributed evenly around the circumference, the object is a hoop, as shown in Figure 13.10, it has the same moment of inertia, mr^2.

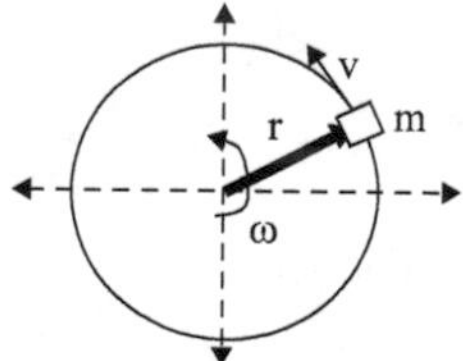

FIGURE 13.9 A mass, m, at a distance, r, from a rotation point and moving with an angular velocity, ω.

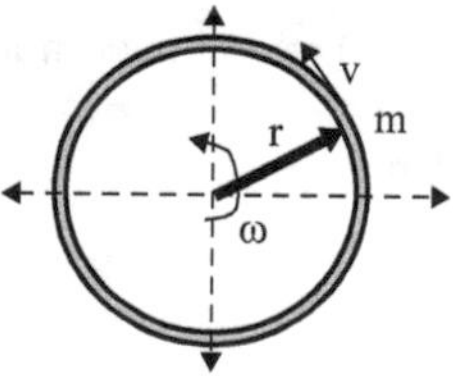

FIGURE 13.10 A thin hoop with a mass, m, and a radius, r, rotating with an angular velocity, ω.

If the mass, m, is distributed evenly across the entire surface of a disk with a radius, r, as shown in Figure 13.11, the moment of inertia is ½ mr^2.

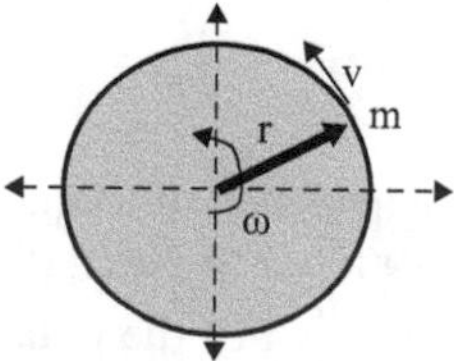

FIGURE 13.11 A uniform disk with a mass, m, and a radius, r, rotating with an angular velocity, ω.

In Table 13.2, the moments of inertia of a few standard objects are listed.

TABLE 13.2
Moments of Inertia

Object	Moment of inertia (I)	Object	Moment of inertia
Hoop	mr^2	Disk	$\frac{1}{2}mr^2$
Thin rod of length L about the center	$\frac{1}{12}mL^2$	Thin rod of length L about an end	$\frac{1}{3}mL^2$
Solid Sphere	$\frac{2}{5}mr^2$	Hollow Sphere	$\frac{2}{3}mr^2$

13.4.1 Rotational Dynamics Examples

Example 13.1

A torque of 0.4 Nm is applied in a clockwise direction to the axle of a wheel. Given that the spokes are negligible, the wheel can be considered a hoop with a radius of 50 cm and a mass of 2 kg. Find the angular displacement of the point P on the outer rim of the wheel if the wheel started from rest and the torque is applied for 8 seconds.

Solution:

Step 1: Create a free-body diagram of the wheel with the torque, angular acceleration, and the angular displacement.

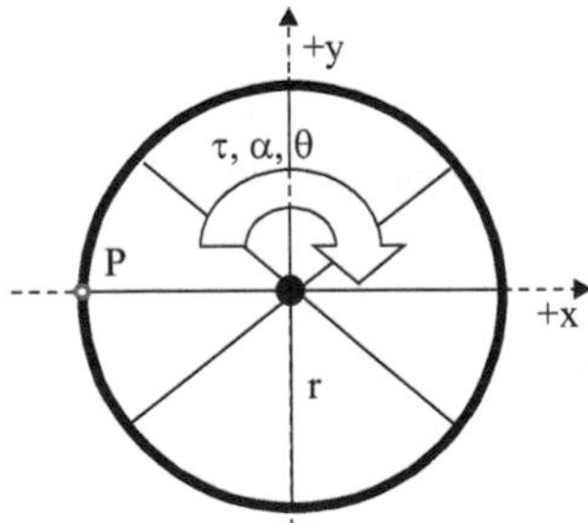

FIGURE 13.12 Free-body diagram for Example 13.1.

Step 2: Compute the z-components of the torque.

$$\tau_z = -0.4 \text{ N} \times \text{m}$$

By inspection, the sense of rotation of the torque (as viewed from above) is clockwise, and from the right-hand rule, that means the torque is directed inward (into of the page) in the –z-direction.

Step 3: Applying Newton's Second Law for rotational motion yields:

$\tau_{Net,\,z}$=I α_z, which can be solved for the angular acceleration as: $\alpha_z = \tau_z/I$.

First, compute the moment of inertia of the hoop:

$$I = mr^2 = (2 \text{ kg})(0.5 \text{ m})^2 = 0.5 \text{ kg m}^2$$

Step 4: Substituting values in the expression of rotational dynamics gives:

$$\alpha_z = \frac{-0.4\ \mathrm{N \cdot m}}{0.5\ \mathrm{kg\ m^2}} = -0.8\frac{\mathrm{radians}}{\mathrm{s^2}}$$

The angular displacement can be found by plugging into the rotational kinematic equation (13.6):

$$\theta_{01z} = 0\ \mathrm{rad} + (0\ \mathrm{rad/s})(8\ \mathrm{s})_+ (1/2)(-0.8\frac{\mathrm{radians}}{\mathrm{s^2}})(8\ \mathrm{s})^2 = 25.6\ \mathrm{rad} = 4.07\ \mathrm{rotations}$$

Example 13.2

A horizontal disk, in Figure 13.12, which has a mass M of 45.0 kg and a radius R = 1.60 m, is mounted on a frictionless vertical axle. A horizontal force of magnitude 142 N and at a constant angle of 90.0° to the radius of the disk and at a distance of r = 1.10 m from the center of the disk is applied for 5 seconds. Find the angular velocity of the disk after the 5 seconds of applied force.

Solution:

Step 1: Generate a free-body diagram of the horizontal forces on the disk.

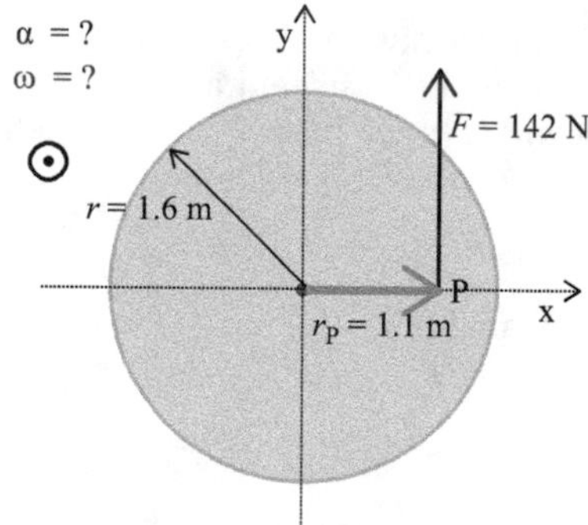

FIGURE 13.13 Free-body diagram for rotational dynamics example.

Step 2: Compute the z-components of the torque:

$$\tau_z = r_P F \,|\sin(\theta_F)| = +1.1\ \mathrm{m}\ (142\ \mathrm{N})\sin(90°) = +156.2\ \mathrm{N \times m}$$

By inspection, the sense of rotation of the torque (as viewed from above) is counterclockwise, and from the right-hand rule, it means the torque is directed upward (out of the page) in the +z-direction (Figure 13.13).

Step 3: Applying Newton's Second Law for rotational motion yields:

$\tau_{\mathrm{Net},\,z} = I\ \alpha_z$, which can be solved for the angular acceleration as: $\alpha_z = \tau_z/I$.

Compute the moment of inertia of the disk:

$$I = \frac{1}{2}mr^2 = \frac{1}{2}(45\ \mathrm{kg})(1.6\ \mathrm{m})^2 = 57.6\ \mathrm{kg\ m^2}$$

Step 4: Substituting values in the expression of rotational dynamics gives:

$$\alpha_z = \frac{156.2\ \mathrm{N \cdot m}}{57.6\ \mathrm{kg\ m^2}} = 2.712\frac{\mathrm{radians}}{\mathrm{s^2}}$$

The angular velocity can be found by plugging into the rotational kinematic equation (13.5):

$$\omega_z = \omega_{oz} + \alpha_z t = 0 + \left(2.712 \frac{\text{radians}}{\text{s}^2}\right)(5 \text{ s}) = 13.56 \frac{\text{rad}}{\text{s}}$$

13.5 ANSWER TO THE CHAPTER QUESTION

The body of the bacterium is the gray oval section in Figure 13.14 and the flagellum is the corkscrew-shaped part of the figure.

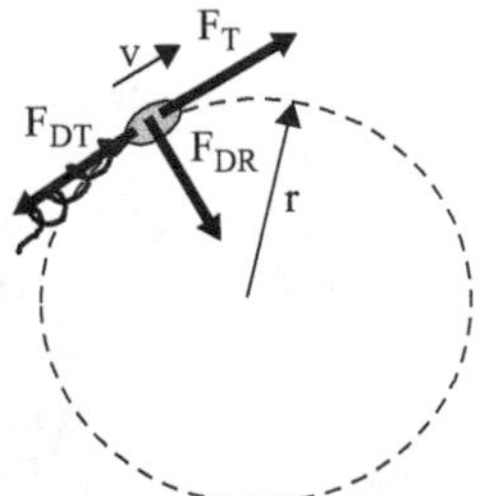

FIGURE 13.14 Free-body diagram for this chapter question.

As the bacteria apply a torque from their body, their flagellum rotates at a constant angular velocity, ω. Since the bacteria generate a forward thrust force, with a magnitude of (F_T), due to the rotating flagellum and they are moving with a constant translational velocity (v), there must be an equal but opposite transverse drag force, given by Stokes Law ($F_{DT} = 6\pi\mu Rv$), due to the viscosity of the water. The bacterium is applying a torque on their flagellum, which has a moment of inertia (I), to produce the forward thrust. Thus, since the body of the bacterium is not rotating there must be a torque on the body of the bacterium. This torque is due to the imbalance of the frictional force on the top and bottom of the body of the bacterium. This imbalance is due to the presence of the microscope slide on the bottom of the bacterium and results in the radial drag force (F_{DR}) that results in the bacterium's circular path. Using the analysis from the previous chapter, the motion of the bacteria can be studied to learn more about these forces.

13.6 Questions and Problems

13.6.1 Multiple-Choice Questions

1. When using a drill bit to drill a hole in a horizontal piece of wood, if the drill bit is rotating clockwise, as viewed from above, about a vertical axis, and is slowing down. Using the right-hand rule, what is the direction of the angular velocity of the drill bit?
   A. upward B. downward C. it has no direction because it is zero

2. When using a drill bit to drill a hole in a horizontal piece of wood, if the drill bit is rotating clockwise, as viewed from above, about a vertical axis, and is slowing down. Using the right-hand rule, what is the direction of the angular acceleration of the drill bit?
   A. upward B. downward C. it has no direction because it is zero

3. The turntable of a microwave oven, located on a horizontal countertop, is rotating at a constant rate, counterclockwise, as viewed from above, about a vertical axis. Using the right-hand rule, what is the direction of the angular velocity of the turntable?
   A. upward B. downward C. it has no direction because it is zero

4. The turntable of a microwave oven, located on a horizontal countertop, is rotating at a constant rate, counterclockwise, as viewed from above, about a vertical axis. What is the direction of the angular acceleration of the turntable?
   A. upward B. downward C. it has no direction because it is zero

5. A wheel, with a radius of 10 m, is rotating with a constant period of 10 seconds. What is the magnitude of the angular velocity for a point on the outer rim of the wheel?
   A. $\frac{\pi}{10}$ rad/s B. $\frac{\pi}{5}$ rad/s C. π rad/s
   D. 2π rad/s E. 20π rad/s

6. A wheel, with a radius of 0.5 m, is rotating with a constant angular velocity of magnitude $\frac{\pi}{2}$ rad/s. What is the translational speed of a point on the outer rim of the wheel?
   A. $\frac{\pi}{4}$ m/s B. $\frac{\pi}{2}$ m/s C. 2π m/s D. 4π m/s

For 7 & 8: A merry-go-round has two rows of horses, the inner row is 2 m from the center of rotation, and the outer row is 3 m from the center of rotation. The merry-go-round is rotating at a rate of 1 revolution every 25 seconds.

7. Is the magnitude of the angular velocity of the outer horses greater than, less than, or equal to the magnitude of the angular velocity of the inner horses?
   A. greater than B. less than C. equal to

8. Is the translational speed of the outer horses greater than, less than, or equal to the translational speed of the inner horses?
   A. greater than B. less than C. equal to

For questions 9 & 10: In Figure 13.15, a string is wrapped around each solid object and 8 N of tension is applied downward by the string. Both objects have a mass of 1 kg. Cylinder A has a radius $r_A = 1$ m and disk B has a radius of $r_B = 2$ m.

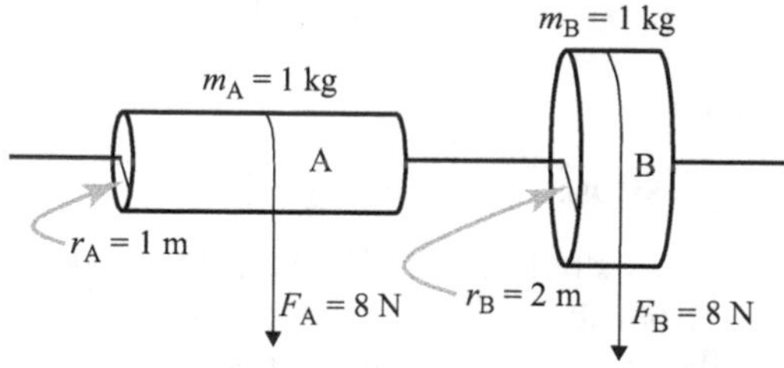

FIGURE 13.15 Diagram for questions 9 and 10.

9. Is the torque on cylinder A greater than, less than, or equal to the torque on disk B?
   A. greater than B. less than C. equal to

10. Is the magnitude of the angular acceleration of cylinder A greater than, less than, or equal to the magnitude of the angular acceleration of disk B?
   A. greater than B. less than C. equal to

13.6.2 Problems

1. A merry-go-round is spinning counterclockwise as viewed from above, but is slowing down at a constant rate. Starting at that instant when it is spinning 25.0 rpm, the merry-go-round completes 16.2 turns before coming to rest. Determine the angular acceleration (axial vector) of the merry-go-round in units of radians per seconds squared.
2. Starting at rest, the shaft of an electric motor experiences a constant angular acceleration of magnitude .180 rad/s^2. How fast is the shaft spinning after 28.2 seconds?
3. The shaft of an electric motor is spinning counterclockwise, as viewed from above, about a vertical axis at a rate that, at time 0, is 4.00 radians per second. The shaft is experiencing an angular acceleration of .222 rad/s^2 counterclockwise, as viewed from above. Starting at time 0, how long does it take for the shaft to achieve a spin rate of 11.0 rad/s?
4. A rotating restaurant at the top of a skyscraper makes one complete 360° of rotation in 20 minutes. Express the maximum value of the magnitude of the angular velocity of the restaurant in radians per second.
5. A windmill that, at time 0, is spinning at a rate of.120 rad/s is experiencing a constant angular acceleration (in the same direction as its initial angular velocity) such that, after 5.20 seconds, the windmill is spinning at a rate of 1.200 radians per second. How many revolutions does the windmill complete during those 5.20 seconds?
6. A disk, with a mass of 200 kg and a moment of inertia (with respect to its axis of symmetry) of 400 kg m^2, starts from rest and experiences an angular acceleration for 5 seconds due to a constant net torque of 1,200 N m about the axis of rotation of the disk. The axis of rotation of the disk is the axis of symmetry of the disk. What is the translational speed of a point on the rim of the disk after 5 seconds of this constant net torque?
7. A horizontal disk of mass $m=2$ kg and radius $r=10$ cm is mounted such that the axis of rotation of the disk is vertical and passes through the center of the circular face of the disk. A piece of string of negligible mass is wound around the outer circumference of the disk. One end of the string extends horizontally away from the disk. As long as any string is in contact with the disk, the string does not slip relative to the disk. The experiment starts when the string is pulled with a constant force of magnitude 2.00 N so that the disk, starting from rest, achieves a maximum angular velocity of 14.80 radians/s when the last of the string is pulled off the disk. The disk then gradually slows down to a stop due to a constant frictional torque of magnitude .0400 Nm (which acts on the disk whenever it is spinning) and the experiment ends. Consider the process to start with the disk at rest at the instant the force starts being applied and end at the instant the disk comes to rest. How many rotations does the disk complete during the entire process and how long does the process last? (Figure 13.16)

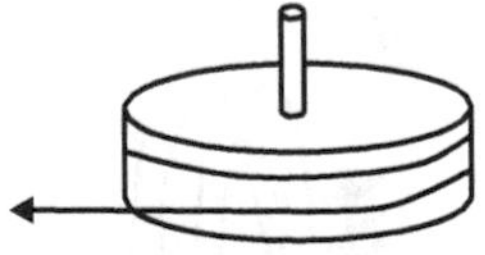

FIGURE 13.16 Diagram for problem 7.

8. As described in Figure 13.17, a large stone disk with a radius $R=50$ cm and a mass of 20 kg is mounted on a frictionless bearing. A smaller, light-weight cylinder, with a radius of $r=2$ cm, is attached to the center of the stone disk so that it can be used as an axle of rotation for the stone disk. A rope is wrapped around the axle and pulled so the tension force in the rope is $F=200$ N. Assuming the stone is at rest at $t=0$ second and the axle has negligible moment of inertia, calculate the rotational rate for the stone disk after 5 seconds of the applied force (F).

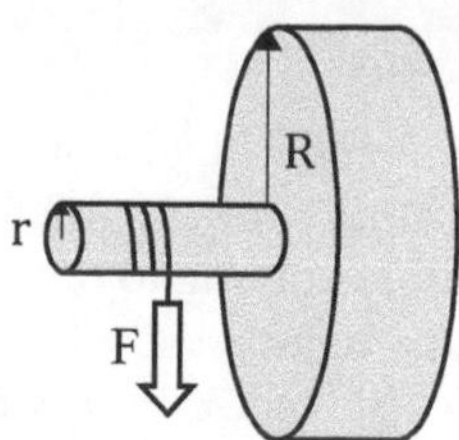

FIGURE 13.17 Diagram for problem 8.

9. In Figure 13.18, a disk, with a mass of 15 kg and a radius of $R = 20\,\text{cm}$, is mounted on an axle so that it spins with negligible frictional torque. A light strong string is wrapped around the outer edge of the disk. The string is pulled with a force of $F = 12\,\text{N}$ for 5 seconds, so that the disk speeds up from rest to a final angular speed, ω. Calculate the angular speed, ω, of the disk after the force is applied to the string for 5 seconds.

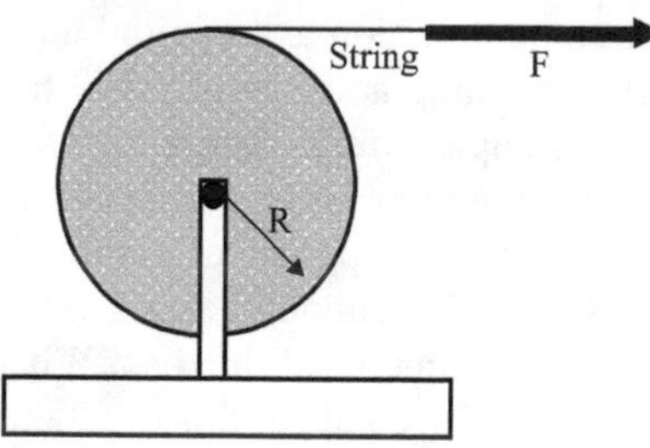

FIGURE 13.18 Diagram for problem 9.

10. A string is wrapped around a disk of radius r and mass m_D, which is mounted with frictionless bearings on a horizontal axle that is collinear with the axis of symmetry of the disk. The disk is held in place by a person while a block of mass m_B is suspended from the end of the string. When the person lets go of the disk, it is observed that the block accelerates downward. Find (for the time interval after the person lets go of the block but before the string runs out and before the block hits anything) the expressions for the magnitude of the acceleration of the block, the magnitude of the angular acceleration of the disk, and the tension in that segment of the string that extends vertically from the disk to the block (Figure 13.19).

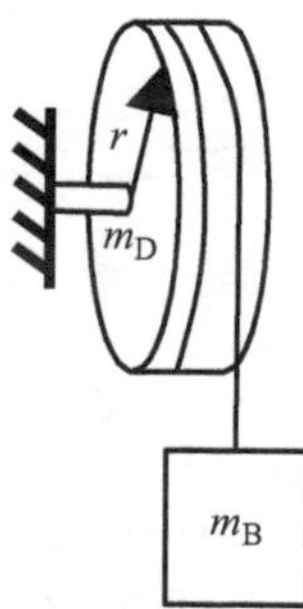

FIGURE 13.19 Diagram for problem 10.

14 Simple Harmonic Motion

14.1 INTRODUCTION

The analysis of an object moving, so that the displacement of the system occurs over and over in the same path, so that it can be expressed with a simple trigonometric function like sine or cosine is known as simple harmonic motion (SHM). Systems with a motion following these functions are characterized by a force that has a form that pulls the system back toward a position; these forces are known as restoring forces because their structure is such that they restore the system to an equilibrium position. Several examples of different restoring forces are presented in the chapter and the motion resulting from these situations are explained.

Chapter question: A pendulum is a simple device with many uses, including its use as the timing mechanism of some clocks. For the case of a small mass, known as a pendulum bob, on the end of a string, what effect does a change in the length of the string or the mass of the pendulum bob have on the time it takes for one complete oscillation and the period of oscillation (back and forth) of the pendulum bob? This question will be answered at the end of the chapter using the techniques developed in this chapter to analyze the motion of an object in SHM.

14.2 SIMPLE HARMONIC MOTION

The key to understanding simple harmonic motion (SHM) is finding the force that pulls the object back toward the equilibrium location. This force is known as the restoring force and can take the form of many different forces such as gravity, a spring force, an electrical force, or even the combination of a few different forces. The key feature of the restoring force is that this restoring force needs to pull the system back toward an equilibrium position. As described in Figure 14.1, if equilibrium of the system is at the origin and the object is displaced a distance x in the positive direction from equilibrium, the force and thus the acceleration of the object in SHM must be in the opposite direction and proportional to the displacement of the object from equilibrium. This is critical to SHM because as the particle moves further away from the equilibrium the force must be greater to bring the object back to equilibrium.

Therefore, from Newton's Second Law for an object undergoing SHM in the x-direction, the relationship between the acceleration (a_x) and the displacement (x) from equilibrium is given by $a_x = -C\ x$, where C is a constant. For a system in SHM, the constant $C = \left(\frac{2\pi}{T}\right)^2$ where T is period of oscillation, which is the time it takes the object to undergo one complete oscillation, and 2π is the

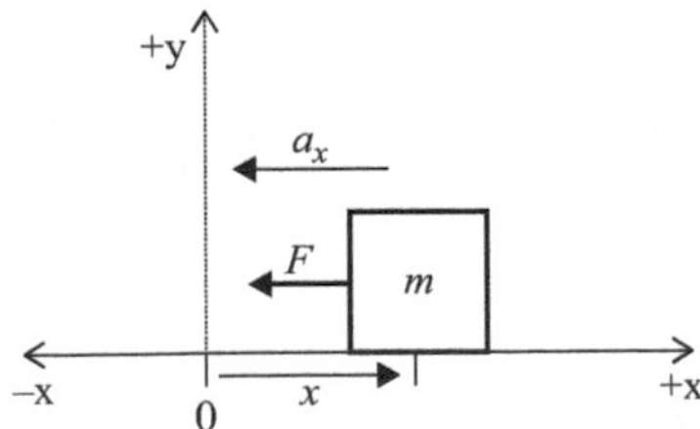

FIGURE 14.1 Restoring force of a system in simple harmonic motion.

DOI: 10.1201/9781003308065-14

number of radians for one complete rotation around a circle. Thus, the characteristic equation of the acceleration of an object in SHM is given in equation (14.1),

$$a_x = -\left(\frac{2\pi}{T}\right)^2 x \tag{14.1}$$

This expression requires calculus to derive it so it will be applied in this chapter without a formal derivation. What is more important than the derivation is the understanding of its physical meaning. For an object with a positive displacement (x) that is increasing, the acceleration (a_x) and thus the restoring force is in the opposite direction and increasing. Therefore, this restoring force will pull the object back toward equilibrium. When the object is on the negative side of equilibrium, the acceleration and thus the force will switch to the positive direction and again pull the object back toward equilibrium. This is why it is called a restoring force; it always is in a direction and has a magnitude that restores the system to equilibrium. It is interesting to note that it is equal to zero at equilibrium, so the inertia of the object carries it through the origin as the restoring force switches direction.

14.3 SPRING FORCE

Since springs themselves or as a representation of other forces play an important role in many systems that exhibit SHM, it is important to first establish the nature of the force associated with a spring. As shown in Figure 14.2, if a spring is hung vertically with nothing attached to the bottom of the spring, it will still stretch to a length labeled here as L_o.

For the particular spring in this example, if a mass, M_1, with a weight of 1 N is hung from the bottom of the spring, the spring will stretch a distance of 10 cm until the force provided by the spring, F_{s1}, is equal and opposite to the weight of M_1. With a mass of M_2 and M_3 hung on the same spring, it will stretch 20 and 30 cm from the zero point, respectively.

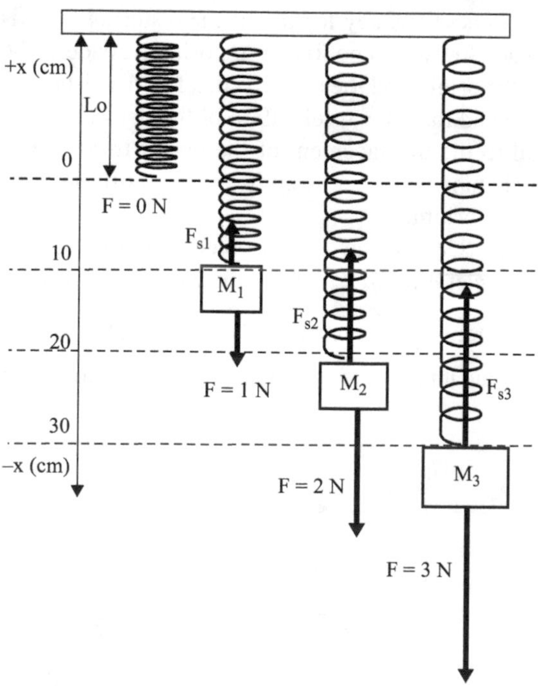

FIGURE 14.2 Springs and hanging mass.

Shown in Figure 14.3 is a graph of the stretch distance, |s|, in meters and the force in Newtons applied by the spring to achieve equilibrium.

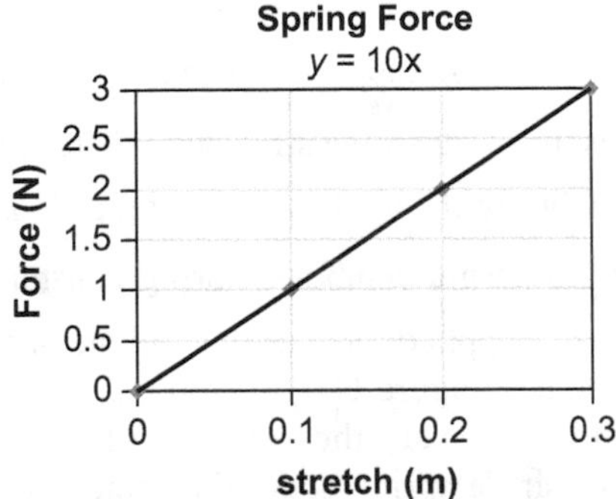

FIGURE 14.3 A graph of the spring force vs stretch distance for a spring.

The line that fits the data is linear for most springs, within a limit. Beyond the limit, the spring is broken and the spring does not behave in the same way. The equation of the line that fits to these data is y = 10x. Therefore, within the elastic limit of the spring the magnitude of this spring force is $F = \left(10\frac{\text{N}}{\text{m}}\right)\text{x}$. An expression of this sort is known as Hooke's Law, which is given in equation (14.2):

$$F_{\text{spring}} = k|s| \tag{14.2}$$

where s is the amount by which the spring is stretched (or compressed) and k is called the force constant of the spring, or just the spring constant, and is the slope of the line in the graph in Figure 14.2. A strong spring has a large value of k and a weak, flimsy spring has a small value of k. Equation (14.2) is valid for both the expansion and the compression of the spring and the sign of the force is dictated by the free-body diagram of the situation; therefore, equation (14.2) gives only the magnitude of the force.

14.3.1 Parallel and Series Spring

In some applications, springs need to be combined to provide the necessary resistance and/or length for a certain task. Springs in series are arranged one after another, whereas springs arranged side by side are in parallel. Please see the Figure 14.4 for a depiction of these arrangements.

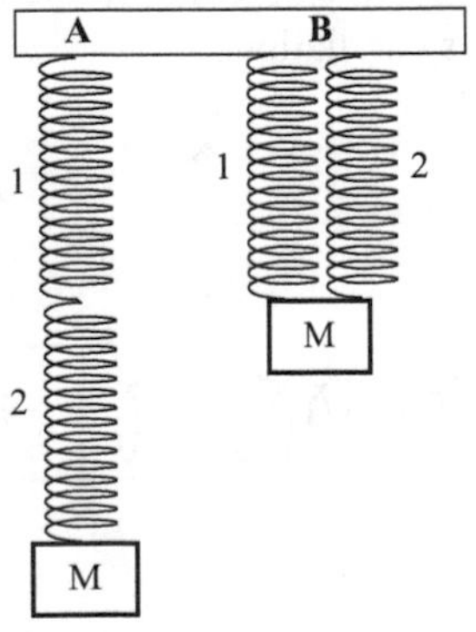

FIGURE 14.4 (a) Series and (b) parallel springs.

For springs in equilibrium and in a series arrangement, labeled **a** in Figure 14.4, the weight of the hanging mass (*M*) is equal in magnitude to the force exerted upward by the bottom spring (2). Since spring (2) pulls downward on the top spring (1) and the springs are in equilibrium all forces are equal and opposite. Since a force equal to the magnitude of the weight is exerted on each spring, each spring will stretch a distance, *s*, to supply the needed force for equilibrium. Thus, the net results of two springs in series is that each spring stretches a distance $s = \frac{F}{k}$ so that the both springs supply the same force as the weight and the total stretch distance is 2*s*. Therefore, two springs, each with a spring constant *k*, in series have a combined spring constant of $\frac{1}{2}k$ since the system will stretch twice as much with the same applied force as that on one of the springs.

For springs in parallel, labeled **b** in Figure 14.4, the force is distributed between the two springs so that each spring supplies half the force and thus stretch half as much. So, springs in parallel will stretch by ½(*s*) as compared to a single spring under the same load. Therefore, two springs, each with a spring constant *k*, in parallel, have a combined spring constant of 2*k* since the system will stretch half as much with the same applied force as that on one of the springs.

14.3.2 Spring-Mass System—Horizontal

One of the simplest systems that exhibit harmonic motion is that of a mass, M, on a flat frictionless surface attached to the end of a horizontal spring. As depicted in Figure 14.5, a spring is attached to a block at rest and the origin of the x-axis is set at the equilibrium location of the block.

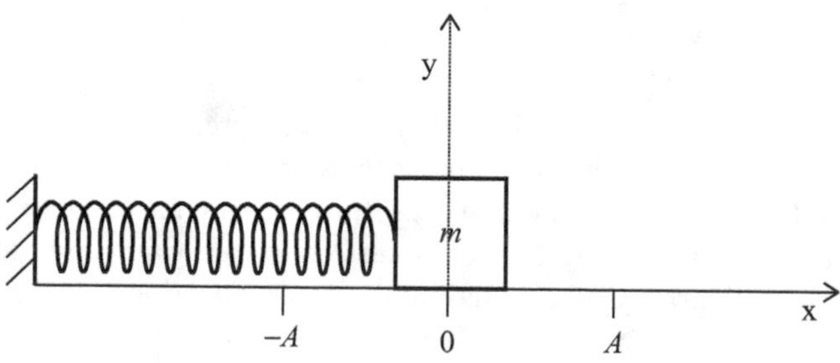

FIGURE 14.5 Block of mass *m* attached to spring on a horizontal frictionless surface at equilibrium.

As depicted in Figure 14.6, if the block is moved a distance *A* directly away from the point on the wall where the spring is attached to the wall, and then released from rest, the block will oscillate about the equilibrium position, back and forth between –*A* and *A*.

For this system, the displacement *A* is then the amplitude of oscillations and is the maximum *x* value achieved by the object during its oscillations.

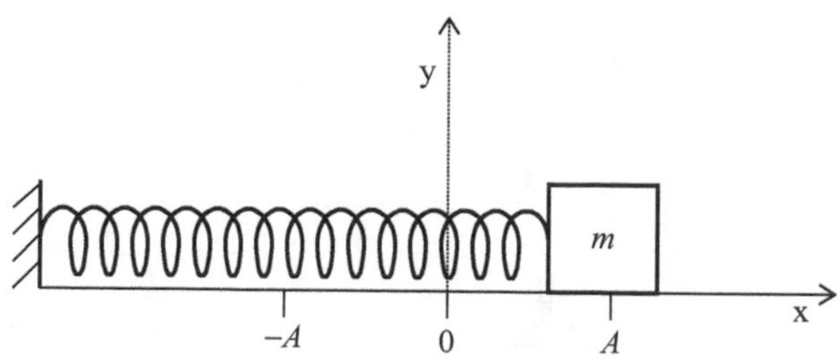

FIGURE 14.6 Block of mass *m* attached to spring on a horizontal frictionless surface at a point *A* displaced from equilibrium.

14.4 EXAMPLE OF DYNAMIC ANALYSIS

The goal of the following dynamic analysis of the block of mass, *m*, is to find an expression for the acceleration of the block as a function of time. With this expression, the kinematic equations can be applied to find expressions of the velocity and displacement of the block as a function of time.

Step 1: Generate a free-body diagram of the block at an instant of time. In Figure 14.7, the point was chosen when the block is to the right of its equilibrium position.

This free-body diagram and the following analysis is for the block to the right of the equilibrium position, but the analysis is good for every position of the block in its oscillation, except for exactly at equilibrium, where all the forces balance. At equilibrium, it is understood that if the block was displaced it will pass through this equilibrium point and continue to compress or expand the spring beyond this point and the analysis continues to work.

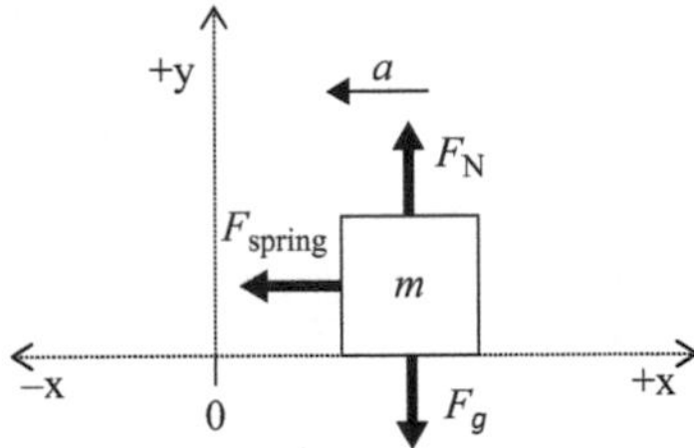

FIGURE 14.7 Free-body diagram of the block of mass *m* attached to spring on a horizontal frictionless surface at a point to the right of equilibrium.

Step 2: The components of the forces are written for the x- and y-directions.

First the y-components:

$$F_{Nx} = F_N \qquad F_{gx} = -F_g \qquad F_{spring,y} = 0$$

Next the x-components:

$$F_{Nx} = 0 \qquad F_{gx} = 0 \qquad F_{spring,x} = -F_{spring} = -k|s| = -kx$$

Step 3: Apply Newton's Second Law for the forces in each coordinate direction separately. The y-components of the vectors in the free-body diagram:

$$\Sigma F_y = ma_y$$

$$F_{Ny} + F_{gy} + F_{spring,y} = 0$$

$$F_N - F_g + 0 = 0$$

$$F_N = F_g = mg.$$

This is the expected result that the normal force is equal in magnitude to the weight of the block.

The x-components of the vectors in the free-body diagram give:

$$\Sigma F_x = ma_x$$

$$F_{Nx} + F_{gx} + F_{spring,x} = m\ a_x$$

$$0 + 0 + -kx = m\ a_x$$

Solving for the x-component of acceleration results in the following expression of the acceleration of the block:

$$a_x = -\frac{k}{m}x$$

This is in the same form as the characteristic equation of the acceleration of an object in SHM. Comparing the expression for a_x from the dynamic analysis of a mass attached to a spring on a horizontal frictionless surface to equation (14.1), the terms in front of the x must be equal, so

$$\left(\frac{2\pi}{T}\right)^2 = \frac{k}{m}$$

Solving for T, the period of oscillations of a block on a horizontal frictionless surface attached to a spring is given in equation (14.3) as:

$$T = 2\pi\sqrt{\frac{m}{k}}. \tag{14.3}$$

If a graph of the displacement of the block as a function of time is generated, it should look like the graph of the displacement vs. time in Figure 14.8.

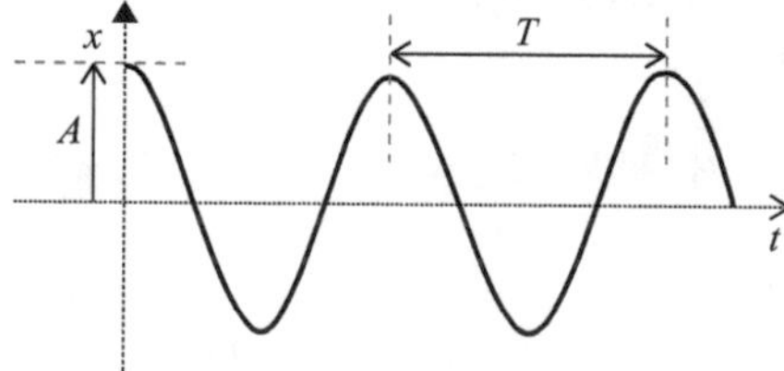

FIGURE 14.8 Displacement as a function of time of a mass attached to a spring oscillating in the horizontal.

It is clear that the displacement should not be linear since the restoring force increases and decreases with time, so the acceleration is not constant. The shape of the displacement graph is the trigonometric function of a cosine. So, the displacement can be expressed as the position x of the object as a function of time t as equation (14.4):

$$x = A\cos\left(2\pi\frac{t}{T}\right) \tag{14.4}$$

where A is the maximum displacement from the equilibrium, which is the amplitude of the motion. The lower-case t is the time that has elapsed since time 0, and the capital T is the period of oscillations, which is the time for one complete oscillation. The 2π is included in the expression since the cosine function repeats itself in 2 π radians. The ratio $\frac{t}{T}$ is the number of oscillations that the block has completed at time t. For example, if the period of oscillation is 4 seconds, the block will have completed one-half of an oscillation in 2 seconds: $\frac{t}{T} = \frac{2 \text{ seconds}}{4 \text{ seconds}} = \frac{1}{2}$. Plugging this into our equation for the position of the block as a function of time yields:

$$x = A\cos\left(2p\left(\tfrac{1}{2}\right)\right) = A\cos(\pi) = -A,$$

which is where you would expect the block to be after half an oscillation starting at $x=A$.

The period is the number of seconds per oscillation. The reciprocal of that is the number of oscillations per second. The number of oscillations per second is called the frequency of oscillations, f, given in equation (14.5) as:

$$f = \frac{1}{T} \tag{14.5}$$

The unit of frequency works out to be 1/*s*, which is defined to be a hertz, abbreviated as Hz. Not only is this system easy to analyze, but the mathematical analysis developed for this system is also used to study electrical circuits, vibrational molecular transitions, and other physical and chemical systems. With the spring attached and the mass at rest, the equilibrium location of the mass is labeled as the zero point. If the spring is stretched by displacing the mass at distance A away from the equilibrium position, 0, the mass will oscillate about the zero point between $+A$ and $-A$.

14.4.1 Vertical Spring

For the case of a block of mass m suspended from a vertical spring with a force constant k, the spring has an initial length L_o before the block is attached. When the block is attached, the spring will stretch to a new equilibrium length of $(L_o + s_e)$, as shown in Figure 14.9.

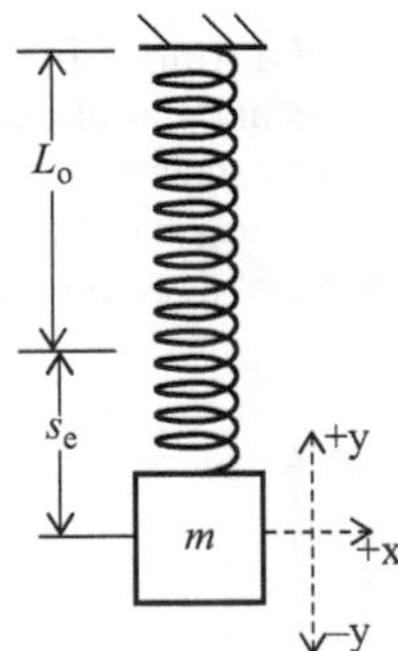

FIGURE 14.9 Block suspended from a vertical spring.

Example

Dynamic analysis of the block on a vertical spring (equilibrium)

An analysis of the static situation reveals a common way of finding the spring constant. If a mass (m) is placed on the end of a spring, the spring will stretch a distance (s_e); therefore, the free-body diagram of the mass in this situation is given in Figure 14.10.

Step 1: Generate a free-body diagram of the situation in Figure 14.10:

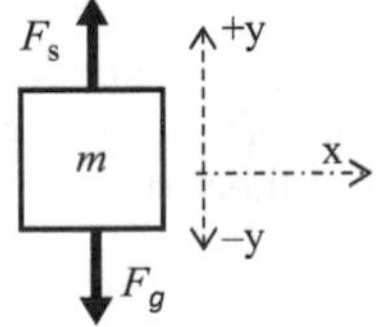

FIGURE 14.10 Free-body diagram of a block suspended from a vertical spring at equilibrium.

Step 2: Write out the y-components of the forces

$$F_{s,y} = ks_e \qquad F_{gy} = -F_g$$

Step 3: Apply the static equation in the y-direction

$$\Sigma F_y = 0$$

$$F_{sp,y} + F_{gy} = 0$$

$$ks_e + (-mg) = 0$$

Step 4: Solve for the spring constant

$$k = \frac{mg}{s_e}$$

where s_e is the amount of stretch for the spring to provide an upward force that balances the weight of the block at equilibrium.

Example

Dynamic analysis of the block on the end of a spring above equilibrium

If the block is then lifted upward at a distance *A* above its equilibrium position and released from rest, find the expression of its period of oscillation.

Step 1: Draw a new free-body diagram, shown in Figure 14.11, for the block when it is at some position, *y*, between its release point, *A*, and the equilibrium position.

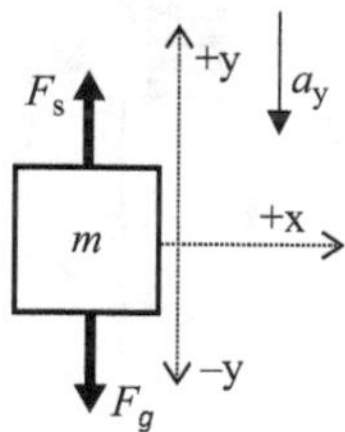

FIGURE 14.11 Free-body diagram of a block on a spring at a point, *y*, moving downward from above the equilibrium position, so the spring is stretched less than it was at equilibrium, s_e. So, the stretch of the spring is $(s_e - y)$.

Step 2: Find the y-components of the forces.

$$F_{s,y} = k(s_e - y) \qquad F_{gy} = -F_g$$

Notice that the value of stretch length is $(s_e - y)$ because the spring stretches an amount s_e to achieve equilibrium, and it is at a point *y* above equilibrium.

Step 3: Apply Newton's Second Law in the y-direction for the block at a position *y* above the equilibrium position.

$$\Sigma F_y = ma_y$$

$$F_{s,y} + F_{gy} = ma_y$$

$$k(s_e - y) + (-mg) = ma_y$$

$$a_y = -\frac{k}{m}y + \frac{(ks_e - mg)}{m}$$

$k\,s_e$ is the magnitude of the force that is exerted on the block when it is at its equilibrium position and that is equal to $m\,g$ so the quantity in parentheses is 0. Therefore,

$$a_y = -\frac{k}{m}y$$

This is the characteristic equation for the acceleration of a system in SHM, equation (14.1). So, a block hanging from a vertical spring undergoes up and down oscillations with the same frequency that the same block at the end of the same spring in a horizontal orientation with the block moving on a frictionless surface.

14.4.2 Mass and Spring Examples

Example 14.1

A mass is hung from the end of a spring, and then it is displaced from equilibrium and it oscillates with a frequency of 4 Hz. What is its period of the oscillations?

Solution: Plug into equation (14.5) to get: $T = \frac{1}{f} = \frac{1}{4\text{ Hz}} = 0.25\text{ second.}$

Example 14.2

A mass hung on the end of a spring, with a constant $k = 100$ N/m, is displaced from equilibrium and oscillates with a frequency of 2 Hz. What is the value of the mass hung from the spring?

Solution: ***Start equation (14.5) to find the period of oscillation:*** $T = \frac{1}{f} = \frac{1}{2\text{Hz}} = \frac{1}{2}\text{ second.}$

Solve equation (14.3) for mass:

$$m = k\left(\frac{T}{2\pi}\right)^2 = \left(100\ \frac{\text{N}}{\text{m}}\right)\left(\frac{\frac{1}{2}\text{second}}{2\pi}\right) = 0.633\text{ kg}$$

Example 14.3

A 50 g mass is hung at the end of a spring and allowed to come to equilibrium. It is then displaced from equilibrium and the mass oscillates with a period of 0.6 second. How far did the spring stretch to achieve equilibrium when the 50 g mass was first hung on the spring?

Solve equation (14.3) for the spring constant:

$$k = m\left(\frac{2\pi}{T}\right)^2 = (0.05\text{ kg})\left(\frac{2\pi}{0.6\text{ second}}\right)^2 = 5.48\ \frac{\text{N}}{\text{m}}$$

Step 1: Generate a free-body diagram such as Figure 14.10.
Step 2: Write out the y-components of the forces

$$F_{s,y} = ks_e \qquad F_{gy} = -F_g$$

Step 3: Apply the static equation in the y-direction

$$\Sigma F_y = 0$$

$$F_{sp,y} + F_{gy} = 0$$

$$ks_e + (-mg) = 0$$

Step 4: Solve for the stretch length for equilibrium

$$s_e = \frac{mg}{k} = \frac{0.05\ \text{kg}\left(9.8\ \frac{\text{N}}{\text{kg}}\right)}{5.48\ \frac{\text{N}}{\text{m}}} = 0.089\ \text{m}$$

14.5 RESONANCE

Another important relationship for all types of vibrating systems is resonance. When an SHM system is driven at its natural frequency, the amplitude of the oscillations will grow larger than if driven at any other frequency. A simple example of this is when you push your friend on a swing. You always try to push your friend at the same frequency that the swing is oscillating. Another common example of this is when a singer hits a note that matches the natural frequency of a crystal glass. As the oscillations grow larger, they become so large that the glass breaks.

14.6 ANSWER TO THE CHAPTER QUESTION

The expression for a pendulum can be found in a similar way to the expression of the marble oscillating in the bowl. In the case of the marble, the normal force of the bowl is perpendicular to the restoring force. In the case of the pendulum in Figure 14.12, the tension is perpendicular to the restoring force, which is the x-component of gravity, $F_{gx} = F_g \cos(270° - \theta)$

Applying Newton's Second Law in the x-direction gives: $ma_x = mg\ \cos(270° - \theta)$
divide both sides by m: $a_x = \mathbf{g}\ \cos(270° - \theta)$
since $\cos(270° - \theta) = -\sin(\theta)$: $a_x = -g\ \sin(\theta)$

With x defined as the horizontal distance, the pendulum bob is away from the equilibrium position and l as the length of the pendulum string, $\sin(\theta) = x/l$. So, the acceleration is:

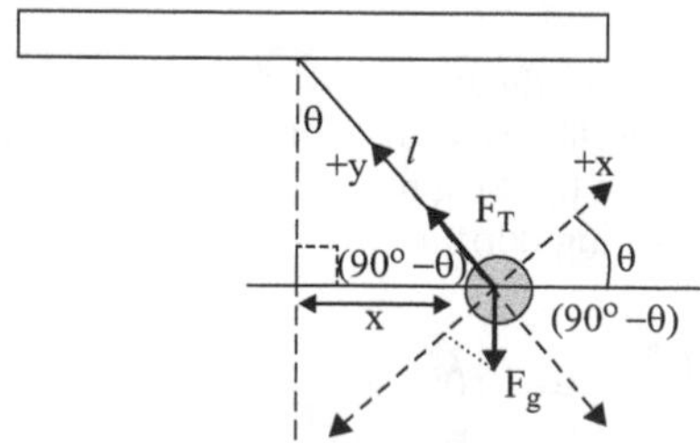

FIGURE 14.12 Free-body diagram of a pendulum bob.

$$a_x = -\frac{mg/L}{m}x$$

The m in the numerator having to do with how strong the gravitational force is, cancels the m in the denominator, which is the inherent resistance of the object to a change in its velocity. Hence, *the result does not depend on the mass of the bob.* The acceleration of the pendulum bob is: $a_x = -\frac{g}{L}x$

Comparing this to the characteristic equation of SHM, equation (14.1), gives: $\left(\frac{2\pi}{T}\right)^2 = \frac{g}{L}$

So, the period of a simple pendulum is given in equation (14.6) as:

$$T = 2\pi\sqrt{\frac{L}{g}} \tag{14.6}$$

where l is the length of the pendulum string and g is the gravitational field strength. So, increasing the length of the pendulum string will make the period of oscillation longer, but a change in the mass of the pendulum bob will have no effect on the period of the pendulum. For a pendulum with more mass, the force of gravity on the bob will be larger, but the inertia of the pendulum will also be larger. The effects will cancel out each other.

14.6.1 Pendulum Examples

Example 14.1

What length would you need to make a pendulum to have a period of 1 second?

$$l = g(T/2\pi)^2 = 9.8 \text{ N/kg } (1 \text{ second}/2\pi)^2 = 0.248 \text{ m} = 25 \text{ cm}$$

Example 14.2

A pendulum oscillates through 20 complete cycles in 10 seconds:

a. What is the period of one oscillation?

$$T = 10 \text{ seconds} / 20 \text{ cycles.} = 0.5 \text{ second.}$$

b. What is the frequency of the pendulum?

$$f = 1/T = 2 \text{ Hz}$$

14.7 QUESTIONS AND PROBLEMS

14.7.1 Multiple-Choice Questions

1. The magnitude of the force being exerted by a stretched spring is given by $F=k|s|$. Is s the amount of stretch of the spring or is it the length of the spring?
   a. s is the amount of stretch of the spring.
   b. s is the length of the spring.

2. In the case of a spring, what is the meaning of a negative value of s?
   a. A negative value of s means that the spring is compressed by an amount equal to the absolute value of s.

b. A negative value of s means that the spring is stretched by an amount equal to the absolute value of s.
c. A negative value of s means that the spring is compressed by an amount $L_0 - s$ where L_0 is the relaxed length of the spring.

3. Do the constant acceleration equations apply to the SHM of a block on a spring?
A. Yes. B. No.

Multiple-Choice Questions 4–7 refer to the position versus time graph in Figure 14.13

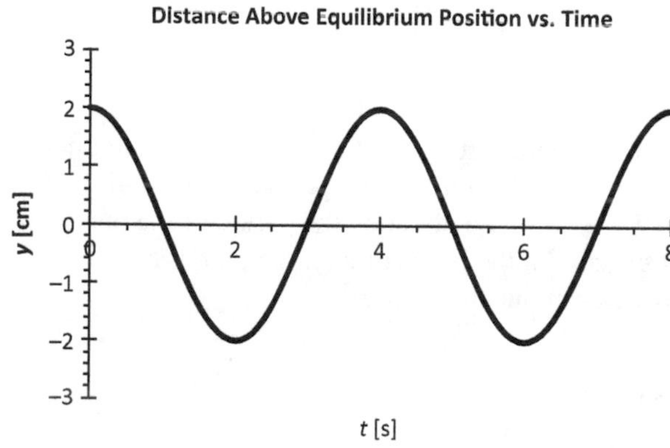

FIGURE 14.13 Position versus time graph for a block on the end of a vertical spring for questions 4–7.

4. The amplitude of the oscillations of the block is:
A. 0 cm B. 2 cm C. 4 cm D. 16 cm

5. In one complete cycle, the block whose motion is characterized by the graph will travel:
A. 2 cm B. 4 cm C. 8 cm D. 16 cm

6. The period of oscillations of the block is:
A. 1 second B. 2 seconds C. 3 seconds D. 4 seconds

7. The frequency of oscillations of the block is:
A. (1/4) Hz B. (1/2) Hz C. 2 Hz D. 4 Hz

8. A vertical spring whose upper end is attached to the ceiling is stretched 10 cm when a block of mass 20 g is attached to the spring and is hanging at rest at its equilibrium position. What will the total stretch of the spring be when a block of mass 80 g is attached to the spring, instead of the 20 g mass, and the 80 g block is hanging at rest at its equilibrium position?
A. 5 cm B. 10 cm C. 20 cm D. 40 cm E. 80 cm

9. Each of two freely hanging springs of one and the same unstretched length has a spring constant of 2 N/m. When they are placed in parallel, as depicted in Figure 14.4b, and a single block that weighs 2 N is attached to them, how much will each of them be stretched when the block is at rest at its equilibrium position:
A. (1/4) m B. (1/2) m C. 1 m D. 2 m E. 4 m

10. Each of two freely hanging springs of the same unstretched length have spring constants of 2 N/m each. When they are placed in series, as depicted in Figure 14.4a, and a single block that weighs 2 N is attached to them, how much will each one be stretched when the block is at rest at its equilibrium position:
    A. (1/4) m B. (1/2) m C. 1 m D. 2 m E. 4 m

14.7.2 Problems

1. A block of mass 250.0 g is on a frictionless surface and is attached to one end of an ideal, massless, horizontal spring. The other end of the spring is attached to a wall. The person sets the block into oscillation by pulling the block away from the wall and releasing it from rest at time $t = 0$. A coordinate system has been established in which the +x-direction is the horizontal direction from the end of the spring that is attached to the wall, toward the end of the string that is attached to the block. The origin of that coordinate system is at the equilibrium position of the block (the position of the block when the spring is neither stretched nor compressed). The block oscillates with an angular frequency of $7.00\,\frac{\text{rad}}{\text{s}}$ and an amplitude of 15 cm. Find the force constant of the spring.
2. A block of mass 150.0 g is on a frictionless horizontal surface and is attached to one end of an ideal, massless, horizontal spring having a force constant of 7.50 N/m. The other end of the spring is attached to a wall. The person sets the block into oscillation by pulling the block away from the wall a distance of 10.0 cm and releasing it from rest at time, $t = 0$. A coordinate system has been established in which the +x-direction is the horizontal direction from the end of the spring that is attached to the wall, toward the end of the string that is attached to the block. The origin of that coordinate system is at the equilibrium position of the block (the position of the block when the spring is neither stretched nor compressed). Find the period of oscillations of the block.
3. A block with a mass of 0.250 kg is attached to a spring of unstretched length 15.0 cm and allowed to hang at rest at its equilibrium position. The mass is then lifted up 10 cm from its equilibrium position and released from rest. If the block oscillates such that its position, relative to the equilibrium position, is characterized by the position vs. time graph in Figure 14.14, how far was the spring stretched when the block was placed on the spring and allowed to come to rest at its equilibrium position?

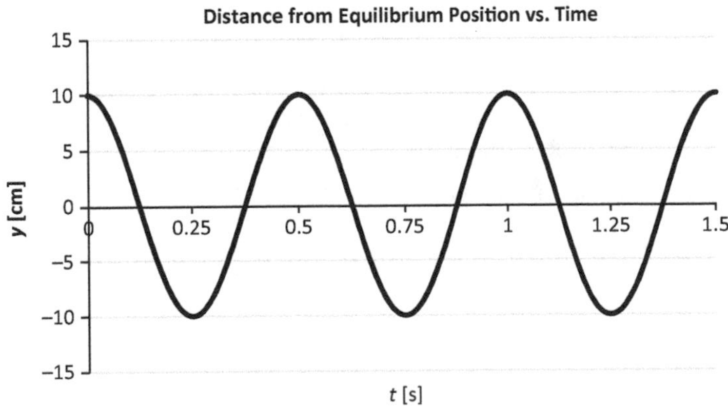

FIGURE 14.14 Position versus time graph for a block on the end of a vertical spring for problem 3.

4. What would the length of a simple pendulum have to be in order for its period of oscillations to be 2.00 seconds given that the pendulum bob is a solid brass sphere of diameter 1.88 cm? (The density of the brass of which the bob is made is 8,700 kg/m³.)
5. A 0.5 kg mass is oscillating up and down on the lower end of a spring that has a relaxed length of 31.0 cm and whose upper end is attached to the ceiling. Compute the force constant of the spring needed for the period of oscillations of the mass to be the same as that of a simple pendulum of length 22.0 cm whose bob has a mass of 200.0 g.
6. An object with a mass of 0.15 kg is attached to the lower end of a vertical spring whose upper end is fixed to the ceiling. The height y of the object above its equilibrium position is given as a function of time as:

$$y = 0.0540\text{m}\cos\left(12\frac{\text{rad}}{\text{s}}\,t\right).$$

For this oscillating object:

a. Determine the amplitude of the oscillations.
b. Determine the period of oscillations.
c. Determine the frequency of oscillations.
d. Determine the acceleration of the object at time 0.10 second.
e. Determine the spring constant.

7. An object with a mass of 0.550 kg hangs at rest from the end of a spring whose spring force constant is 21.5 N/m and whose relaxed length is 16.0 cm. A person lifts the object a distance of 0.200 m above its equilibrium position and releases it from rest.
   a. Find the time it takes, subsequent to the object's release, for the object to arrive (the first time) at a position 0.100 m above its equilibrium position.
   b. Find the acceleration of the object at the time found in part a of this problem.

8. A 500 g block is suspended at equilibrium from the end of a spring, which has a constant of 2 N/m. The mass is displaced slightly from equilibrium and it oscillates in SHM. Compute the period and frequency of the oscillation.
9. A 400 g block is hung from a spring of negligible mass and when it is displaced a small amount from equilibrium, it vibrates with a natural frequency of 0.55 Hz.
   a. What is the spring constant of the spring?
   b. How far did the spring stretch when the block was first placed on the spring?

10. A spring is hung from a stand and 200 g mass is attached to the bottom of the spring and the spring stretches to an equilibrium position. The mass is then pulled down a small distance from equilibrium and released from rest, so that the mass oscillates with a frequency of $\frac{4}{3}$ Hz. Compute the spring constant.

15 Waves

15.1 INTRODUCTION

The ability to generate, analyze, and manipulate waves is important to the success of modern society. In this chapter, the terminology employed to represent waves is presented along with the mechanism required to support the propagation (movement) of a wave through a material (medium). In addition, the analysis techniques applied to the study of waves as they combine are presented along with the situations that result from these interacting waves.

Chapter question: Ultrasound is used extensively in medical applications, including maternity, cardiology, and diagnosing other soft tissue and organ ailments. The question is what is meant by the term ultrasound? This question will be answered at the end of this chapter after investigating waves, including sound waves.

15.2 MECHANICAL WAVES

A wave is a periodic disturbance that travels, or propagates, from one location to another. Mechanical waves, such as sound waves, water waves, or waves on a string, require a material or medium for the wave to move through. There must also be an initial force applied to disturb the medium, like when displacing a mass-on-a-spring from its equilibrium position. Like the mass-on-a-spring, there must also be a restoring force to sustain the material. For a wave on a guitar string, the tension in the string provides the restoring force while the pluck of the string provides the displacement force. For a more complicated wave, like a sound wave, it is the interaction between the molecules in the air, summarized in quantities such as the bulk modulus, to provide the restoring force, while a person may provide the disturbance from pushing the air with the air coming from their mouth. For water waves, the restoring forces are gravity and buoyancy, and the disturbance forces are from the wind at the air–water interface.

15.3 SPEED OF THE WAVE

The speed at which a wave travels through a medium is known as the velocity of propagation, v, and it depends on properties of the medium. This velocity of propagation is the speed the wave moves from one point to another, like a sound wave coming from the radio to your ear. It is not the speed of the individual particles of the medium; this speed is associated with the temperature of the material and not the motion of the wave. For sound waves, the speed of propagation (v) depends on the compressibility, measured as the bulk modulus β, and the density, ρ, of the material in the format given in equation (15.1) as:

$$v = \sqrt{\frac{\beta}{\rho}}. \tag{15.1}$$

This relationship works well for air: $v = \sqrt{\frac{\beta}{\rho}} = \sqrt{\frac{1.42 \times 10^5 \frac{\text{N}}{\text{m}^2}}{1.204 \frac{\text{kg}}{\text{m}^3}}} = 343.4 \ \frac{\text{m}}{\text{s}}$

DOI: 10.1201/9781003308065-15

TABLE 15.1
Speeds of Sound Waves Passing through Some Materials

Gases (0°C)	v (m/s)	Liquids (25°C)	v (m/s)	Solids	v (m/s)
Hydrogen	1,286	Water	1,493	Iron	5,130
Helium	972	Sea water	1,533	Aluminum	5,100
Air	331	Kerosene	1,324	Pyrex glass	5,640
Air (20°C)	343	Methyl alcohol	1,143	Gold	3,240
Air (average)	340	Mercury	1,450	Diamond	12,000

and for water: $v = \sqrt{\frac{\beta}{\rho}} = \sqrt{\frac{2.2\times 10^9 \frac{\text{N}}{\text{m}^2}}{1{,}000 \frac{\text{kg}}{\text{m}^3}}} = 1{,}483.23 \ \frac{\text{m}}{\text{s}}.$

The speeds of sound for some materials are given in Table 15.1.

For waves traveling on a string under tension, like on a piano or guitar string, the speed of propagation of the wave (v) depends on the tension (T) in the string and the mass per length (μ) of the string, as given in equation (15.2) as:

$$v = \sqrt{\frac{F_\text{T}}{\mu}}. \tag{15.2}$$

For example, the speed of a wave traveling down a guitar string with a mass per length of 2.5 g/m = 0.0025 kg/m under a tension of 80 N is

$$v = \sqrt{\frac{F_\text{T}}{\mu}} = \sqrt{\frac{80 \text{ N}}{0.0025 \ \frac{\text{kg}}{\text{m}}}} = 178.9 \ \frac{\text{m}}{\text{s}}.$$

That's about 400 mph. It is important to note that the properties of the medium in which the wave is propagating sets the speed of the wave and the characteristics of the wave change to match the properties of the material, not the other way around.

15.4 WAVE CHARACTERISTICS

In determining the characteristics of a wave, it is important to first determine the origin of the situation, denoted as the *equilibrium position*. This equilibrium position is the location of the medium before any disturbing force acts on the material. For example, the calm flat surface of a lake before a rock is tossed into the water and waves begin to propagate out in a circular pattern with the location at which the rock entered the water at their center or the location of the straight guitar string before it is plucked by a person's finger and it begins to vibrate.

Once the disturbance of the equilibrium position of the medium occurs, a wave can be measured in two independent ways. The first way in which a wave can be observed, is by remaining at one fixed location and observing the wave passing by the observer who monitors the wave as a function of time. For example, if you watch a buoy in the water attached to the heavy, dense object and monitor its motion with a stopwatch. For this type of observation, a graph of the vertical displacement vs. time of the buoy similar to Figure 15.1 can be produced.

For the wave in Figure 15.1, the amplitude is denoted with an **A** and it is the maximum displacement from equilibrium. In Figure 15.1, it has a value of $A = 0.6\,\text{m}$.

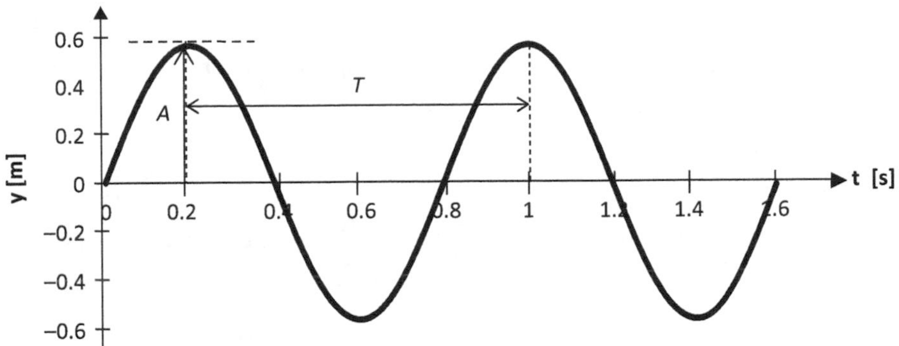

FIGURE 15.1 The displacement of a wave measured as a function of time.

For the wave in Figure 15.1, the period of the wave is denoted with a **T** and it is the time it takes the wave to repeat, in this case from peak to peak. In Figure 15.1, it has a value of $T=0.8$ second.

Another way to observe a wave is to take a snapshot of the wave at one instant of time. For example, if the picture of a water wave is taken in reference to a boat of known length on the water. If such an observation is made, a graph of the vertical components of waves' displacement could be produced vs. the horizontal distance. For this type of observation, a graph of displacement vs. time similar to Figure 15.2 can be produced.

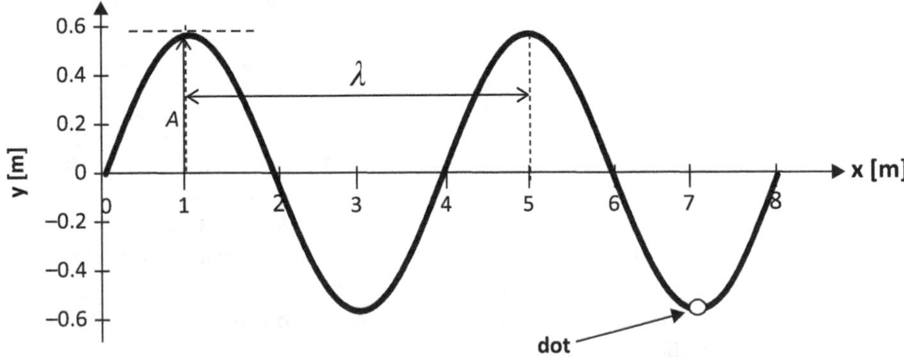

FIGURE 15.2 The displacement of a wave measured as a function of distance.

The amplitude of the wave, **A**, is the maximum displacement from equilibrium. For the wave in Figure 15.1, the amplitude is $A=0.6\,\text{m}$. The wavelength of the wave, λ, is the distance along the direction of the wave velocity it takes the wave to repeat, in this case from one peak of the wave to the next. For the wave in Figure 15.1, the wavelength is from 1 to 5 m, so $\lambda=4\,\text{m}$.

Note that the wavelength is a distance and it is measured along the length of the medium.

The speed of the wave, which is determined by the properties of the medium, is also related to the wave characteristics. Since the wave moves forward a distance of one wavelength (λ) in an amount of time equal to one period (T) and the magnitude of velocity is the distance traveled over time for that travel and the wavelength and period of the length and time between the points on the wave that repeat, then the speed of a wave (v) is equivalent to the ratio of the wavelength and period of a wave, as given in equation (15.3) as:

$$v=\frac{\lambda}{T} \tag{15.3}$$

As described in Chapter 14, the relationship between period T and frequency is reciprocal in nature, that is, $f = \frac{1}{T}$. So, the speed of a wave can be expressed as the product of wavelength and frequency as given in equation (15.4) as:

$$v = \lambda f \tag{15.4}$$

This is the expression that is commonly applied when relating the velocity of the wave to the wave characteristics.

Example

A sound wave in air has a frequency of an A_3 note 220 Hz, what is the wavelength of this wave?

Solution:

From Table 15.1, the average speed of sound in air is 340 m/s, so $v = 340$ m/s.

Solving equation (15.4) for wavelength gives $\lambda = \frac{v}{f} = \frac{340 \frac{\text{m}}{\text{s}}}{220 \text{ Hz}} = \frac{340 \frac{\text{m}}{\text{s}}}{220 \frac{1}{\text{s}}} = 1.55 \text{ m}$. Notice that the unit of a $\text{Hz} = \frac{1}{\text{s}}$, so the final units of the wavelength are meters. It is surprising that sound waves are quite long. This frequency is one that humans can easily hear and it is moving fast and vibrating quickly, but its length is about 5 ft.

15.5 TYPES OF WAVE

Throughout the chapter up until this point, waves have been described, as shown in Figure 15.2. These are *transverse waves*, since the particles of the medium are displaced in a direction that is perpendicular to the direction of the wave propagation. If a dot was made on a piece of string, like the one at the 7 m mark in Figure 15.2, and the string is disturbed so that a wave propagated down the string from left to right, like the one in Figure 15.2, the dot would remain at the same horizontal position, the 7 m mark in this case, as the wave moved from left to right. The wave moves from left to right, but the particles of the string only move up and down.

The other type of wave is the *longitudinal wave* in which the particles move back and forth in the same direction that the wave propagates. Sound waves in air represent an example of longitudinal waves. For the sound wave produced by a speaker in Figure 15.3, the sound wave moves from left to right and the molecules of air move back and forth in the direction of the x-axis.

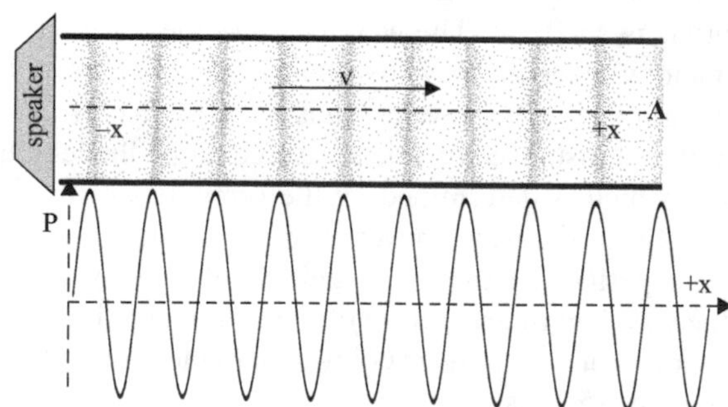

FIGURE 15.3 Longitudinal sound wave in air moving from left to right and the graph of pressure vs. x-location of the wave.

The actual air molecules at the location of the speaker do not race from left to right, from the speaker to point **A**, but the wave does. The air molecules are oscillating back and forth to the right and to the left. The restoring force is provided by air and is summarized as the bulk modulus of the air. As the wave travels through the air, the air molecules tend to form regions of high density (high pressure) and low density (low pressure) along the line along which the wave is traveling. It is common to represent these locations of high and low pressures on a plot, as shown in Figure 15.3, below the depiction of the actual longitudinal wave. This graph provides a way to visualize the longitudinal wave in the same way that a transverse wave is analyzed.

It is interesting to note that waves traveling along the surface of a body of water, are both transverse and the longitudinal waves since the water rolls in both a direction parallel to and perpendicular to the velocity of the wave. Light is not a mechanical wave, but it consists of oscillating electric and magnetic fields, which are perpendicular to the direction in which the light is traveling, so light waves are transverse waves.

15.6 FUNCTION OF WAVE DISPLACEMENT

An expression of the displacement in the *y-direction* of a particle of the medium, through which a wave is traveling as a function of time, is known as the wavefunction and is given in equation (15.5) as:

$$y = A\cos(\omega\, t + \phi) \tag{15.5}$$

In that expression:

A is the amplitude of the wave,
t is the clock reading, how much time has elapsed since the start of observations,
ϕ is phase of the wave.
ω is the angular frequency that equals $\frac{2\pi}{T}$, where T is the period of the wave.

Consider the wave in Figure 15.1; the wave has a period of T=0.8 second, so the angular frequency is $\omega = \frac{2\pi}{0.8\text{ second}} = 2.5\pi$ Hz; it is a good idea to keep the π as a value since it can be entered in your calculator directly and also it may remind you that the cosine needs to be calculated in Radians. Since the wave starts with a displacement of zero and not at the maximum displacement associated with a cosine, the wave is $\frac{1}{4}$ ahead of where it should be, so the phase is $\phi = -2\pi\left(\frac{1}{4}\right) = -\frac{\pi}{2}$. So, for this wave, the expression of the y-displacement of the wave as a function of time is

$$y = A\cos\left((2.5\ \pi\text{Hz})t - \frac{\pi}{2}\right).$$

As a test of this expression, plug in an amplitude of A=0.6 m and a time of t=1 second into the expression. With these inputs, the value for y is 0.6 m, which matches the location of the y-displacement at this time in Figure 15.1.

15.7 PHASE

As introduced in the previous section, the phase of a wave indicates the starting point of the wave as compared to a standard. This phase of a wave can also be used to compare individual waves to each other, setting one of the waves as the standard and finding the relative phase of the other wave. As shown in Figure 15.4, two waves that are in-phase have a phase difference, ϕ, of zero and the peaks are in-step with the peaks of other waves.

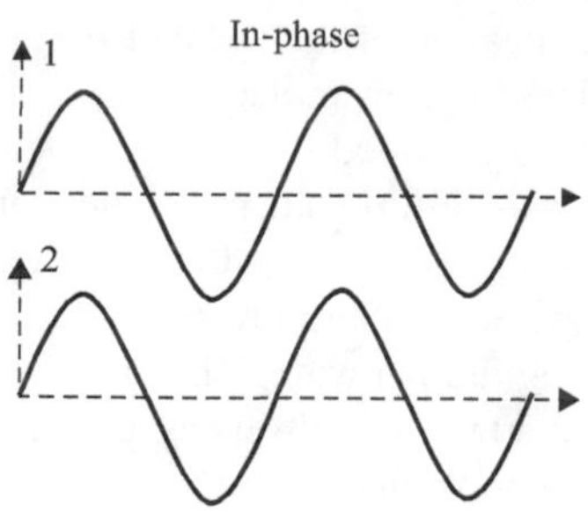

FIGURE 15.4 Two waves that are in-phase with each other.

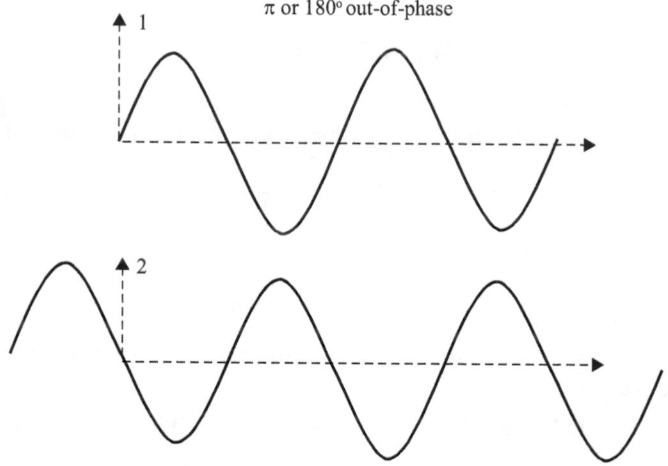

FIGURE 15.5 Two waves that are out-of-phase with each other.

As shown in Figure 15.5, if two waves are out-of-phase, the phase difference, ϕ, between them is π Rad or 180° and the peaks lineup with the troughs of the two waves.

In general, the phase difference, ϕ, between waves can be any angle from 0 to 2π Rad (0° – 360°) but these two extremes of in-phase and out-of-phase are critical to the understanding of the phenomena of interference.

15.8 INTERFERENCE

One of the most important concepts regarding waves that is related to phase and is important to the understanding of many phenomena is *interference*. Interference is the addition and subtraction of the amplitude of waves as they overlap. This overlapping of waves, called *superposition*, can occur at one instant of time as waves pass by each other or along an entire wave if the waves are traveling in the same direction. Constructive-interference occurs when waves that are in-phase overlap to produce a wave with a larger amplitude. If the two waves in Figure 15.4, that are in-phase, overlap, they will produce a wave with an amplitude equal to the sum of the amplitudes of the two waves and the wave will have period and wavelength equal to that of the original waves. This wave would look like the single wave in Figure 15.6.

Destructive-interference occurs when waves that are 180° out-of-phase overlap and produce a single wave with a smaller amplitude. If the two waves in Figure 15.5, which are out-of-phase, overlap, they will produce a wave with an amplitude equal to the difference of the amplitudes, which in

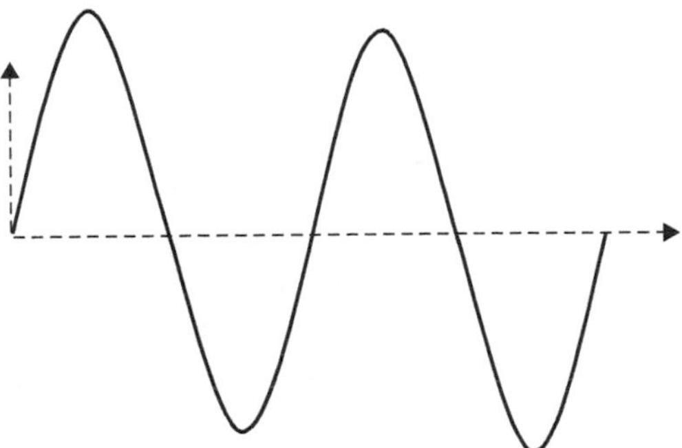

FIGURE 15.6 Single wave produced by the superposition of two waves in Figure 15.4 that are in-phase with each other.

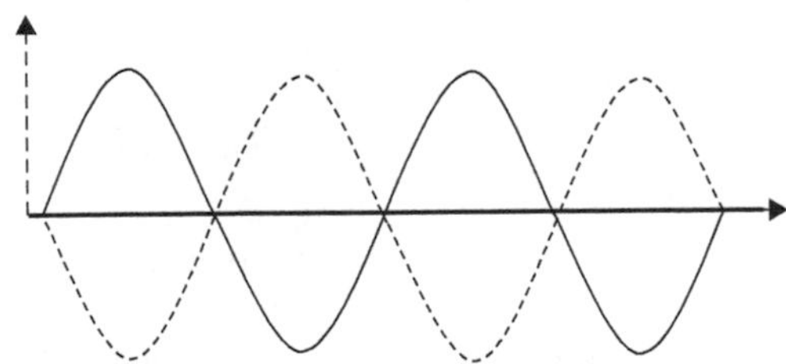

FIGURE 15.7 Superposition of two waves in Figure 15.5 that are out-of-phase with each other.

this case is zero. This wave would look like the single flat line shown in Figure 15.7, the transparent waves in Figure 15.7 are not visible—they are simply a representation of the two waves that are destructively interfering.

It is important to note that when waves reflect off a surface, like a water wave off a wall, a wave on a string off the end, and a light wave off a mirror, the waves undergo an π Rad or 180° phase-shift, in a sense they flip over. So, *reflection* is a natural cause of a π Rad or 180° phase shifts, thus it plays a part in creating situations in which interference occurs.

15.9 STANDING WAVES

If waves are confined along a specific length, *L*, whether it is along a string, between two parallel mirrors, or in an enclosure of any kind, waves that pass by with an amplitude in the same direction, like waves 1 and 2 in Figure 15.8, will constructively interfere.

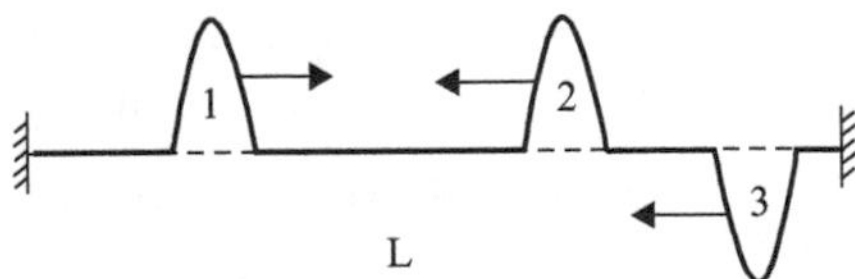

FIGURE 15.8 Waves traveling along a line.

If the waves have amplitudes in the opposite direction, like waves 1 and 3 in Figure 15.8, these waves will destructively interfere. If the timing is just right, the locations of constructive and destructive will remain fixed as the waves pass through each other and a standing wave will result. This is a bit of a misnomer, because the waves are traveling across the medium but are moving at just the right speed so that they overlap constructively and destructively at the same locations every time. The fixed locations of minimum and maximum displacement are called *nodes* and *antinodes*, respectively.

As on a guitar or in a piano, a string with a mass per length of μ is stretched between two posts a distance L apart, as shown in Figure 15.9, and tightened to just the right tension, F_T.

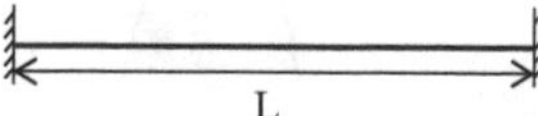

FIGURE 15.9 String of length L.

The string is disturbed, plucked, or struck, and a wave is produced with a speed given by equation (15.2). If the tension is set to just the right value, the speed will be such that the waves traveling back and forth across the string interfere, and produce a standing wave that is twice as long as the string, as shown in Figure 15.10.

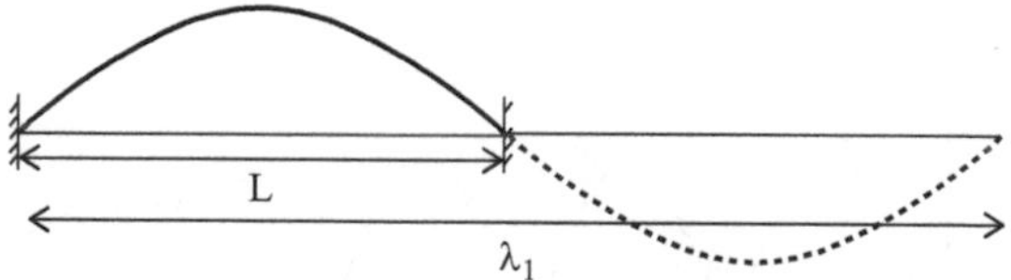

FIGURE 15.10 String of length L with a standing wave that is twice as long as the string.

The boundary condition is that there must be a node at each end of the string, since it is attached at these locations, and in this case, there is one antinode in the middle. Notice that the subscript on the symbol of the wavelength is a 1, λ_1. This is called the first harmonic or the fundamental standing wave and it is one-half of a wavelength long. The next possible standing wave that could satisfy the boundary condition of nodes at the ends of the string is a wave that is exactly as long as the string, as shown in Figure 15.11.

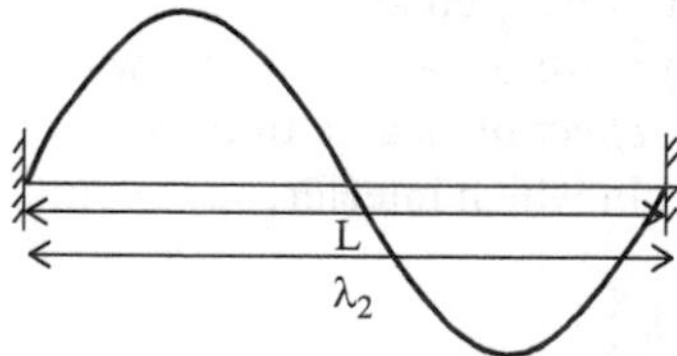

FIGURE 15.11 String of length L with a standing wave that is the same length as the string.

This wave has a node in the middle along with the nodes at each end. Notice that the subscript on the wavelength is a 2, λ_2. This is called the second harmonic and it is two-halves of a wavelength long. The next possible standing wave that could satisfy the boundary condition of nodes at the ends of the string is a wave that has $\frac{3}{2}$ of a wave on the string, as shown in Figure 15.12.

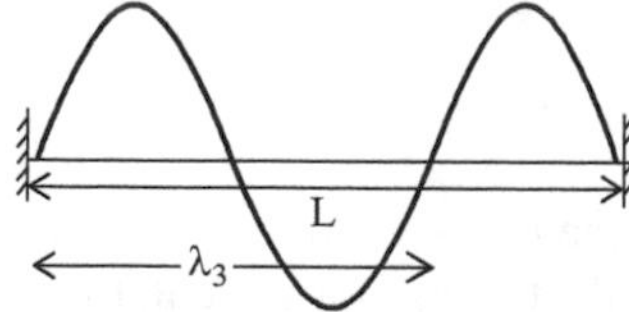

FIGURE 15.12 String of length L with a standing wave that is (3/2) the length of the string.

This wave has two nodes across the string along with the nodes at each end. Notice that the subscript on the wavelength is a 3, λ_3. This is called the third harmonic and it is three-halves of a wavelength long. Thus, higher harmonics of the standing waves have wavelengths that are multiples of ½. Thus, the condition for a standing wave starts with a wave that is twice the length of the string L, then a wave that is the same length as the string, then a wave that is $\frac{2}{3}$ the length L, then $\frac{2}{4}$ the length of the string, and so on; this condition can be expressed as equation (15.5) and is the condition for a standing wave on a string of length L, which is given by equation (15.6) as:

$$\lambda_n = \frac{2}{n}L \tag{15.6}$$

where n is an integer, $n=1, 2, 3, 4, \ldots$.

Because the speed of the wave is set by conditions of the medium in which the waves travel, the tension and the mass per length of the string in equation (15.2), the condition set by equation (15.6), determines the frequency of waves, as expressed in equation (15.7) as:

$$f_n = \frac{v}{\lambda_n} = n\frac{v}{2L} \tag{15.7}$$

where n is an integer, $n=1, 2, 3, 4, \ldots,$. Since the first frequency, f_1, of a standing wave that can be achieved is called the fundamental frequency, the expression of frequencies that can be produced by standing waves are often given in the form of equation (15.8) as:

$$f_n = n\ f_1,\ \text{where}\ f_1 = \frac{v}{2L} \tag{15.8}$$

All waves in a string travel with the same speed, so these waves with different wavelengths have different frequencies as shown. These different situations that result in standing waves are known as the modes of vibration. The fundamental frequency (f_1) is the first mode and the nth mode has frequency n times that of the fundamental.

Remember the waves that produced these standing waves are not stationary, but the locations of constructive (antinodes) and destructive (nodes) interference are fixed. Obviously, this is a special condition that takes just the right timing, but by no means is this event rare. In fact, standing waves are all around us. Important examples of these standing waves in music are the waves on the strings of instruments like guitars, pianos, and violins. They are also in the form of sound waves in musical instruments such as flutes, trumpets, and organs. In addition to musical instruments, standing waves play an important role in a laser, which is a topic of study in Volume 2.

15.10 ANSWER TO THE CHAPTER QUESTION

The ultrasound systems used in medical applications generate sound waves that are sent into the body and the reflected waves are measured and turned into an image. Since the speed of the sound waves in the body is set by the medium, it is not surprising that the speed of sound waves in the body is 1,540 m/s, which is very close to the speed of sound in sea water, quoted in Table 15.1. The speed of the sound waves is not ultra; it is close to the speed of sound of a whale call. It is the frequency that is ultra-high. Humans can hear sounds with frequencies between approximately 20 and 20,000 Hz, with some difference between individuals and this range decreases with age. On the other hand, most ultrasound devices operate at around three megahertz (3 MHz), which is 3×10^6 Hz. In the chapter, an example was given showing that a sound wave with a frequency of 220 Hz in air has a wavelength of 1.55 m. For an ultrasound wave at 3 MHz, the wavelength in the body is about ½ mm, $\lambda = \frac{v}{f} = \frac{1540\ \frac{\text{m}}{\text{s}}}{3{,}000{,}000\ \text{Hz}} = 5.13\times10^{-4}\ \text{m} = 0.5\ \text{mm}$. So, these waves are small, but not exceptionally small, so it is definitely the frequency that is associated with the ultra-part of the name.

15.11 QUESTIONS AND PROBLEMS

15.11.1 Multiple-Choice Questions

1. The number of oscillations of a wave per second is defined as the wave's:
   A. period B. frequency C. amplitude D. wavelength

2. The distance along the direction of the wave velocity it takes the wave to repeat is defined as the wave's:
   A. period B. frequency C. amplitude D. wavelength

3. The time for one complete oscillation of the wave is defined as the wave's:
   A. period B. frequency C. amplitude D. wavelength

4. The maximum displacement of the wave from equilibrium is defined as the wave's:
   A. period B. frequency C. amplitude D. wavelength

5. What is the wavelength of the standing wave in Figure 15.13?
   A. 1 m B. 2 m C. 4 m D. 10 cm

6. What is the amplitude of the standing wave in Figure 15.13?
   A. 5 cm B. 10 cm C. 20 cm D. 4 m

7. What is the speed of the traveling waves that produce the standing wave in Figure 15.13?
   A. 2 m/s B. 4 m/s C. 8 m/s
   D. 12 m/s E. 16 m/s

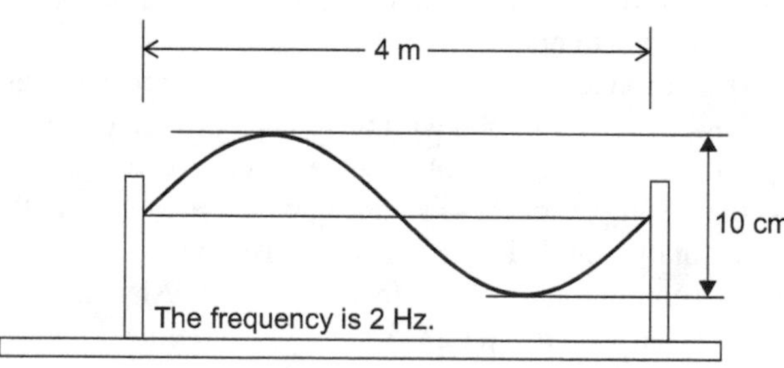

FIGURE 15.13 Multiple-choice questions 5–7.

8. What is the frequency of the standing wave in Figure 15.14?
   A. 3 Hz B. 5 Hz
   C. 6 Hz D. 20 Hz
   E. 50 Hz F. 100 Hz

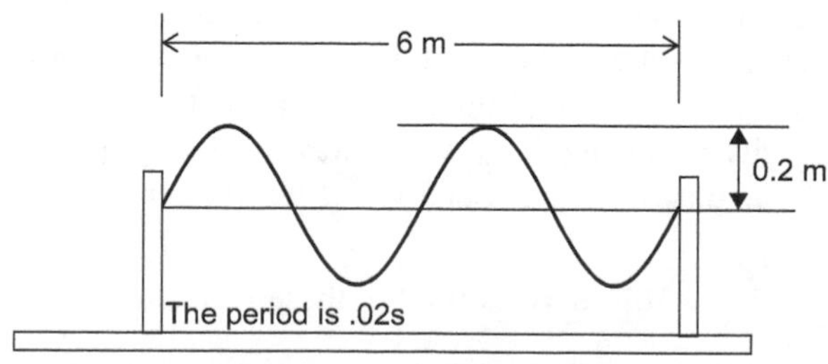

FIGURE 15.14 Multiple-choice questions 8–10.

9. What is the wavelength of the standing wave in Figure 15.14?
   A. 1 m B. 2 m C. 3 m
   D. 4 m E. 5 m F. 6 m

10. What is the speed of the traveling waves that produce the standing wave in Figure 15.14?
    A. 15 m/s B. 36 m/s C. 50 m/s D. 150 m/s E. 300 m/s

15.11.2 Problems

1. Find the speed of a wave traveling in a string when the frequency of the wave oscillations is 77.0 Hz and the wavelength is 0.650 m.
2. Find the speed of a wave traveling in a string whose linear mass density is 0.0360 kg/m when the tension in the string is 14.4 N.
3. Find the wavelength of a wave in a string of linear density 15.0 g/m and tension 9.70 N when the frequency of the wave is 66.0 Hz.
4. A string stretched with a linear density of 0.01 kg/m is stretched across posts that are 1 m apart. The string is displaced and a standing wave in Figure 15.15 is produced that oscillates with a frequency of 200 Hz. Find the tension in the string in Figure 15.15.

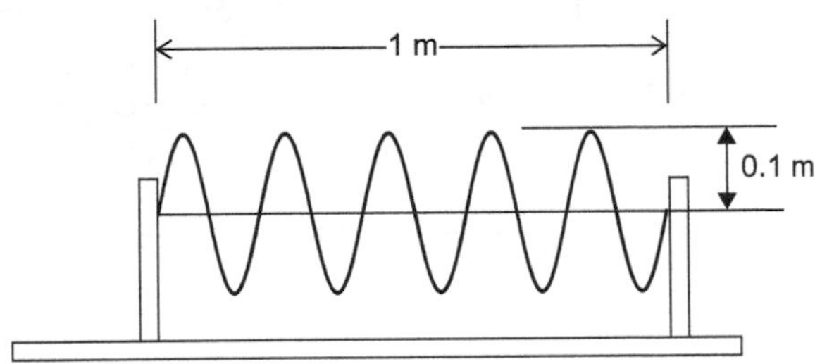

FIGURE 15.15 Problem 4.

5. A wave on a string of linear density 0.0032 kg/m and a wavelength of 0.9 m is governed by the equation

$$y = 0.28\text{m} \cos\left[\left(58\frac{\text{rad}}{\text{s}}\right)t - \pi \text{ rad}\right]$$

   a. What is the period of the wave?
   b. What is the frequency of the wave?
   c. What is the wave speed?
   d. What is the tension in the string?
   e. What is the amplitude of the wave?
6. The tension in a string of linear density 0.00250 kg/m is 7.80 N. The string oscillates at 94.0 Hz producing a wave in the string. The maximum displacement of any point on the string, from its equilibrium position, at any point in time, is 0.310 m. Consider the positive direction for x to be away from the oscillator. (Assume the string to be so long that during the period of observations under consideration here, the wave produced by the mechanical oscillator does not make it to the other end of the string.)
   a. What is the wave speed?
   b. What is the wavelength of the wave?
   c. What is the period of the wave?
   d. What is the amplitude of the wave?

7. One end of a 1.5 m long string is attached to the end of a vibrating arm and the other end is attached to a block of mass M and hung over a pulley, as depicted in Figure 15.16.

   The vibrating arm is set into motion at 120 Hz and a standing wave in Figure 15.16 is produced on the string. Find the velocity of the waves in this string that produce this standing wave if $L=0.9$ m and $M=1.0$ kg. Assume that the vibrations are so small that, even though the vibrating arm is producing the waves, the end of the string that is tied to the vibrating arm can be considered to be a node of the standing wave.

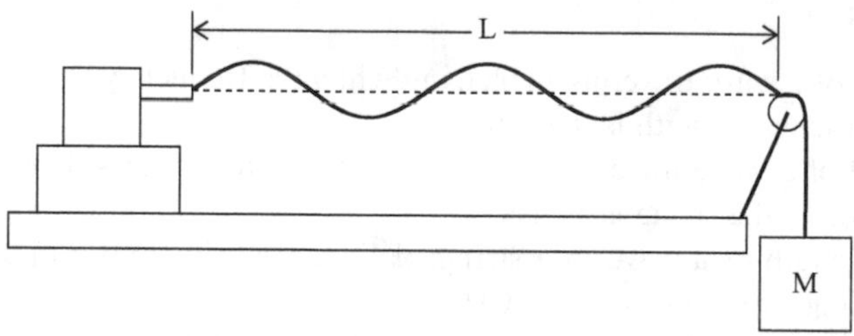

FIGURE 15.16 Problem 7.

8. One end of a string that is 1.350 m long and has a mass of 4.022 g is attached to an oscillator, as depicted in Figure 15.17, and the other end is strung over a pulley and attached to a mass that weighs 65.000 N. The oscillator is turned on and a standing wave is produced on the string, as depicted in Figure 15.17. If $L=0.9$ m what is the frequency of the wave?

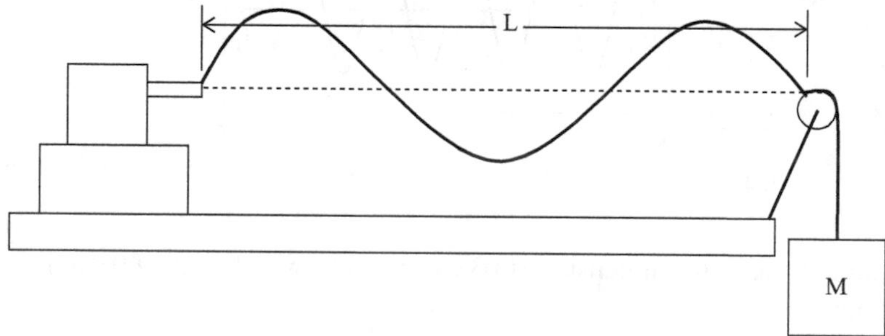

FIGURE 15.17 Problem 8.

9. a. Determine the wavelength of the third harmonic for a string of length 1.00 m, which is fixed at both ends. Include sketches of a snapshot of the string undergoing the standing wave motion associated with the first, second, and third harmonics as parts of your solution (three sketches). Make sure your solution is consistent with your third sketch.
   b. Suppose that the mass of the 1.00 m length of string is 4.60 g and that the tension in the string is 1.25 N. Determine the frequency of the third harmonic.
10. How tightly must a guitar string of linear mass density 3.45 g/m be strung for its fundamental frequency to be 110 Hz (the frequency of the note *A*) when the fixed ends of the string are .668 m apart from each other.

Index

Printed in the USA
CPSIA information can be obtained
at www.ICGtesting.com
LVHW082031070224
771122LV00006B/325

9 781032 311067